Progress in Inflammation Research

Series Editor

Advisory Board

Forthcoming titles:

Inflammatory Cardiomyopathy (DCMi) – Pathogenesis and Therapy, H.-P. Schultheiß, M. Noutsias (Editors), 2009
Occupational Asthma, T. Sigsgaard, D. Heederick (Editors), 2009
Endothelial Dysfunction and Inflammation, A. Karsan, S. Dauphinee (Editors), 2010

(Already published titles see last page.)

The Hygiene Hypothesis and Darwinian Medicine

Graham A.W. Rook

Editor

Birkhäuser
Basel · Boston · Berlin

Editor

Graham A.W. Rook
Centre for Infectious Diseases and International Health
Windeyer Institute for Medical Sciences
University College London
46 Cleveland Street
London W1T 4JF
United Kingdom

Library of Congress Control Number: 2009927790

Bibliographic information published by Die Deutsche Bibliothek
Die Deutsche Bibliothek lists this publication in the Deutsche Nationalbibliografie;
detailed bibliographic data is available in the internet at http://dnb.ddb.de

ISBN 978-3-7643-8902-4 Birkhäuser Verlag AG, Basel – Boston – Berlin

The publisher and editor can give no guarantee for the information on drug dosage and administration contained in this publication. The respective user must check its accuracy by consulting other sources of reference in each individual case. The use of registered names, trademarks etc. in this publication, even if not identified as such, does not imply that they are exempt from the relevant protective laws and regulations or free for general use.

Basel · Boston · Berlin
P.O. Box 133, CH-4010 Basel, Switzerland
Part of Springer Science+Business Media
Printed on acid-free paper produced from chlorine-free pulp. TCF ∞
Cover design: Markus Etterich, Basel
Cover illustration: by Anne Rook, with friendly permission
Printed in Germany
ISBN 978-3-7643-8902-4
9 8 7 6 5 4 3 2 1

e-ISBN 978-3-7643-8903-1
www.birkhauser.ch

Contents

List of contributors

Hafid Ait-Oufella, Paris Cardiovascular Research Center, INSERM and Assistance Publique-Hôpitaux de Paris, Hôpital Européen Georges Pompidou, 75015 Paris, France; e-mail: hafid.aitoufella@inserm.fr

George J. Armelagos, Goodrich C. White Professor of Anthropology, Department of Anthropology, Emory University, Atlanta, Georgia 30320, USA; e-mail: antga@learnlink.emory.edu

Anne Cooke, Department of Pathology, University of Cambridge, Tennis Court Rd, Cambridge, CB2 1QP, UK; e-mail: ac@mole.bio.cam.ac.uk

Jorge Correale, Dr. Raúl Carrea Institute for Neurological Research, FLENI, Montañeses 2325, 1428 Buenos Aires, Argentina; e-mail: jcorreale@fleni.org.ar; bairesla@fibertel.com.ar

Rosemarie H. DeKruyff, Harvard Medical School, Children's Hospital Boston, One Blackfan Circle, Boston, MA 02115, USA; e-mail: rosemarie.dekruyff@childrens.harvard.edu

David E. Elliott, Division of Gastroenterology (4611 JCP), University of Iowa College of Medicine, 200 Hawkins Drive, Iowa City, IA 52242-1009, USA; e-mail: david-elliott@uiowa.edu

Martin Feelisch, Clinical Sciences Research Institute, Warwick Medical School, The University of Warwick, Gibbet Hill Road, Coventry, CV4 7AL, UK; e-mail: mf@warwick.ac.uk

Mel Greaves, Section of Haemato-Oncology, The Institute of Cancer Research, Brookes Lawley Building, 15 Cotswold Road, Sutton, Surrey SM2 5NG, UK; e-mail: mel.greaves@icr.ac.uk

Sue Griffin, Donald W. Reynolds Institute on Aging and Department of Geriatrics, University of Arkansas for Medical Sciences, 629 Jack Stephens Drive and the Geriatric Research Education Clinical Center, Central Arkansas Veterans Healthcare System, 4300 West Seventh Street, Little Rock, AR 72205, USA; e-mail: griffinsuet@uams.edu

Eckard Hamelmann, Department of Paediatrics, Ruhr-University Bochum, Alexandrinenstraße 5, 44791 Bochum, Germany; e-mail: eckard.hamelmann@klinikum-bochum.de

Leonard C. Harrison, Autoimmunity & Transplantation Division, Walter and Eliza Hall Institute of Medical Research, 1 G Royal Parade, Parkville, Victoria, Australia 3052; e-mail: harrison@wehi.edu.au

Margo C. Honeyman, Walter and Eliza Hall Institute of Medical Research, Parkville, Victoria, Australia 3052; e-mail: honeyman@wehi.edu.au

Christopher A. Lowry, Department of Integrative Physiology, University of Colorado, Boulder, CO 80309-0354, USA; e-mail: christopher.lowry@colorado.edu

Rick M. Maizels, Institute of Immunology and Infection Research, University of Edinburgh, West Mains Road, Edinburgh EH9 3JT, UK; e-mail: rick.maizels@ed.ac.uk

Ziad Mallat, Paris Cardiovascular Research Center, INSERM and Assistance Publique-Hôpitaux de Paris, Hôpital Européen Georges Pompidou, 75015 Paris, France; e-mail: ziad.mallat@inserm.fr

Paolo M. Matricardi, Pediatric Pneumology and Immunology, Charité University Medicine, Berlin, Germany

Robert E. Mrak, Department of Pathology, University of Toledo College of Medicine, Toledo, OH 43614, USA; e-mail: robert.mrak@utoledo.edu

Graham A.W. Rook, Centre for Infectious Diseases and International Health, Windeyer Institute of Medical Sciences, University College London, 46 Cleveland Street, London W1T 4JF, UK; e-mail: g.rook@ucl.ac.uk

Fergus Shanahan, Alimentary Pharmabiotic Centre, Department of Medicine, Clinical Science Building, Cork University Hospital, Cork, Ireland; e-mail: f.shanahan@ucc.ie

Alain Tedgui, Paris Cardiovascular Research Center, INSERM and Assistance Publique-Hôpitaux de Paris, Hôpital Européen Georges Pompidou, 75015 Paris, France; e-mail: alain.tedgui@inserm.fr

Dale T. Umetsu, Harvard Medical School, Children's Hospital Boston, One Blackfan Circle, Boston, MA 02115, USA; e-mail: dale.umetsu@childrens.harvard.edu

Joel V. Weinstock, Division of Gastroenterology and Hepatology, Department of Internal Medicine, Tufts Medical Center, Boston, MA, USA; e-mail: jweinstock2@tuftsmedicalcenter.org

David R. Whitlock, Nitroceutic LLC, Dover, MA 02030, USA; e-mail: dwhitlock@nitroceutic.com

Ursula Wiedermann, Department of Specific Prophylaxis and Tropical Medicine, Medical University of Vienna, Vienna, Austria; e-mail: ursula.wiedermann@meduniwien.ac.at

Preface

The hygiene hypothesis hit the headlines in 1989 when David Strachan observed that hay fever is less common in children with older siblings, suggesting that multiple childhood cross-infections from dirty older brothers might protect from allergic disorders. The implication was that elimination of these infections by modern hygiene might be responsible for the increased prevalence of allergies in developed countries. However the notion that mankind's changing lifestyle is contributing to the increased incidences of various chronic inflammatory disorders was not new. In 1873 Blackley had observed that farmers rarely had hay fever, and in 1966 Leibowitz and colleagues noted that, in Israel, the incidence of multiple sclerosis correlated with the level of sanitation. Similarly, inflammatory bowel disease, which was almost non-existent before the 20th century, was already known to be increasing in Northern, urban, educated white-collar populations. Meanwhile animal models studied in the 1970s had shown that different manipulations of the microbial flora of the gut could either increase or decrease susceptibility to autoimmune arthritis.

The findings in Strachan's 1989 paper were initially considered narrowly in relation to allergic disorders, and this led research in two unhelpful directions. First, there was emphasis on the notion that protection was provided by the burden of recognisable symptomatic childhood infections: subsequent epidemiological surveys have consistently failed to support this hypothesis. Secondly it was thought that the role of these infections might be to drive Th1 cells able to inhibit development of the Th2 cells that mediate allergic disorders: this view ignored the simultaneous increases in Th1- and Th17-mediated disorders, and the increasing evidence for the importance of helminths and of orofaecally transmitted organisms. The latter were often subclinical, so had been identified by antibody studies rather than by documenting manifest clinical infections.

Therefore by the late 1990s several authors were suggesting an inclusive hypothesis that could account for the simultaneous rises in allergies, autoimmunity and inflammatory bowel diseases in rich developed countries, while at the same time incorporating all the different epidemiological clues, in addition to the protective effects of older siblings. These clues included orofaecal transmission, living on a farm, economic deprivation, keeping a dog, and having helminth infections.

This reworking of the hygiene hypothesis has three components. First, it suggests that the increases in chronic inflammatory disorders are partly due to an imbalance between effector cells and regulatory cells, rather than an imbalance between Th1 and Th2. Secondly the new synthesis suggests that the most relevant organisms will turn out to be those with very long associations with the mammalian immune system, often as commensals, environmental 'pseudocommensals', subclinical infections or asymptomatic carrier states. (This long association is necessary for the development of evolved dependence and has led us to use the term 'old friends' hypothesis.) Thirdly, it suggests that the most relevant lifestyle changes are those that deprive us of contact with these particular organisms (less contact with animals, soil, and faeces), rather than the details of modern hygiene technology.

This book seeks to summarise the mass of data from epidemiological studies, laboratory experiments, animal models and clinical trials that lead to this view of the hygiene hypothesis. It also seeks to establish the broader significance of the hypothesis, and its boundaries. We first present a Darwinian view of the role of microorganisms in setting up the correct balances within the immune system. Then we consider the history of human interactions with microorganisms in order to analyse how microbial exposures have changed from Paleolithic to modern times. This is followed by chapters on the mechanisms involved in the priming of immunoregulatory circuits by microorganisms, including sections devoted specifically to gut microbiota and skin flora. There are then chapters considering the evidence for *and against* the importance of the hygiene hypothesis as a mechanism contributing to increases in allergies, multiple sclerosis, inflammatory bowel disease, Type 1 diabetes, depression, atherosclerosis, cancer and neurodegenerative disorders. Finally, there is an important chapter describing other aspects of the modern western lifestyle that might provide alternatives to the hygiene hypothesis, or might be interacting with, and exacerbating, the effects of our changing microbial environment.

This book is therefore unusually interdisciplinary, with implications for those working in essentially all areas of medicine. The year 2009 is the 150th anniversary of the publication of *On the Origin of Species* (24 November 1859) and the 200th anniversary of Darwin's birth (12 February 1809). We hope that this book will provoke debate and research on the hygiene hypothesis, which should now be seen as a crucial and expanding aspect of Darwinian Medicine.

April 2009 Graham A.W. Rook

Introduction: The changing microbial environment, Darwinian medicine and the hygiene hypothesis

Graham A.W. Rook

Centre for Infectious Diseases and International Health, Windeyer Institute of Medical Sciences, University College London, UK

Abstract

Man has undergone rapid cultural and technological evolution, with little simultaneous genetic evolution. Thus our physiology is adapted to the microbial exposures that prevailed in the hunter-gatherer environment, rather than to the clean living conditions of the rich industrialised countries. There is increasing evidence that lack of exposure to organisms that were part of mammalian evolutionary history is leading to disordered regulation of the immune system, and hence to increases in several chronic inflammatory disorders. The concept began with the allergic disorders, but there are now good reasons for extending it to autoimmunity, inflammatory bowel disease, neuroinflammatory disorders, atherosclerosis, depression associated with raised inflammatory cytokines, and some cancers. We discuss these possibilities in the context of Darwinian medicine. This approach enables one to identify some of the organisms that are important for the 'hygiene' or 'old friends' hypothesis, and to point to their potential exploitation in novel prophylactics and treatments, with applications in several branches of medicine.

The early history of the hygiene hypothesis

Several categories of chronic inflammatory disorder have become much more prevalent in developed countries [1]. The 'hygiene hypothesis', based originally on epidemiology, but now supported by experimental work *in vivo* and *in vitro*, and by clinical trials, suggests that some of this increased prevalence is due to a malfunction of the immune system caused by the altered patterns of exposure to microorganisms that result from economic development and changing lifestyles.

Although clearly applicable to many types of chronic inflammatory disorder, it is in the field of allergic disorders that these ideas have the longest history. In 1873 Charles Harrison Blackley noted that hay fever was associated with exposure to pollen, but also stated that "*farmers rarely experience the condition*". Indeed hay fever began to be regarded as a mark of wealth, education and sophistication. In the 1880s Morell Mackenzie, a British physician, went so far as to state "*As, therefore, summer sneezing goes hand-in-hand with culture, we may, perhaps infer that*

The Hygiene Hypothesis and Darwinian Medicine, edited by Graham A.W. Rook

the higher we rise in the intellectual scale, the more is the tendency developed." He added the entertaining remark that "*...the tendency to sneeze at the sight of a rose will become a test, surer than the letter h, for the separation of the elect from the common herd.*"

During the 1980s and 1990s both Strachan and Matricardi and colleagues observed that having many siblings, especially older ones, correlated with a diminished risk of hay fever. These findings were considered consistent with a protective influence of postnatal infection, which might be lost in the presence of modern hygiene [2–4]. So the 'hygiene hypothesis' was born. Subsequent studies rediscovered the protective effect of the farming environment, particularly exposure to cowsheds [5, 6]. However this did not provide an immediately obvious microbiological common factor. What, if any, is the link between exposure to older siblings and exposure to cowsheds? You do not pick up the common infections of childhood from cows!

The situation was complicated further by epidemiological or experimental studies showing protection from allergies attributable to exposure to endotoxin [7] helminths [8], lactobacilli [9], mycobacteria [10, 11], gut microbiota [12, 13], and organisms transmitted by the faeco-oral route [14–16]. These observations consolidated the view that microorganisms or their components were a crucial factor, but failed initially to provide a unifying hypothesis. Thus the concept was initially vague and lacked mechanistic explanations, so it is not surprising that in the 20 years since Strachan's original study a multitude of different, sometimes overlapping, often mutually exclusive versions of this hypothesis have been considered. From time to time this has led to the 'disproving' of hypotheses or mechanisms that few had intended to propose in the first place.

Recently we have approached the issue from a different angle; Darwinian Medicine. It was Theodosius Dobzhansky who first made the now famous statement that "*Nothing in biology makes sense except in the light of evolution*" [17]. This type of thinking is rapidly enhancing our understanding of all branches of medicine and is particularly relevant to the hygiene hypothesis [18].

The evolutionary view of the hygiene hypothesis

Environment of evolutionary adaptedness (EEA)

The term EEA was first used in 1969 by John Bowlby, who was concerned that those aspects of human behaviour that are genetically determined (such as instincts) might be adapted to the hunter-gatherer existence rather than to modern city life [19]. The basis for this was the view that since the start of agriculture and pastoralism about 10,000 years ago, human adaptation to new environments has been cultural and technological rather than genetic. Interestingly human genetic diversity appears to be increasing more rapidly than ever before, which might seem discordant with this view [20]. However this increase is due to the population explosion, rather than to adaptation to specific environments. For example, we have not adapted genetically to living in cold places: we have learnt to make fur coats. Humans easily detect problems within the physical environment and invent appropriate technological adaptations. We deal with excessive heat, cold, light, dark, water, drought etc., by technology. However there are two physiological systems where we lack conscious awareness that there is something wrong, so we fail to seek technological solutions. For example, only since Bowlby have we been wondering if the brain is fully adapted to the modern social environment. But the immune system, like the brain, is a learning system that requires the inputs that it has evolved to receive. Only since the hygiene hypothesis appeared have we been wondering if the immune systems of people living in clean modern cities are receiving the appropriate inputs.

The human EEA is the hunter-gatherer environment. Does this allow us to define the microbial inputs that our immune systems have evolved to 'expect'? The hunter-gatherer lifestyle was in fact many different lifestyles, in many different environments, so this is a complex issue [21, 22]. Nevertheless one can identify types of organism that will inevitably have been abundant in all manifestations of the human EEA.

First, there were commensal organisms on the skin and in the gut microbiota. The latter constitute an 'organ' with the metabolic activity and importance at least equal to that of the liver [23], and profound effects on the immune system that are outlined later [24]. Modern lifestyles and antibiotic use are changing the intestinal microbiota (see the chapters by Fergus Shanahan and by David Elliott and Joel Weinstock in this volume). It is also possible that commensal skin organisms had important immunoregulatory roles, and these are discussed by David Whitlock and Martin Feelisch.

Secondly, there were harmless environmental organisms present throughout hominid (and indeed mammalian) evolution. These can be considered as 'pseudo-commensals' (i.e., always present, and consumed every single day, but not actually replicating in the host) associated with rotting vegetation, soil and untreated water. The latter include lactobacilli, and many actinomycetes including saprophytic mycobacteria.

Thirdly there were helminths [21]. Helminth infections were universal throughout hominid evolution. Only recently, and only in developed countries, has infestation with these organisms become rare.

These issues, and the history of man's encounters with microorganisms, are considered in the chapter by George J. Armelagos. Interestingly, these groups of organisms (commensal microbiota, environmental saprophytes and helminths) that man and his ancestors encountered continuously and in large quantities for millions of years, are among those that are changed or depleted from the modern environment and have been shown to be relevant to immunoregulation, as discussed in detail later [25]. Our need for these organisms is an example of 'evolved dependence'.

Evolved dependence

This concept refers to situations where an organism has become adapted to the presence of a partner, and can no longer perform well without that partner [26]. It was originally used to describe endosymbioses. A classical example was seen in the laboratory environment when an amoeba (*Amoeba discoides*) became infected with a bacterium [27]. Initially this infection severely compromised the growth of the amoebae. However, after 5 years the relationship between the two species had changed and neither organism could survive without the other. This indicates genetic changes leading to dependence. For instance, an enzyme that is encoded in the genome of both species might be dropped from the genome of one of them. Access to that gene is now 'entrusted' to the other species. This idea is at first somewhat alien to immunologists, but it is in fact rather commonplace. For instance, most mammals can synthesise vitamin C, but large primates and guinea-pigs have lost the relevant pathways. In effect man and guinea-pig are now in a state of evolved dependence on fruit and vegetables. Of course the same is true for many other genes involved in the synthesis of vitamins and other essential nutrients that we have to consume after other organisms have created them for us.

Evolved dependence of the immune system on commensal organisms

The immune systems of germ-free animals fail to develop correctly, and are functionally distorted. There is lack of cellularity, and lack of effective immunoregulation. In 2005 Sarkis Mazmanian and colleagues showed that a single polysaccharide from an intestinal commensal, *Bacteroides fragilis*, could partly correct these developmental abnormalities [12]. More recently they have shown, using three different models of intestinal inflammation that the same polysaccharide, given by mouth, can turn on crucial immunoregulatory pathways [28]. In the discussion of the latter paper they state:

> *We propose that the mammalian genome does not encode for all functions required for immunological development but rather that mammals depend on critical interactions with their microbiome (the collective genomes of the microbiota) for health.*

To put it even more simply, some genes needed for the development and regulation of the mammalian immune system might have been 'entrusted' to microorganisms; a clear example of 'evolved dependence'. It is obvious that these organisms have to be those with which mammals have co-evolved for a very long time, and that were always present. They cannot be organisms that merely cause sporadic infections. The latter can modify the human genome by elimination of susceptible genotypes, but they cannot play the role of supplying genes and functions that we need. These concepts are expanded later in relation to the 'old friends' hypothesis.

Microorganisms potentially involved in the hygiene hypothesis

Against this evolutionary background we can take a closer look at the organisms that have been implicated as inducers or inhibitors of the chronic inflammatory disorders that are increasing in the developed countries.

Changed exposure to microorganisms that provoke chronic inflammatory disorders

Is it possible that in the rich countries we are seeing consequences of *increased* exposure to organisms that *provoke* chronic inflammatory disorders? Some authors believe that chronic infections have a causal role in many chronic inflammatory disorders, either because they persist as cryptic infections, or because they disturb immune function [29]. Likely mechanisms for the latter include molecular mimicry, where a microbial structure evokes a response that is cross-reactive with a human component. Other organisms might exert adjuvant effects that overcome down-regulation of responses to self, or cause more subtle disturbances of inflammation, immunoregulation or innate immunity [30]. Thus some authors implicate *Chlamydia* in atherosclerosis [31] or mycobacteria in Crohn's disease [32]. The viruses discussed in this context include herpesviruses [33], rotavirus [34] and hepatitis B virus [35], and the list of disorders in which these authors tentatively implicate them is long (systemic lupus erythematosus, aplastic anaemia, antiphospholipid syndrome, polyarteritis nodosa, rheumatoid arthritis, Type 1diabetes, multiple sclerosis, thyroid disease, and uveitis).

Are these points relevant to the hygiene hypothesis? First, the truth of these findings remains uncertain, and it might be that none of them is a true causal association. Nevertheless, we cannot dismiss these possibilities. Changing lifestyles might be resulting in changes in the frequency of exposure to these organisms, or changes in the *timing* of exposure. The latter is an interesting concept. Modern living could cause delayed infection, so that the child encounters the virus after levels of antibody derived from the mother have fallen to an ineffective level, or when the immune system has achieved a different stage of development. This might cause significant changes in pathology. Some of these issues are discussed in the chapters by Mel Greaves and by Margo C. Honeyman and Leonard C. Harrison.

Most recognised childhood infections do not protect from chronic inflammatory disorders

It was the observations on family size and birth order that gave impetus to the hygiene hypothesis [2–4], so childhood infections were initially considered to be likely candidates. Perhaps these infections are protective, but less frequent in the developed countries?

Children are exposed to many types of infection. However when considered in the light of the evolutionary considerations outlined above, it becomes clear that the childhood *virus* infections are extraordinarily unlikely to be important in this context, and this prediction has been substantiated in several studies.

Domesticated and peri-domesticated animals are said to harbor approximately 184 different zoonotic diseases, so from about 10,000 years ago man will have increasingly encountered these, particularly viruses. Diseases that emerged during this period include influenza, measles, mumps, and smallpox [21]. However until recently human populations were too small to sustain these as endemic infections, so it is theoretically improbable that they became physiological necessities.

In short, the history of our interactions with these organisms establishes the fact that they were not part of the human EEA, and in any case they were not appropriate for the setting up of a relationship such as evolved dependence.

Thus as might be expected, the common infections of childhood do not, with notable exceptions considered in a later section, protect from allergic disorders [14, 36]. Children in daycare centres do not have an increased risk of atopy if they wash more often and reduce their infection rate [37].

Although the studies are less numerous, similar conclusions are being drawn from investigation of the relationship between childhood infections and susceptibility to other members of the groups of chronic inflammatory disorders that are increasing in the rich countries. Thus no evidence could be found that the childhood infections exert protective effects against Type 1 diabetes [38], or inflammatory bowel disease [39–41].

Microorganisms that protect from chronic inflammatory disorders

Despite the apparent lack of relevance of the common childhood infections, the protective effects of large family size [2–4] and of exposure to the farming environment [5, 6] that had been noted previously, are rather consistently reproduced in the more recent studies [36, 40]. Other work reveals a role for organisms that are transmitted by the faeco-oral route [14, 15] in addition to helminths [8, 42, 43], Bacillus Calmette et Guérin (BCG) [44, 45], latent tuberculosis (or large tuberculin skin test responses) [46, 47] environmental saprophytes (both environmental mycobacteria [11, 48] and other bacterial groups) [49], and gut commensals [28, 50] and probiotics [51–53]. The evidence comes from epidemiology, animal experiments, *in vitro* studies, and in some cases, clinical trials. Decreased exposure to these organisms, all of which have the long history of association with mammals required by the evolutionary viewpoint, readily provides an explanation for the epidemiology linking modern living with increased chronic inflammatory disorders, and they are considered in greater detail below.

Organisms transmitted by the faeco-oral route

There are several reports that protection from allergies is associated with evidence of infection with organisms that are transmitted by the faeco-oral route [14, 15]. These organisms include *Helicobacter pylori*, *Toxoplasma gondii*, and Hepatitis A virus (HAV) [16]. The receptor for HAV on human lymphocytes is TIM-1, which is involved in the regulation of T cell subsets, including Treg and Th2 [54]. Exposure to HAV might selectively remove Th2, or alter the balance of T cell effector and regulatory subsets [16, 54], particularly in individuals with an insertion polymorphism of TIM-1 [55]. Before 1975 the incidence of infection with HAV approached 100% in the general population, but it has declined rapidly over the past three decades [55]. Thus HAV might explain in part the original observations of Strachan [2], showing protective effects of larger family size, and in particular, older siblings. This issue is discussed in detail by Dale T. Umetsu and Rosemarie H. DeKruyff.

Helminths

Individuals infected by helminths are less likely to have allergic sensitisation, or allergic disorders, and treating the infection tends to increase allergic sensitisation [8, 42]. This negative association has been documented again in a recent study. A total of 1,601 children age 6 to 18 participated. Sensitisation to dust mites was present in 14.4% and to cockroach in 27.6% of children. The risk of sensitisation to dust mites was reduced in those with higher hookworm burden and with Ascaris infection. Nevertheless the complexity of the situation was also revealed by this study. Sensitisation to house dust mite was also increased in those using flush toilets. On the other hand, sensitisation to cockroach was not independently related to helminth infection but was increased in those regularly drinking piped or well water rather than from a stream [43]. It is likely that the stream water also supplied other regulation-inducing organisms such as the saprophytic mycobacteria discussed below.

The protective effect of helminths might also depend on the parasite load. High loads may drive regulatory circuits, while lower loads act as Th2 adjuvants, and enhance allergic sensitisation. For instance in a Costa Rican population with low prevalence of parasitic infection but high prevalence of parasitic *exposure*, sensitisation to *Ascaris lumbricoides* was associated with increased severity and morbidity of asthma [56]. On the other hand, antibody to *Ascaris lumbricoides*, indicating exposure to this or to cross-reactive parasites, could not account for the protective effect of the farming environment in Europe [57], but nor was it associated with increased risk.

Animal models have confirmed the hypothesis that helminths can oppose allergic manifestations by driving immunoregulation. One of the first reports used *Strongyloides stercoralis* [58]. More recently mice were infected with the gastrointestinal nematode *Heligmosomoides polygyrus* to test its influence on experimentally induced airway allergy to ovalbumin and to the house dust mite allergen Der p 1 [59]. Inflammatory cell infiltrates in the lung were suppressed in infected mice compared with uninfected controls. Suppression was reversed in mice treated with antibodies to CD25 and could be transferred with mesenteric lymph node cells. The protective cell populations contained elevated numbers of $CD4^+$ $CD25^+$ $Foxp3^+$ T cells, and expressed TGF-β and IL-10. Interestingly, the regulatory cells could be taken from IL-10-deficient animals so IL-10 produced by the Treg themselves was not required. Other workers, using the same parasite found IL-10 to be important [60]. Similar effects have been achieved using components of helminths, rather than living infections. *S. japonicum* egg antigens were active in a murine model of asthma. This treatment increased the number and suppressive activity of $CD4^+$ $CD25^+$ T cells, which made IL-10. These cells were associated with decreased expression of Th2 cytokines and diminished antigen-induced airway inflammation [61].

Similar results have been obtained in a model of autoimmunity. In the Non-Obese Diabetic (NOD) mouse a spontaneous Th1-mediated autoimmune response destroys the β cells in the pancreas, leading to a diabetic state analogous to human Type 1diabetes. Infection with the gastrointestinal helminths *Trichinella spiralis* or *Heligmosomoides polygyrus* was able to inhibit the development of diabetes [62].

Since the parasites used in the animal models are highly species specific, they are not appropriate for use in man. Two parasites are undergoing trials. *Trichuris suis*, the pig whipworm, is undergoing trials in inflammatory bowel disease, because it fails to complete its life cycle in man, and is considered safe [63, 64]. This issue is discussed in detail by David E. Elliott and Joel V. Weinstock . Other trials are using *Necator americanus* because although this is a human parasite it is considered safe if a well standardised low dose infection (10 larvae) is used. In a pilot study it was well-tolerated [65].

Meanwhile a fascinating experiment of nature has encouraged research in this area. Patients in Argentina suffering from multiple sclerosis (MS) were followed up for 4.6 years. It was found that those who developed parasite infections (which were not treated) had significantly fewer exacerbations than those who did not [66]. Moreover, they also developed specific Treg that responded to myelin basic protein by releasing IL-10 and TGF-β. In other words, the presence of the parasite appeared to drive the development of Treg that recognised the autoantigen, and inhibited the disease process. This is discussed further in the chapter by Jorge Correale.

Mycobacteria; Mycobacterium tuberculosis *(and BCG)*

There has been confusion in the literature about the relationship between exposure to mycobacteria, and protection from allergic disorders. The confusion arises from the erroneous assumption that responses evoked by vaccination with BCG, progressive tuberculosis, latent tuberculosis, or contact with environmental mycobacteria are all functionally equivalent, and that all that matters is the diameter of the tuberculin skin test response. Numerous authors have tried to relate incidence or severity of allergic disorders to tuberculin test reaction size. In view of the multiple immunologically disparate causes of tuberculin positivity and the multiple patterns of immunological reactivity that the tuberculin response represents, this is misleading, as extensively reviewed elsewhere [67]. However, it is clear that individuals with definite latent tuberculosis (i.e., *resistant* to the disease, despite infection) have reduced allergic disease and reduced allergic sensitisation [10, 46, 47, 67, 68], and increased expression of IL-4d2 [69, 70], an inhibitory splice variant of IL-4. In striking contrast, individuals with progressive tuberculosis (i.e., *susceptible* to the disease), particularly if living in a developing country, have increased expression of IL-4 [71] and correspondingly *increased* IgE and allergic sensitisation [67, 72, 73]. Thus although the direction of causality is uncertain, *latent* TB might protect from allergic disorders, but is clearly not a therapeutic option. Progressive TB, by contrast, is associated with *increased* allergic sensitisation.

The effect of BCG vaccination on the subsequent incidence of allergic disorders is variable [44, 45, 74–76]. This is to be anticipated since it is also variable in its effects on susceptibility to tuberculosis [77]. Effects of BCG are unlikely to be detectable in countries where a large proportion of the population has latent TB, since the latter will exert a stronger protective effect [47]. In animal models BCG can inhibit subsequent induction of an allergic response, though this appeared to be largely due to establishment of a Th1 pattern that inhibited subsequent induction of Th2 [78].

Saprophytic mycobacteria, and other abundant environmental organisms; 'pseudocommensals'

The saprophytic mycobacteria, and other *Actinomycetales*, are enormously abundant in untreated water supplies and mud. Throughout mammalian evolution it will have been normal to consume milligrams (10^9 or more) of these organisms every day. For this reason we have coined the term 'pseudocommensals' to designate organisms that were always present at mucosal surfaces throughout evolution, even if not actually replicating in the host.

The environmental mycobacterium that has received most attention in animal models is *M. vaccae*, an environmental saprophyte. It is effective in mouse models of allergy, in therapeutic and well as preventative protocols [11, 48, 79–83]. *M. vaccae* works by inducing Treg that inhibit Th2, rather than by inducing a Th1 bias [11, 48]. It is encouraging that *M. vaccae* was equally effective in the mouse model when administered orally since this is a practical route of administration that avoids the scar at the injection site [84].

More recently *M. vaccae* was tested in a formal clinical trial in dogs with eczema. A single intradermal injection was found to be very effective on all disease parameters for several months, and further studies are now in progress in this species, with a view to eventual commercialisation [85].

M. vaccae is derived, as its name suggests (*vacca* is latin for cow) from the environment of cows. In view of the striking ability of exposure to the cowshed to protect from subsequent allergy it makes sense to see what other relevant microbes might be present in that environment [5, 6]. Two of the most abundant organisms in these cowsheds have been identified and isolated. These were *Acinetobacter lwoffii* F78 (Gram-negative) and *Lactococcus lactis* G121 (Gram-positive; a fermenting organism used in the manufacture of many brands of cheese). Both were found to reduce allergic reactions in mice [49]. Unlike *M. vaccae*, which works by inducing Treg rather than by polarising the response to Th1, these two organisms appeared to bias the response to Th1 [49]. Thus their effects in models of Th1-mediated autoimmunity or IBD cannot be predicted.

This is likely to be the tip of an iceberg. There are many more relatively harmless environmental organisms that are part of the evolutionary heritage of mammals because they are abundant in soil, mud and untreated water, but that are depleted from the environment in rich western countries.

Gut commensals (intestinal microbiota)

The idea that modulating the intestinal microbiota can modulate chronic inflammatory disorders has a long history. In 1985 Kohashi and colleagues observed that the susceptibility of rats to adjuvant arthritis depended on the nature of the gut flora [50]. It was also established 10 years ago that the intestinal microbiota are required for successful induction, by the oral route, of tolerance to ovalbumin [86]. The recent work of Mazmanian and colleagues, discussed earlier, has shown that some immunoregulatory function can be restored to germ-free mice by administering a single polysaccharide from *Bacteroides fragilis* [28]. Interestingly, there is some variation in time of appearance of *Bacteroides fragilis* in the microbiota (delayed by caesarean section, for example), but this does not relate to allergic manifestations [87], and ultimately it is always present, so while this work illustrates the principle, we do not know if this particular organism is important in the context of the current increases in chronic inflammatory disorders.

Equally striking is a recent study showing that the altered gut microbiota that develops in MyD88 knockout mice is inherently able to inhibit development of Type 1diabetes, even in non-genetically-modified animals [13].

Some of the reasons for the importance of the microbiota are beginning to emerge. The expression of IL-17 (often implicated in autoimmune pathology) in the gut depends on the nature of the gut microbiota. Genetically identical inbred mice from different suppliers had different ratios of expression of Th17 cells and Treg. This turned out to be attributable to the presence of different microbiota, particularly the relative abundance of *Bacteroidetes*. The latter seem to drive Th17 cells, and the striking immunological differences between the mice waned when they were housed together, or when microbiota were transferred [88].

A critical factor might be secretion of ATP. This causes DC to secrete IL-23, which maintains Th17 cells. Enteric nervous system function is also driven *via* ATP sensors [89]. Some members of the microbiota may induce IL-25 in intestinal epithelial cells, and this downregulates IL-23, so secondarily reducing maintenance of Th17 cells [90]. TLR9 is another relevant sensor of microbiota. TLR9-/- mice have more Foxp3$^+$ cells, and less Th17 and Th1 cells in intestine-associated lymphoid tissue. This suggests that TLR9 inhibits Treg development in the gut, and CpG or DNA extracted from gut bacteria inhibit Treg development. If microbiota are depleted by antibiotic treatment, development of Th1 and Th17 cells is reduced, but can be restored by administering DNA [91]. Thus the DNA of microbiota acts as an 'adjuvant' for local homoeostatic inflammatory responses by limiting induction of Treg. Thus, administering DNA from gut microbiota to animals that had been rendered more susceptible to *Encephalitozoon cuniculi* by treatment with antibiotics partially restored protection [91].

In view of all these findings it is perhaps not surprising that there is evidence that antibiotic treatment can predispose to allergies [92]. Antibiotics have profound effects on the nature and diversity of the intestinal microbiota [93]. Thus changes in the resident flora are taking place in the developed countries, both as a result of antibiotics and of changing diet and lifestyle [93, 94]. Allergic children have diminished numbers of colonising lactobacilli in their intestinal microbiota [9]. It is likely that modern life and frequent consumption of antibiotics are altering the nature of the human gut microbiota in ways that affect our immune systems.

Probiotics

Further evidence for the potential immunomodulatory role of intestinal microbiota comes from studies using some of these strains, or related non-colonising fermenting *Lactobacillus* strains, as probiotics in models of chronic inflammatory disorders. Thus oral administration of these organisms can inhibit systemic inflammatory states.

Probiotics were protective in a model of dextran sulphate induced colitis [95]. Similarly *Lactobacillus salivarius* protected against the inflammatory bowel disease that occurs in IL-10 KO mice [53]. Interestingly, the probiotic was effective by both the oral and subcutaneous route [53].

Similar results have been achieved in allergy models, using oral treatment with various live *Lactobacilli* [96, 97] or *Bifidobacteria* [52, 98, 99], and in models of autoimmunity such as type II collagen (CII)-induced arthritis (CIA) in DBA/1 mice [100], and autoimmune beta cell destruction in NOD mice (Type 1diabetes) [51].

Not all authors have studied the mechanism of the protective effects of probiotics in these animal models of chronic inflammatory disorders. Nevertheless efficacy in both Th1- and Th2-mediated disorders suggests that Th1/Th2 balance is not involved. Indeed, where the issue has been studied it seems that these probiotic effects are due to induction of regulatory T cells [28, 52, 53]. For instance feeding *Bifidobacterium infantis* to mice results in increased numbers of $CD4^+$ $CD25^+$ $Foxp3^+$ T cell numbers in the mucosa and spleen [101].

Conclusion

Man's interactions with the organisms described above extends back into the Palaeolithic, with the exception of BCG, though other mycobacteria were always abundant. We designate these organisms 'old friends', to emphasise the duration of these interactions, and their evolutionary significance. Importantly, the 'old friends' all satisfy the following criteria (reviewed in [102]):

1. Abundant during mammalian evolution (i.e., going back much further than the ~500 generations since the start of agriculture)
2. Virtually absent, and increasingly so over the last century, from the modern environment
3. Proven to have therapeutic effects in animal models of chronic inflammatory disorders
4. Some proven to have therapeutic effects in human clinical trials

How do microorganisms protect from chronic inflammatory disorders?

Why might the heterogeneous groups of microorganisms discussed above be essential for the correct functioning of the immune system? Most of the material discussed in the previous section points to their ability to induce Treg, and this is expanded below. First however we must consider the adjuvant effects of microbial components, and also the effects on the balance between Th1 and Th2 that was initially thought to explain the epidemiology underlying the increase in allergic disorders.

Adjuvant effects

One important aspect of microbe-dependent immunoregulation is adjuvanticity leading to enhanced responses rather than downregulation. Clearly these two facets of immunoregulation are closely related. The interplay between the two is particularly clear in relation to cancer. Microbial components such as endotoxin (LPS) are powerful adjuvants that can enhance immune responses. Workers in the cotton industry in Shanghai are heavily exposed to LPS, and they are reported to have reduced incidences of cancers of the oesophagus, stomach [103], breast [104] and lung [105]. Similarly dairy workers, who are exposed to a multitude of microbial components including LPS, are also protected from lung cancer [106]. In the 1970s a number of authors claimed that BCG vaccination protected against leukaemia and other haematological tumours, though the validity of the findings remains uncertain [107]. More recently it has been suggested that BCG or smallpox vaccine can protect against melanoma [108]. This issue is not discussed further in this review, which is focussed on situations where the problem may be failure to terminate inflammation, rather than failure to initiate it. Cancer is discussed by Mel Greaves.

Th1/Th2 balance; relevant at the level of the individual

It was thought that lack of infections driving Th1 was leading to overproduction of Th2 cells. This was never a strong theory. First, Th1 cytokines such as IFN-γ are present in large quantities both in asthma [109] and in established atopic dermatitis [110]. Secondly, profound defects in the IL-12 or IFN-γ (Th1) pathways do not lead to an increased incidence or severity of allergic disorders, implying that in man Th1 is not a physiological regulator of Th2 responses [111]. Thirdly, superimposing polarised Th1 cells onto a Th2-mediated inflammatory site can lead to synergistic inflammation rather than to downregulation of immunopathology [112], and other examples of synergy between Th1 and IL-4 responses have been reported [113, 114]. It may be that when IL-4 is associated with immunosuppression it is coming from Treg that simultaneously release other regulatory mediators [115].

In reality the Th1/Th2 balance hypothesis has been untenable since as early as 1998 [116] by which time it had been well-documented that there was a simultaneous increase in Th1-mediated (or perhaps Th17-mediated) chronic inflammatory diseases (Type 1 diabetes, multiple sclerosis, inflammatory bowel disease) [1], occurring in the same countries as the increases in allergic disorders [117].

Interestingly, *at the level of the individual*, developing a Th1-mediated disorder may be associated with a reduced likelihood of developing a Th2-mediated disorder and *vice versa* [118, 119]. Thus at the level of the individual the Th1/Th2 balance might be important. Th1- and Th2-mediated chronic inflammatory diseases are increasing in parallel in developed countries but whether an *individual* living in such a country develops Th1- or Th2-mediated disorder (or no disorder) will depend on his genetic constitution and the Th1/Th2 balance evoked by his previous immunological history. This is represented by the 'double see-saw' (Fig. 1).

Moreover individuals infected by helminths, which enhance Th2 responses, are paradoxically *less* likely to have allergic sensitisation or allergic disorders, and treating the infection leads to increased allergic sensitisation [8]. It has been argued that the helminths merely cause secretion of IL-10 and so suppress other responses. However we now know that the situation is more complex and interesting; the

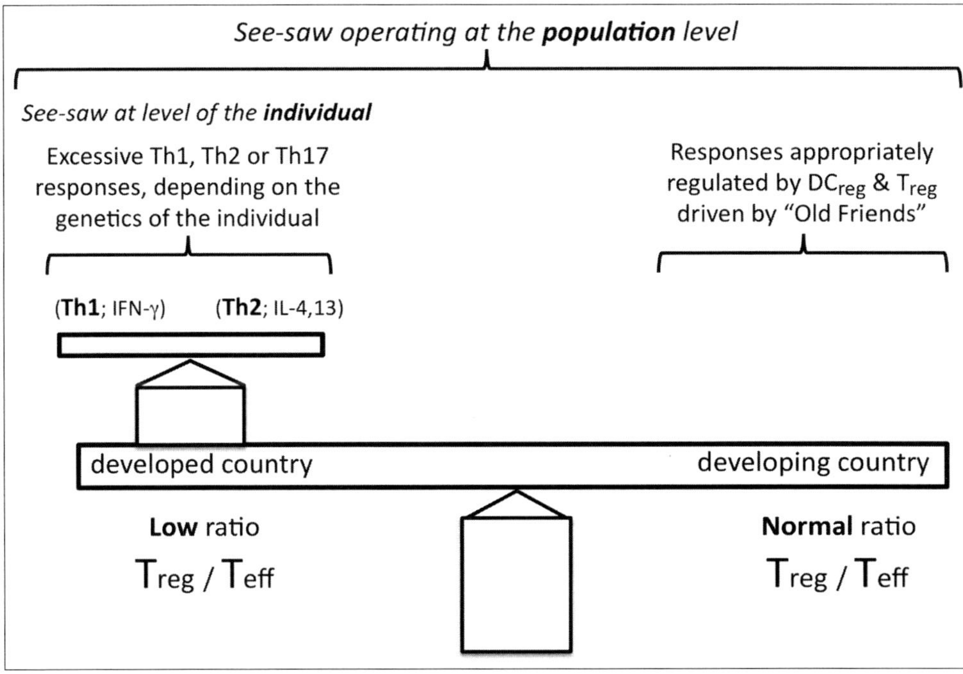

Figure 1
The double see-saw. Our evolved dependence on the 'old friends' operates via many mechanisms. One important one is illustrated here. In populations adequately exposed to 'old friends', priming of regulatory cells is sufficient to suppress inappropriate inflammation (right side of large see-saw). When the priming of regulatory cells is too low, the population is at risk of syndromes attributable to inadequate termination of inappropriate inflammatory responses. Some individuals have a genetic background and immunological history that makes their Th1 mechanisms more likely to become pathologically dysregulated, and these people are at risk of Th1-mediated conditions (Type 1diabetes, multiple sclerosis, Crohn's disease). In other individuals it is the Th2 response that is most liable to lack of control, resulting in allergic disorders. A further category of individuals suffering from failure to terminate inflammation does not develop any gross pathology, but is susceptible to CNS effects of chronic cytokine exposure, such as depression or anxiety (see the chapter by Rook and Lowry). It is relatively common however for patients to have both a chronic inflammatory disorder and a psychiatric disturbance.

helminths can act as 'Treg adjuvants', and so lead to the development of circulating Treg that specifically recognise the autoantigen, rather than the helminth itself. Thus multiple sclerosis patients who develop helminth infections also develop circulating T cells that release IL-10 or TGF-β in response to myelin peptides *in vitro* [66].

Finally, virtually all the 'old friends' discussed in this chapter have been shown to be protective and/or therapeutic in experimental models of both Th2-mediated and Th1-mediated disorders. The mechanism must therefore be distinct from modulation of Th1/Th2 balance, and with few exceptions, the 'old friends' have been shown to induce regulatory T cells.

Induction of immunoregulatory circuits; evolved dependence on the 'old friends'

The component of the hygiene hypothesis that implicates faulty induction of immunoregulation has been designated 'the old friends' hypothesis. We introduced this term in 2003 when it became clear that Th1/Th2 balance cannot explain the facts [102, 120]. The 'old friends' hypothesis' suggests that the gut microbiota, and environmental saprophytes, needed to be tolerated by the immune system because they were harmless and always present in large numbers in food and water (i.e., 'pseudocommensals'). Similarly the helminthic parasites needed to be tolerated because, although not always harmless, once they were established in the host any effort by the immune system to eliminate them was likely to cause tissue damage. For instance, a futile effort to destroy *Brugia malayi* microfilariae results in lymphatic blockage and elephantiasis [121]. So over millions of years a state of Evolved Dependence might have developed, where the induction of appropriate levels of immunoregulation by the 'old friends' has become a physiological necessity. In other words, some genes involved in setting up appropriate levels of immunoregulation are located in microbial rather than mammalian genomes. Reduced exposure to the old friends should therefore lead to a failure to terminate inflammatory episodes, and to a range of chronic inflammatory disorders. We know that a generalised dysfunction of immunoregulatory mechanisms can lead to simultaneous increases in diverse types of pathology, because genetic defects of Foxp3, a transcription factor that plays a crucial role in the development and function of regulatory T cells (Treg), leads to a syndrome known as X-linked autoimmunity-allergic dysregulation syndrome (XLAAD) that includes aspects of allergy, autoimmunity and enteropathy [122]. In support of this concept immunoregulation has been shown to be faulty in individuals suffering from allergic disorders [123] and some autoimmune diseases [124, 125] and probably in IBD too [126, 127]. Further evidence for and against the view that the increases in these and other chronic inflammatory diseases that are occurring in the rich developed countries are indeed due to failing immunoregulation, is discussed in other chapters of this book.

The cellular and molecular mechanisms that enable microorganisms and their components to drive immunoregulation are discussed in detail by Rick M. Maizels and Ursula Wiedermann. The host–parasite relationship evolved so that rather than

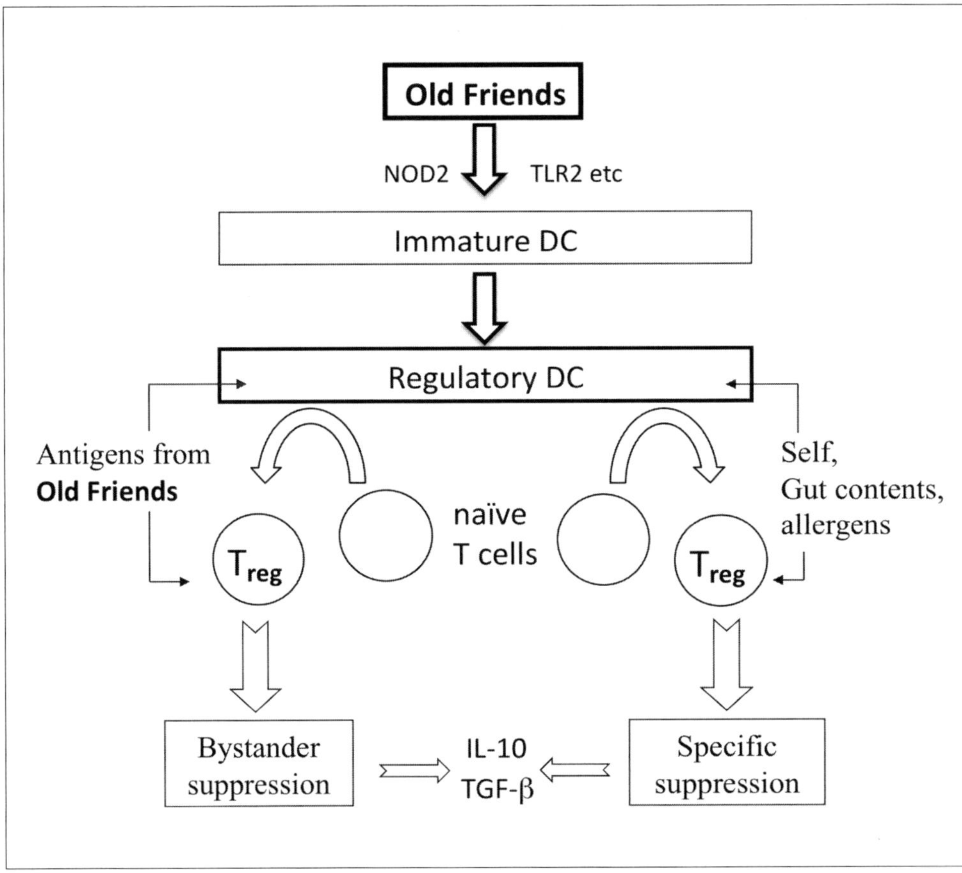

Figure 2

The old friends' hypothesis. Organisms such as helminths and environmental saprophytes, that are part of mammalian evolutionary history ('old friends') and must be tolerated, are detected by pattern recognition receptors such as TLR2 and CARD15 (NOD2) on DC. The DC mature into regulatory DC that drive regulatory T cell (Treg) responses to the antigens of these organisms. The continuing presence of these antigens in the gut flora, in food, or resident as parasites such as microfilariae, leads to continuous background release of regulatory cytokines from these Treg, exerting bystander suppression of other responses, as shown on the left. Meanwhile the increased numbers of regulatory DC lead to increased processing by such DC of self antigens, gut content antigens and allergens, as shown on the right. Therefore the numbers of Treg specifically triggered by these antigens are also increased, specifically downregulating autoimmunity, IBD and allergies respectively.

Clearly, other mechanisms are also involved. For instance microorganisms also provide adjuvants (such as inhaled endotoxin) that might enhance development and polarisation of responses to some allergens and tumour antigens (see main text).

provoking needless, damaging aggressive immune responses, these organisms cause a pattern of maturation of dendritic cells (DC) such that these drive Treg rather than Th1 or Th2 effector cells [128, 129]. This in turn leads to two mechanisms that help to control inappropriate inflammation (Fig. 2). First, the constitutive presence of the 'old friends' causes continuous background activation of the DCreg and of Treg specific for the old friends themselves, resulting in constant background bystander suppression of inflammatory responses, partly attributable to release of IL-10 and TGF-β. Second, these DCreg inevitably sample self, gut contents and allergens, and so induce Treg specific for the illicit target antigens of these three groups of chronic inflammatory disorder [11, 66]. A fascinating example of this mechanism occurring in man is described by Jorge Correale.

The validity of this hypothetical model is further supported by clinical trials and experimental models in which exposure to microorganisms that were ubiquitous during mammalian evolutionary history, but are currently 'missing' from the environment in rich countries (or from animal units with Specific Pathogen-Free facilities) will treat allergy [11, 59, 85], autoimmunity [66, 130] or intestinal inflammation [64].

Conclusions

We conclude therefore that a variety of organisms that are harmless (commensals, environmental saprophytes) or must be tolerated (helminths), and that were part of the environment throughout mammalian evolution (the old friends), evolved roles in the priming of immunoregulation. We have summarised the evidence that diminished exposure to these old friends is contributing to the increases in disorders of immunoregulation that we see in rich developed countries. This introductory chapter provides a broad overview of the evolutionary aspects, and of the range of microorganisms implicated. George J. Armelagos discusses in his chapter the microbiology of human evolution. Rick M. Maizels and Ursula Wiedermann in their chapter go in greater detail into the mechanisms within the immune system that allow microorganisms to modulate immunoregulatory pathways.

The hygiene hypothesis originally focussed on allergic disorders, and then broadened to consider IBD and some autoimmune disorders (Type 1diabetes and MS). If there is indeed a relative failure of immunoregulatory pathways in the developed countries, might this be relevant to other chronic inflammatory disorders that are increasing, in addition to these three groups [131]? Figure 3 illustrates other possible relevant disorders. The following chapters 4 to 14 are written by authors who have been asked to consider inflammatory disorders within their own particular field of expertise in the light of the 'old friends' hypothesis.

Finally, Margo C. Honeyman and Leonard C. Harrison explain some of the other changes in the modern lifestyle that could be contributing to the rise in chronic

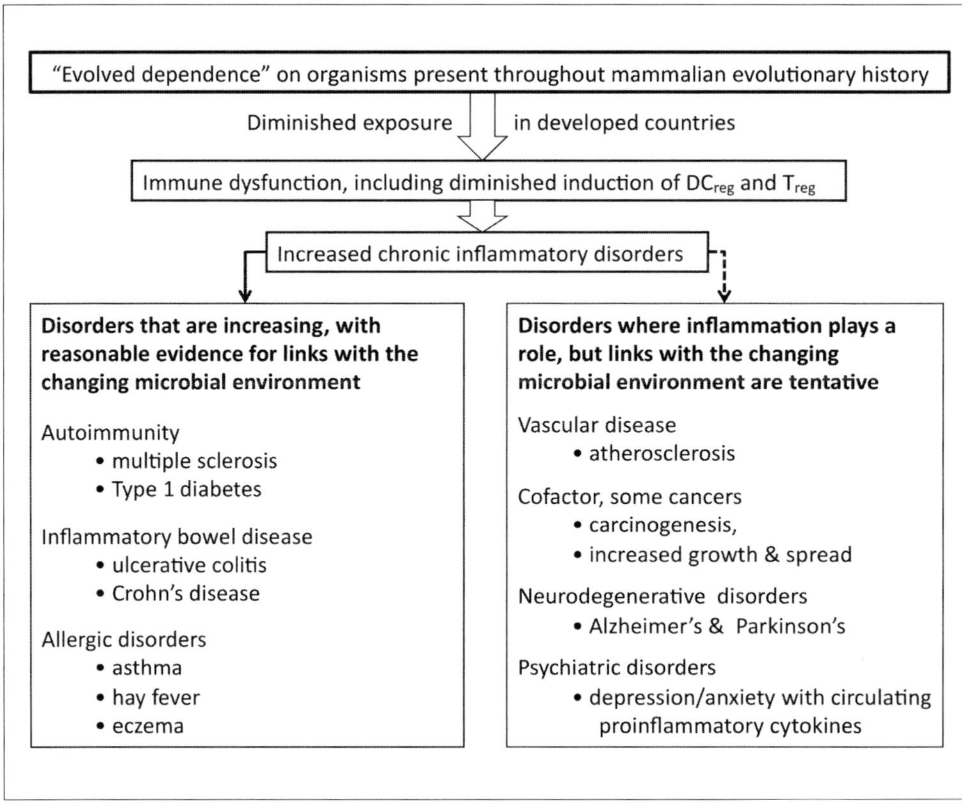

Figure 3
Pathologies that are increasing, and that might be partly attributable to defective immunoregulation. Human evolution and physiology were shaped by the hunter-gatherer way of life, which is regarded as the human 'Environment of Evolutionary Adaptedness'. Despite increasing human genetic diversity, most human adaptation to novel environments in the last few millennia has been cultural and technological rather than genetic, so a gene-environment misfit may be occurring, particularly in the immune system, which is not linked to a conscious sensory modality that can warn us of problems. On the left we list chronic inflammatory disorders where the link to the changing microbial environment is supported by epidemiology, animal models, in vitro *models, and in some cases, direct observation in man. On the right we list conditions where such links seem possible, and are discussed in depth in subsequent chapters in this volume. All of the disorders listed are considered in detail in the appropriate chapters in this volume.*

inflammatory disorders. It is not the purpose of this book to suggest that all of the increase in chronic inflammatory disorders is due to diminished contact with the old friends.

References

1 Bach JF (2002) The effect of infections on susceptibility to autoimmune and allergic diseases. *N Engl J Med* 347: 911–920

2 Strachan DP (1989) Hay fever, hygiene, and household size. *Brit Med J* 299: 1259–1260

3 Strachan DP, Taylor EM, Carpenter RG (1996) Family structure, neonatal infection, and hay fever in adolescence. *Arch Dis Child* 74: 422–426

4 Matricardi PM, Franzinelli F, Franco A, Caprio G, Murru F, Cioffi D, Ferrigno L, Palermo A, Ciccarelli N, Rosmini F (1998) Sibship size, birth order, and atopy in 11,371 Italian young men. *J Allergy Clin Immunol* 101: 439–444

5 von Ehrenstein OS, von Mutius E, Illi S, Baumann L, Bohm O, von Kries R (2000) Reduced risk of hay fever and asthma among children of farmers. *Clin Exp Allergy* 30: 187–193

6 Riedler J, Braun-Fahrlander C, Eder W, Schreuer M, Waser M, Maisch S, Carr D, Schierl R, Nowak D, von Mutius E (2001) Exposure to farming in early life and development of asthma and allergy: a cross-sectional survey. *Lancet* 358: 1129–1133

7 Braun-Fahrlander C, Riedler J, Herz U, Eder W, Waser M, Grize L, Maisch S, Carr D, Gerlach F, Bufe A et al (2002) Environmental exposure to endotoxin and its relation to asthma in school-age children. *N Engl J Med* 347: 869–877

8 Yazdanbakhsh M, Kremsner PG, van Ree R (2002) Allergy, parasites, and the hygiene hypothesis. *Science* 296: 490–494

9 Bjorksten B, Naaber P, Sepp E, Mikelsaar M (1999) The intestinal microflora in allergic Estonian and Swedish 2-year-old children. *Clin Exp Allergy* 29: 342–346

10 Shirakawa T, Enomoto T, Shimazu S, Hopkin JM (1996) The inverse association between tuberculin responses and atopic disorder. *Science* 275: 77–79

11 Zuany-Amorim C, Sawicka E, Manlius C, Le Moine A, Brunet LR, Kemeny DM, Bowen G, Rook G, Walker C (2002) Suppression of airway eosinophilia by killed *Mycobacterium vaccae*-induced allergen-specific regulatory T-cells. *Nat Med* 8: 625–629

12 Mazmanian SK, Liu CH, Tzianabos AO, Kasper DL (2005) An immunomodulatory molecule of symbiotic bacteria directs maturation of the host immune system. *Cell* 122: 107–118

13 Wen L, Ley RE, Volchkov PY, Stranges PB, Avanesyan L, Stonebraker AC, Hu C, Wong FS, Szot GL, Bluestone JA et al (2008) Innate immunity and intestinal microbiota in the development of Type 1 diabetes. *Nature* 455: 1109–1113

14 Matricardi PM, Rosmini F, Riondino S, Fortini M, Ferrigno L, Rapicetta M, Bonini S (2000) Exposure to foodborne and orofecal microbes versus airborne viruses in relation to atopy and allergic asthma; epidemiological study. *Brit Med J* 320: 412–417

15 Matricardi PM, Rosmini F, Panetta V, Ferrigno L, Bonini S (2002) Hay fever and asthma in relation to markers of infection in the United States. *J Allergy Clin Immunol* 110: 381–387

16 McIntire JJ, Umetsu SE, Macaubas C, Hoyte EG, Cinnioglu C, Cavalli-Sforza LL, Barsh

GS, Hallmayer JF, Underhill PA, Risch NJ et al (2003) Immunology: hepatitis A virus link to atopic disease. *Nature* 425: 576
17 Dobzhansky T (1964) Biology, molecular and organismic. *American Zoologist* 4: 443–452
18 (Audiovisual presentations by multiple authors) edited by Nesse R (2007) Evolution and Medicine. *Henry Stewart Talks* http: //www.hstalks.com/evomed/
19 Bowlby J (1971 (first published by Hogarth press in 1969)) *Attachment and Loss, Volume 1: Attachment*. Penguin, Harmondsworth, Middlesex, England
20 Hawks J, Wang ET, Cochran GM, Harpending HC, Moyzis RK (2007) Recent acceleration of human adaptive evolution. *Proc Natl Acad Sci USA* 104: 20753–20758
21 Armelagos GJ, Harper KN (2005) Genomics at the origins of agriculture, Part Two; evolutionary anthropology. *Evol Anthropol* 14: 109–121
22 Hoberg EP (2006) Phylogeny of Taenia: Species definitions and origins of human parasites. *Parasitol Int* 55 (Suppl): S23–30
23 Wikoff WR, Anfora AT, Liu J, Schultz PG, Lesley SA, Peters EC, Siuzdak G (2009) Metabolomics analysis reveals large effects of gut microflora on mammalian blood metabolites. *Proc Natl Acad Sci USA* 106: 3698–3703
24 Chow J, Mazmanian SK (2009) Getting the bugs out of the immune system: do bacterial microbiota "fix" intestinal T cell responses? *Cell Host Microbe* 5: 8–12
25 Rook GA, Adams V, Hunt J, Palmer R, Martinelli R, Brunet LR (2004) Mycobacteria and other environmental organisms as immunomodulators for immunoregulatory disorders. *Springer Semin Immunopathol* 25: 237–255
26 de Mazancourt C, Loreau M, Dieckmann U (2005) Understanding mutualism when there is adaptation to the partner. *J Ecology* 93: 305–314
27 Jeon KW (1972) Development of cellular dependence on infective organisms: micrurgical studies in amoebas. *Science* 176: 1122–1123
28 Mazmanian SK, Round JL, Kasper DL (2008) A microbial symbiosis factor prevents intestinal inflammatory disease. *Nature* 453: 620–625
29 Cochran GM, Ewald PW, Cochran KD (2000) Infectious causation of disease: an evolutionary perspective. *Perspect Biol Med* 43: 406–448
30 Duffy DL (2007) Genetic determinants of diabetes are similarly associated with other immune-mediated diseases. *Curr Opin Allergy Clin Immunol* 7: 468–474
31 Hoymans VY, Bosmans JM, Ieven MM, Vrints CJ (2007) *Chlamydia pneumoniae*-based atherosclerosis: a smoking gun. *Acta Cardiol* 62: 565–571
32 Behr MA, Kapur V (2008) The evidence for *Mycobacterium paratuberculosis* in Crohn's disease. *Curr Opin Gastroenterol* 24: 17–21
33 Posnett DN (2008) Herpesviruses and autoimmunity. *Curr Opin Investig Drugs* 9: 505–514
34 Honeyman M (2005) How robust is the evidence for viruses in the induction of type 1 diabetes? *Curr Opin Immunol* 17: 616–623
35 Maya R, Gershwin ME, Shoenfeld Y (2008) Hepatitis B virus (HBV) and autoimmune disease. *Clin Rev Allergy Immunol* 34: 85–102

36 Benn CS, Melbye M, Wohlfahrt J, Bjorksten B, Aaby P (2004) Cohort study of sibling effect, infectious diseases, and risk of atopic dermatitis during first 18 months of life. *BMJ* 328: 1223
37 Dunder T, Tapiainen T, Pokka T, Uhari M (2007) Infections in child day care centers and later development of asthma, allergic rhinitis, and atopic dermatitis: prospective follow-up survey 12 years after controlled randomized hygiene intervention. *Arch Pediatr Adolesc Med* 161: 972–977
38 Cardwell CR, Carson DJ, Yarnell J, Shields MD, Patterson CC (2008) Atopy, home environment and the risk of childhood-onset type 1 diabetes: a population-based case-control study. *Pediatr Diabetes* 9: 191–196
39 Amre DK, Lambrette P, Law L, Krupoves A, Chotard V, Costea F, Grimard G, Israel D, Mack D, Seidman EG (2006) Investigating the hygiene hypothesis as a risk factor in pediatric onset Crohn's disease: a case-control study. *Am J Gastroenterol* 101: 1005–1011
40 Bernstein CN, Rawsthorne P, Cheang M, Blanchard JF (2006) A population-based case control study of potential risk factors for IBD. *Am J Gastroenterol* 101: 993–1002
41 Koloski NA, Bret L, Radford-Smith G (2008) Hygiene hypothesis in inflammatory bowel disease: a critical review of the literature. *World J Gastroenterol* 14: 165–173
42 Yazdanbakhsh M, Matricardi PM (2004) Parasites and the hygiene hypothesis: regulating the immune system? *Clin Rev Allergy Immunol* 26: 15–24
43 Flohr C, Tuyen LN, Lewis S, Quinnell R, Minh TT, Liem HT, Campbell J, Pritchard D, Hien TT, Farrar J et al (2006) Poor sanitation and helminth infection protect against skin sensitization in Vietnamese children: A cross-sectional study. *J Allergy Clin Immunol* 118: 1305–1311
44 Garcia-Marcos L, Suarez-Varela MM, Canflanca IM, Garrido JB, Quiros AB, Lopez-Silvarrey Varela A, Hernandez GG, Guillen-Grima F, Diaz CG, Gonzalez IH et al (2005) BCG immunization at birth and atopic diseases in a homogeneous population of Spanish schoolchildren. *Int Arch Allergy Immunol* 137: 303–309
45 Linehan MF, Frank TL, Hazell ML, Francis HC, Morris JA, Baxter DN, Niven RM (2007) Is the prevalence of wheeze in children altered by neonatal BCG vaccination? *J Allergy Clin Immunol* 119: 1079–1085
46 Miyake Y, Arakawa M, Tanaka K, Sasaki S, Ohya Y (2008) Tuberculin reactivity and allergic disorders in schoolchildren, Okinawa, Japan. *Clin Exp Allergy* 38: 486–492
47 Obihara CC, Kimpen JL, Gie RP, Lill SW, Hoekstra MO, Marais BJ, Schaaf HS, Lawrence K, Potter PC, Bateman ED et al (2006) *Mycobacterium tuberculosis* infection may protect against allergy in a tuberculosis endemic area. *Clin Exp Allergy* 36: 70–76
48 Zuany-Amorim C, Manlius C, Trifilieff A, Brunet LR, Rook G, Bowen G, Pay G, Walker C (2002) Long-term protective and antigen-specific effect of heat-killed *Mycobacterium vaccae* in a murine model of allergic pulmonary inflammation. *J Immunol* 169: 1492
49 Debarry J, Garn H, Hanuszkiewicz A, Dickgreber N, Blumer N, von Mutius E, Bufe A, Gatermann S, Renz H, Holst O et al (2007) *Acinetobacter lwoffii* and *Lactococcus*

lactis strains isolated from farm cowsheds possess strong allergy-protective properties. *J Allergy Clin Immunol* 119: 1514–1521

50 Kohashi O, Kohashi Y, Takahashi T, Ozawa A, Shigematsu N (1985) Reverse effect of gram-positive bacteria vs. gram-negative bacteria on adjuvant-induced arthritis in germfree rats. *Microbiol Immunol* 29: 487–497

51 Calcinaro F, Dionisi S, Marinaro M, Candeloro P, Bonato V, Marzotti S, Corneli RB, Ferretti E, Gulino A, Grasso F et al (2005) Oral probiotic administration induces interleukin-10 production and prevents spontaneous autoimmune diabetes in the non-obese diabetic mouse. *Diabetologia* 48: 1565–1575

52 Feleszko W, Jaworska J, Rha RD, Steinhausen S, Avagyan A, Jaudszus A, Ahrens B, Groneberg DA, Wahn U, Hamelmann E (2007) Probiotic-induced suppression of allergic sensitization and airway inflammation is associated with an increase of T regulatory-dependent mechanisms in a murine model of asthma. *Clin Exp Allergy* 37: 498–505

53 Sheil B, McCarthy J, O'Mahony L, Bennett MW, Ryan P, Fitzgibbon JJ, Kiely B, Collins JK, Shanahan F (2004) Is the mucosal route of administration essential for probiotic function? Subcutaneous administration is associated with attenuation of murine colitis and arthritis. *Gut* 53: 694–700

54 Degauque N, Mariat C, Kenny J, Zhang D, Gao W, Vu MD, Alexopoulos S, Oukka M, Umetsu DT, Dekruyff RH et al (2007) Immunostimulatory Tim-1-specific antibody deprograms Tregs and prevents transplant tolerance in mice. *J Clin Invest* 84: S12–16

55 Kuchroo VK, Meyers JH, Umetsu DT, DeKruyff RH (2006) TIM family of genes in immunity and tolerance. *Adv Immunol* 91: 227–249

56 Hunninghake GM, Soto-Quiros ME, Avila L, Ly NP, Liang C, Sylvia JS, Klanderman BJ, Silverman EK, Celedon JC (2007) Sensitization to *Ascaris lumbricoides* and severity of childhood asthma in Costa Rica. *J Allergy Clin Immunol* 119: 654–661

57 Karadag B, Ege M, Bradley JE, Braun-Fahrlander C, Riedler J, Nowak D, von Mutius E (2006) The role of parasitic infections in atopic diseases in rural schoolchildren. *Allergy* 61: 996–1001

58 Wang CC, Nolan TJ, Schad GA, Abraham D (2001) Infection of mice with the helminth Strongyloides stercoralis suppresses pulmonary allergic responses to ovalbumin. *Clin Exp Allergy* 31: 495–503

59 Wilson MS, Taylor MD, Balic A, Finney CA, Lamb JR, Maizels RM (2005) Suppression of allergic airway inflammation by helminth-induced regulatory T cells. *J Exp Med* 202: 1199–1212

60 Kitagaki K, Businga TR, Racila D, Elliott DE, Weinstock JV, Kline JN (2006) Intestinal helminths protect in a murine model of asthma. *J Immunol* 177: 1628–1635

61 Yang J, Zhao J, Yang Y, Zhang L, Yang X, Zhu X, Ji M, Sun N, Su C (2007) Schistosoma japonicum egg antigens stimulate CD4 CD25 T cells and modulate airway inflammation in a murine model of asthma. *Immunology* 120: 8–18

62 Saunders KA, Raine T, Cooke A, Lawrence CE (2007) Inhibition of autoimmune type 1 diabetes by gastrointestinal helminth infection. *Infect Immun* 75: 397–407

63 Summers RW, Elliott DE, Urban JF, Jr., Thompson RA, Weinstock JV (2005) *Trichuris*

suis therapy for active ulcerative colitis: a randomized controlled trial. *Gastroenterology* 128: 825–832

64 Summers RW, Elliott DE, Urban JF, Jr., Thompson R, Weinstock JV (2005) *Trichuris suis* therapy in Crohn's disease. *Gut* 54: 87–90

65 Mortimer K, Brown A, Feary J, Jagger C, Lewis S, Antoniak M, Pritchard D, Britton J (2006) Dose-ranging study for trials of therapeutic infection with *Necator americanus* in humans. *Am J Trop Med Hyg* 75: 914–920

66 Correale J, Farez M (2007) Association between parasite infection and immune responses in multiple sclerosis. *Ann Neurol* 61: 97–108

67 Rook GA, Hamelmann E, Rosa Brunet L (2007) Mycobacteria and allergies. *Immunobiology* 212: 461–473

68 Obihara CC, Beyers N, Gie RP, Potter PC, Marais BJ, Lombard CJ, Enarson DA, Kimpen JL (2005) Inverse association between *Mycobacterium tuberculosis* infection and atopic rhinitis in children. *Allergy* 60: 1121–1125

69 Demissie A, Abebe M, Aseffa A, Rook G, Fletcher H, Zumla A, Weldingh K, Brock I, Andersen P, Doherty TM (2004) Healthy individuals that control a latent infection with *Mycobacterium tuberculosis* express high levels of Th1 cytokines and the IL-4 antagonist IL-4delta2. *J Immunol* 172: 6938

70 Fletcher HA, Owiafe P, Jeffries D, Hill P, Rook GA, Zumla A, Doherty TM, Brookes RH (2004) Increased expression of mRNA encoding interleukin (IL)-4 and its splice variant IL-4delta2 in cells from contacts of *Mycobacterium tuberculosis*, in the absence of *in vitro* stimulation. *Immunology* 112: 669

71 Rook GA (2007) Th2 cytokines in susceptibility to tuberculosis. *Curr Mol Med* 7: 327–337

72 Ellertsen LK, Wiker HG, Egeberg NT, Hetland G (2005) Allergic sensitisation in tuberculosis and leprosy patients. *Int Arch Allergy Immunol* 138: 217–224

73 Suzuki N, Kudo K, Sano Y, Ito K (2001) Can *Mycobacterium tuberculosis* infection prevent asthma and other allergic disorders? *Int Arch Allergy Immunol* 124: 113–116

74 Alm JS, Lilja G, Pershagen G, Scheynius A (1998) BCG vaccination does not seem to prevent atopy in children with atopic heredity. *Allergy* 53: 537

75 Aaby P, Shaheen SO, Heyes CB, Goudiaby A, Hall AJ, Shiell AW, Jensen H, Marchant A (2000) Early BCG vaccination and reduction in atopy in Guinea-Bissau. *Clin Exp Allergy* 30: 644–650

76 Gruber C, Meinlschmidt G, Bergmann R, Wahn U, Stark K (2002) Is early BCG vaccination associated with less atopic disease? An epidemiological study in German preschool children with different ethnic backgrounds. *Pediatr Allergy Immunol* 13: 177–181

77 Fine PE (1995) Variation in protection by BCG: implications of and for heterologous immunity. *Lancet* 346: 1339–1345

78 Herz U, Gerhold K, Gruber C, Braun A, Wahn U, Renz H, Paul K (1998) BCG infection suppresses allergic sensitization and development of increased airway reactivity in an animal model. *J Allergy Clin Immunol* 102: 867–874

79 Wang CC, Rook GAW (1998) Inhibition of an established allergic response to ovalbumin in Balb/c mice by killed *Mycobacterium vaccae. Immunology* 93: 307–313

80 Hopfenspirger MT, Parr SK, Hopp RJ, Townley RG, Agrawal DK (2001) Mycobacterial antigens attenuate late phase response, airway hyperresponsiveness, and bronchoalveolar lavage eosinophilia in a mouse model of bronchial asthma. *Int Immunopharmacol* 1: 1743–1751

81 Ozdemir C, Akkoc T, Bahceciler NN, Kucukercan D, Barlan IB, Basaran MM (2003) Impact of *Mycobacterium vaccae* immunization on lung histopathology in a murine model of chronic asthma. *Clin Exp Allergy* 33: 266–270

82 Smit JJ, Van Loveren H, Hoekstra MO, Schijf MA, Folkerts G, Nijkamp FP (2003) *Mycobacterium vaccae* administration during allergen sensitization or challenge suppresses asthmatic features. *Clin Exp Allergy* 33: 1083–1089

83 Arkwright PD, Fujisawa C, Tanaka A, Matsuda H (2005) *Mycobacterium vaccae* reduces scratching behavior but not the rash in NC mice with eczema: a randomized, blinded, placebo-controlled trial. *J Invest Dermatol* 124: 140–143

84 Hunt JR, Martinelli R, Adams VC, Rook GAW, Rosa Brunet L (2005) Intragastric administration of *Mycobacterium vaccae* inhibits severe pulmonary allergic inflammation in a mouse model. *Clin Exp Allergy* 35: 685–690

85 Ricklin-Gutzwiller ME, Reist M, Peel JE, Seewald W, Brunet LR, Roosje PJ (2007) Intradermal injection of heat-killed *Mycobacterium vaccae* in dogs with atopic dermatitis: a multicentre pilot study. *Vet Dermatol* 18: 87–93

86 Sudo N, Sawamura S, Tanaka K, Aiba Y, Kubo C, Koga Y (1997) The requirement of intestinal bacterial flora for the development of an IgE production system fully susceptible to oral tolerance induction. *J Immunol* 159: 1739–1754

87 Adlerberth I, Strachan DP, Matricardi PM, Ahrne S, Orfei L, Aberg N, Perkin MR, Tripodi S, Hesselmar B, Saalman R et al (2007) Gut microbiota and development of atopic eczema in 3 European birth cohorts. *J Allergy Clin Immunol* 120: 343–350

88 Ivanov, II, Frutos Rde L, Manel N, Yoshinaga K, Rifkin DB, Sartor RB, Finlay BB, Littman DR (2008) Specific microbiota direct the differentiation of IL-17-producing T-helper cells in the mucosa of the small intestine. *Cell Host Microbe* 4: 337–349

89 Atarashi K, Nishimura J, Shima T, Umesaki Y, Yamamoto M, Onoue M, Yagita H, Ishii N, Evans R, Honda K et al (2008) ATP drives lamina propria T(H)17 cell differentiation. *Nature* 455: 808–812

90 Zaph C, Du Y, Saenz SA, Nair MG, Perrigoue JG, Taylor BC, Troy AE, Kobuley DE, Kastelein RA, Cua DJ et al (2008) Commensal-dependent expression of IL-25 regulates the IL-23–IL-17 axis in the intestine. *J Exp Med* 205: 2191–2198

91 Hall JA, Bouladoux N, Sun CM, Wohlfert EA, Blank RB, Zhu Q, Grigg ME, Berzofsky JA, Belkaid Y (2008) Commensal DNA limits regulatory T cell conversion and is a natural adjuvant of intestinal immune responses. *Immunity* 29: 637–649

92 Farooqi IS, Hopkin JM (1998) Early childhood infection and atopic disorder. *Thorax* 53: 927–932

93 Dethlefsen L, Huse S, Sogin ML, Relman DA (2008) The pervasive effects of an antibi-

otic on the human gut microbiota, as revealed by Deep 16S rRNA Sequencingsequencing. *PLoS Biol* 6: e280

94 Guarner F, Bourdet-Sicard R, Brandtzaeg P, Gill HS, McGuirk P, van Eden W, Versalovic J, Weinstock JV, Rook GA (2006) Mechanisms of disease: the hygiene hypothesis revisited. *Nat Clin Pract Gastroenterol Hepatol* 3: 275–284

95 Rachmilewitz D, Katakura K, Karmeli F, Hayashi T, Reinus C, Rudensky B, Akira S, Takeda K, Lee J, Takabayashi K et al (2004) Toll-like receptor 9 signaling mediates the anti-inflammatory effects of probiotics in murine experimental colitis. *Gastroenterology* 126: 520–528

96 Forsythe P, Inman MD, Bienenstock J (2007) Oral treatment with live Lactobacillus reuteri inhibits the allergic airway response in mice. *Am J Respir Crit Care Med* 175: 561–569

97 Ogawa T, Hashikawa S, Asai Y, Sakamoto H, Yasuda K, Makimura Y (2006) A new synbiotic, *Lactobacillus casei* subsp. *casei* together with dextran, reduces murine and human allergic reaction. *FEMS Immunol Med Microbiol* 46: 400–409

98 Kim H, Kwack K, Kim DY, Ji GE (2005) Oral probiotic bacterial administration suppressed allergic responses in an ovalbumin-induced allergy mouse model. *FEMS Immunol Med Microbiol* 45: 259–267

99 Kim H, Lee SY, Ji GE (2005) Timing of bifidobacterium administration influences the development of allergy to ovalbumin in mice. *Biotechnol Lett* 27: 1361–1367

100 Kato I, Endo-Tanaka K, Yokokura T (1998) Suppressive effects of the oral administration of *Lactobacillus casei* on type II collagen-induced arthritis in DBA/1 mice. *Life Sci* 63: 635–644

101 O'Mahony C, Scully P, O'Mahony D, Murphy S, O'Brien F, Lyons A, Sherlock G, MacSharry J, Kiely B, Shanahan F et al (2008) Commensal-induced regulatory T cells mediate protection against pathogen-stimulated NF-kappaB activation. *PLoS Pathog* 4: e1000112

102 Rook GA (2007) The hygiene hypothesis and the increasing prevalence of chronic inflammatory disorders. *Trans R Soc Trop Med Hyg* 101: 1072–1074

103 Wernli KJ, Fitzgibbons ED, Ray RM, Gao DL, Li W, Seixas NS, Camp JE, Astrakianakis G, Feng Z, Thomas DB et al (2006) Occupational risk factors for esophageal and stomach cancers among female textile workers in Shanghai, China. *Am J Epidemiol* 163: 717–725

104 Ray RM, Gao DL, Li W, Wernli KJ, Astrakianakis G, Seixas NS, Camp JE, Fitzgibbons ED, Feng Z, Thomas DB et al (2007) Occupational exposures and breast cancer among women textile workers in Shanghai. *Epidemiology* 18: 383–392

105 Astrakianakis G, Seixas NS, Ray R, Camp JE, Gao DL, Feng Z, Li W, Wernli KJ, Fitzgibbons ED, Thomas DB et al (2007) Lung cancer risk among female textile workers exposed to endotoxin. *J Natl Cancer Inst* 99: 357–364

106 Mastrangelo G, Grange JM, Fadda E, Fedeli U, Buja A, Lange JH (2005) Lung cancer risk: effect of dairy farming and the consequence of removing that occupational exposure. *Am J Epidemiol* 161: 1037–1046

107 Hoover RN (1976) Bacillus Calmette-Guerin vaccination and cancer prevention: a critical review of the human experience. *Cancer Res* 36: 652–654

108 Krone B, Kolmel KF, Henz BM, Grange JM (2005) Protection against melanoma by vaccination with Bacille Calmette-Guerin (BCG) and/or vaccinia: an epidemiology-based hypothesis on the nature of a melanoma risk factor and its immunological control. *Eur J Cancer* 41: 104–117

109 Krug N, Madden J, Redington AE, Lackie P, Djukanovic R, Schauer U, Holgate ST, Frew AJ, Howarth PH (1996) T-cell cytokine profile evaluated at the single cell level in BAL and blood in allergic asthma. *Am J Respir Cell Mol Biol* 14: 319–326

110 Klunker S, Trautmann A, Akdis M, Verhagen J, Schmid-Grendelmeier P, Blaser K, Akdis CA (2003) A second step of chemotaxis after transendothelial migration: keratinocytes undergoing apoptosis release IFN-gamma-inducible protein 10, monokine induced by IFN-gamma, and IFN-gamma-inducible alpha-chemoattractant for T cell chemotaxis toward epidermis in atopic dermatitis. *J Immunol* 171: 1078–1084

111 Lammas DA, Casanova JL, Kumararatne DS (2000) Clinical consequences of defects in the IL-12-dependent interferon-gamma (IFN-gamma) pathway. *Clin Exp Immunol* 121: 417–425

112 Hansen G, Berry G, DeKruyff RH, Umetsu DT (1999) Allergen-specific Th1 cells fail to counterbalance Th2 cell-induced airway hyperreactivity but cause severe airway inflammation. *J Clin Invest* 103: 175–183

113 Lawrence CE, Paterson JC, Higgins LM, MacDonald TT, Kennedy MW, Garside P (1998) IL-4–regulated enteropathy in an intestinal nematode infection. *Eur J Immunol* 28: 2672

114 Van Kampen C, Gauldie J, Collins SM (2005) Proinflammatory properties of IL-4 in the intestinal microenvironment. *Am J Physiol Gastrointest Liver Physiol* 288: G111–117

115 Tiemessen MM, Jagger AL, Evans HG, van Herwijnen MJ, John S, Taams LS (2007) $CD4^+CD25^+Foxp3^+$ regulatory T cells induce alternative activation of human monocytes/macrophages. *Proc Natl Acad Sci USA* 104: 19446–19451

116 Rook GAW, Stanford JL (1998) Give us this day our daily germs. *Immunol Today* 19: 113–116

117 Stene LC, Nafstad P (2001) Relation between occurrence of type 1 diabetes and asthma. *Lancet* 357: 607

118 Tremlett HL, Evans J, Wiles CM, Luscombe DK (2002) Asthma and multiple sclerosis: an inverse association in a case-control general practice population. *Qjm* 95: 753–756

119 Tirosh A, Mandel D, Mimouni FB, Zimlichman E, Shochat T, Kochba I (2006) Autoimmune diseases in asthma. *Ann Intern Med* 144: 877–883

120 Rook GA, Martinelli R, Brunet LR (2003) Innate immune responses to mycobacteria and the downregulation of atopic responses. *Curr Opin Allergy Clin Immunol* 3: 337–342

121 Babu S, Blauvelt CP, Kumaraswami V, Nutman TB (2006) Regulatory networks induced by live parasites impair both Th1 and Th2 pathways in patent lymphatic filariasis: implications for parasite persistence. *J Immunol* 176: 3248–3256

122 Wildin RS, Smyk-Pearson S, Filipovich AH (2002) Clinical and molecular features of the immunodysregulation, polyendocrinopathy, enteropathy, X linked (IPEX) syndrome. *J Med Genet* 39: 537–545
123 Akdis M, Verhagen J, Taylor A, Karamloo F, Karagiannidis C, Crameri R, Thunberg S, Deniz G, Valenta R, Fiebig H et al (2004) Immune responses in healthy and allergic individuals are characterized by a fine balance between allergen-specific T Regulatory 1 and T Helper 2 Cells. *J Exp Med* 199: 1567–1575
124 Viglietta V, Baecher-Allan C, Weiner HL, Hafler DA (2004) Loss of functional suppression by $CD4^+CD25^+$ regulatory T cells in patients with multiple sclerosis. *J Exp Med* 199: 971–979
125 Kriegel MA, Lohmann T, Gabler C, Blank N, Kalden JR, Lorenz HM (2004) Defective suppressor function of human $CD4^+$ $CD25^+$ regulatory T cells in autoimmune polyglandular syndrome type II. *J Exp Med* 199: 1285–1291
126 Kraus TA, Toy L, Chan L, Childs J, Mayer L (2004) Failure to induce oral tolerance to a soluble protein in patients with inflammatory bowel disease. *Gastroenterology* 126: 1771–1778
127 Powrie F, Read S, Mottet C, Uhlig H, Maloy K (2003) Control of immune pathology by regulatory T cells. *Novartis Found Symp* 252: 92–98
128 van der Kleij D, Latz E, Brouwers JF, Kruize YC, Schmitz M, Kurt-Jones EA, Espevik T, de Jong EC, Kapsenberg ML, Golenbock DT et al (2002) A novel host-parasite lipid cross-talk. Schistosomal lyso-phosphatidylserine activates Toll-like receptor 2 and affects immune polarization. *J Biol Chem* 277: 48122–48129
129 Smits HH, Engering A, van der Kleij D, de Jong EC, Schipper K, van Capel TM, Zaat BA, Yazdanbakhsh M, Wierenga EA, van Kooyk Y et al (2005) Selective probiotic bacteria induce IL-10-producing regulatory T cells *in vitro* by modulating dendritic cell function through dendritic cell-specific intercellular adhesion molecule 3–grabbing nonintegrin. *J Allergy Clin Immunol* 115: 1260–1267
130 Zaccone P, Fehervari Z, Jones FM, Sidobre S, Kronenberg M, Dunne DW, Cooke A (2003) *Schistosoma mansoni* antigens modulate the activity of the innate immune response and prevent onset of type 1 diabetes. *Eur J Immunol* 33: 1439–1449
131 Rook GAW (2009) The broader implications of the hygiene hypothesis. *Immunology* 126: 3–11

The paleolithic disease-scape, the hygiene hypothesis, and the second epidemiological transition

George J. Armelagos

Goodrich C. White Professor of Anthropology, Department of Anthropology, Emory University, Atlanta, Georgia 30320, USA

Abstract

The hygiene hypothesis [1, 2] argues that in developed nations the lack of childhood exposure to infectious pathogens, parasites, and symbiotic microorganisms increases susceptibility to allergy and other chronic diseases in adulthood. A modified hygiene hypothesis, ('the old friends hypothesis' proposed by G. A. Rook [3]) excludes childhood diseases as a requisite factor and focuses on organisms such as lactobacilli, a variety of saprophytic mycobacteria and helminthic parasites that are tolerated by the immune system and are absent from the pathogen load of developed nations. The exposure to these ubiquitous agents is postulated to help in the development of the T regulatory response that when absent in industrialized nations results later in the manifestation of allergies and an array of autoimmune diseases such as inflammatory bowel disease, multiple sclerosis, and Type 1 diabetes [3]. To evaluate this hypothesis, I will examine the pattern of human diseases using a model of epidemiological transition modified from A. R. Omran's [4–7] original formulation. The epidemiological transition provides a means of understanding the changing relationship between humans, pathogens and other disease insults from the Paleolithic period to the present [8]. The adaptation of hominid populations in the Paleolithic created a disease ecology that minimized the impact of infectious disease but exposed the foragers to many saprophytic mycobacteria in the soil and decaying plant matter. The shift to agriculture about 10,000 years ago heralded the first epidemiological transition characterized by the continued exposure to helminths and saprophytic environmental organisms, plus the emergence of additional infectious and nutritional diseases that continue to the present. The acceleration of urbanization and social inequality increases the spread of infectious disease. Within the last century, some populations have undergone the second epidemiological transition in which public health measures, improved nutrition and medicine resulted in declines in infectious disease and a rise in non-infectious, chronic and degenerative diseases. This phase with the control of infectious disease and the development of a sanitized water supply and sewer system has played a role in the modified hygiene or 'old friends' hypothesis. It is a period in which 'cleanliness' [9] removes us from contact with 'dirt' [10, 11].

Introduction

My objective is to provide an evaluation of the role that the Paleolithic and Neolithic disease patterns played in evaluating the hygiene hypothesis [12–17]. There are four

The Hygiene Hypothesis and Darwinian Medicine, edited by Graham A.W. Rook

issues that I will consider in unraveling the disease pattern in the Paleolithic; the first is to broaden the concept of epidemiological transition into a model that defines a number of dramatic shifts in disease patterns [18–23]. Secondly, to describe the disease profiles from the perspective of the three epidemiological transitions [8]. While a distinct pattern of disease emerged as our sparsely populated Paleolithic ancestors moved into new ecological niches [24], their mobility, small population size and low density precluded contagious infectious disease from being a factor in the evolution of these populations. I will pay special attention to the omnipresent pathogens included in the disease-scape of these populations. Third, to show that while emerging diseases have been a characteristic of human adaptation, following the shift to primary food production, there has been an acceleration of the trend of emerging disease through the process of urbanization and industrialization. Finally, to demonstrate in the second epidemiological transition that began about 150 years ago there was a decline of infectious disease and a rise in chronic and degenerative disease in some populations. The second epidemiological transition is a key to the hygiene and 'old friends' hypothesis.

Paleolithic baseline

There are misconceptions about the adaptation of our ancestors who lived in Paleolithic period. Thomas Hobbes [25] describes our ancestors living in "*...continual fear*" with "*...a danger of violent death and a life that was ... solitary, poor, nasty, brutish, and short*" (*Leviathan*, i. xiii.9). Paleolithic populations appear to have been relatively healthy and well nourished.

The reconstruction of Paleolithic disease ecology requires the use of methods from a number of fields of inquiry. Paleopathology of archeological populations that have shifted from a subsistence pattern of foraging to the primary food production of agriculture resulted in a novel pattern of disease that can help us understand the Paleolithic pattern of disease as a baseline [26]. Archeological analysis of foraging populations provides direct evidence of their patterns of morbidity and mortality. The genomic diversity of pathogens and parasites provides clues to the phylogenetic relationships and patterns of adaptation to their hosts [27, 28]. Armelagos and Harper [27, 28] analyzed the genomic patterns in the domesticates, the pathogen and the humans to understand factors in the origin of agriculture. Applying a molecular clock to the differentiation of species allows scientists to determine when the pathogen began to parasitize their host. Molecular analysis by Eric Hoberg [29, 30] of the modern taenid tapeworms that parasitize humans and were assumed to have been transmitted to humans during the Neolithic are now thought to have originated as human parasites in the Paleolithic. The molecular clock suggests that human parasite sister species (*T. saginata* and *T. asiatica*) differentiated about 160,000 ago at the time that humans migrated out of Africa. The data suggests that hominids and carnivores preyed on the similar animals. This supposes that the humans transmitted the tape worms to the domesticated cattle and swine. Hoberg suggest that the *T. solium* parasitized hominids in the Paleolithic when they were consuming dog or engaged in cannibalism.

The disease ecology of contemporary gatherer-hunters provides a model for the diseases that would have afflicted Paleolithic foragers. We can speculate that these foragers would have been exposed to the many saprophytic mycobacteria that were present in the soil and decaying organic matter. Stig Bengmark comments,

> *"The most dramatic difference between Paleolithic and modern Western food is that the diet of our forefathers contained at least a billion times more non-pathogenic health-promoting bacteria, mainly of the* Lactobacillus *species. Our ancestors' diet was rich with these bacteria because they consumed unprocessed natural food and they stored most of their food in the soil, where it became rich in fiber-fermenting lactobacilli"* [31: p.612].

He notes that the human body consists of 10 times more bacterial cells (10^{14}) than eukaryotic cells (10^{13}) and that in developing countries humans' commensal flora weighs 2 kg, while in developed countries the commensal flora weighs less than 1.3 kg [31].

On firmer ground, Sprent [32, 33] describes two classes of parasites that would have afflicted gatherer-hunters. 'Heirloom species' are a class of parasites that have had a long-standing relationship with our anthropoid and hominoid ancestors and that continued to infect them as they evolved to hominids. Head and body lice (*Pediculus humanus*), pinworms (*Enterobius vermicularis*), and possibly yaws, malaria are examples of heirloom species. Lice have been ectoparasites since the Oligocene afflicting a variety of species since then [34]. Heirloom species that had longstanding relationship with our ape and hominid ancestors include most of the internal protozoa found in modern humans and also such bacteria as *Salmonella typhi*, and staphylococci [35, 36]. In contrast, 'souvenir' species, those that are 'picked up' along the way as hominids carry out their daily activity. Souvenir species are usually zoonotic pathogens whose primary hosts are non-human animals and only incidentally infect humans through contact.

Zoonoses are passed on to humans through insect bites, preparing and consuming contaminated flesh, from animal bites and contact with urine and feces of infected animals. Sleeping sickness (trypanosomiasis), tetanus caused by toxin produced by *Clostridium tetani*, scrub typhus (*Orientia tsutsugamushi*), relapsing fever caused the spirochete *Borrelia*, trichinosis from the roundworm *Trichinella spiralis*, tularemia (*Francisella tularensis*), avian or ichthyic tuberculosis, leptospirosis (from the spirochete *Leptospira* spp, and schistosomiasis are among the zoonotic diseases that likely afflicted earlier gatherer-hunters [37]. Small population size would have precluded infections of many bacteria and viruses. Forager's synanthropic relationships with the vectors serve to maintain such human host-specific diseases as yellow fever and louse-borne relapsing fever [34]. *Aedes aegypti* is the source of yellow fever and *Pediculus humanus* that carry *Rickettsia prowazekii* the source for typhus. *Anopheles*, the vector that transmits malaria has been in existence since the Miocene era adapting to the canopy environment. The antiquity and ecology would suggest that this vector would have been present in the Paleolithic. However, the threat of malaria in early hominids was unlikely because of the small population size of the foragers and their adaptation to the savanna, an environment that would not have included mosquitoes that carry the malaria plasmodium [38]. Recent reconstruction of the early hominid environment suggests that they inhabited grassy woodland that could have included the *Anopheles* mosquitoes. If malaria was contracted, it would have been as isolated incidents. Recent analysis of the genetic structure of variants of glucose-6-phosphate dehydrogenase confirms that malaria is a recent selective force in human populations [39]. The independent 'A' and 'Med' mutations in glucose-6-phosphate dehydrogenase suggest that this polymorphism originated at least 10,000 years ago.

The range of the earliest hominids was probably restricted to the tropical grassy woodland savannah, limiting the variety of pathogens that could be potential disease agents. As Dicke [40] and Lambrecht [41–44] note, hominids would have found extensive areas of Africa uninhabitable because of tsetse flies and the trypanosomes that they carry. However, Lambrecht argues that as human species moved into new ecological niches the pattern of trypanosome infection would have evolved. As populations moved out Africa 180,000 years ago, there was expansion of the trypanosomes into temperate and tundra habitats that would have changed their disease ecology.

The diseases that are missing from the pantheon of Paleolithic pathogens are very informative. The contagious diseases such as influenza, measles, mumps, and smallpox would not have been present. There would have been few viruses infecting early hominids [45]. Cockburn [35] has argued that some viral diseases found in non-human primates could have been easily transmitted to early hominids. Regardless, these intereactions would be unlikely to have affected the regulation of the immune system.

The first epidemiological transition: disease in agricultural populations

The earliest evidence of primary food production in the Old World dates from about 10,000 years ago. Agriculture began in seven independent areas of cultivation in Mesopotamia (based on barley and wheat), Sub-Saharan Africa (based on millets and plantains), Southeast Asia (based on rice), northern China (based on millet) and southern China (based on rice). Centers in the New World originate later based on the domestication of maize (Mesoamerica) and potatoes (South America). Within 7,000 years, there were major settlements in the Tigris-Euphrates area, and a thousand years later centralized governments rose in response to the need to control vast irrigation systems. These developments created social classes with differential access to resources. Social inequality increases with urbanization. In the Valley of Mexico, there were well-established settlements by 3500 BP and by 1 CE these settlements show extensive social hierarchies.

Ecological changes increase the potential for disease load following the shift to primary food production. Sedentary villages increase parasitic disease infection by increasing contact with human waste that accumulates as settlement size increases. While sedentarism could and did occur prior to the Neolithic period in those areas with abundant resources such as salmon and other marine products, the shift to agriculture necessitated sedentary living. In sedentary popula-tions, the living area where human waste is deposited is often near the source of water that can be easily contaminated. Gatherer-hunters frequently move their home base and as they forage away from the base camp their contact with their human waste would decrease.

The world is rife with potential pathogens. A total of 1,407 species are recognized as human pathogens and 58% of these are of zoonotic origin [46]. Woolhouse and Gowtage-Sequeria [46] claim that 177 species are considered emerging and reemerging pathogens. Animal domestication and husbandry increase the frequency of contact with animals and thus with a steady supply of disease vectors. Zoonotic infections contracted from domesticated animals, such as goats, sheep, cattle, pigs, and fowl, as well as the peri-domestic animals such as rodents and sparrows that develop permanent habitats in and near human dwellings. Domesticated animals products such as milk, hair, and skin, as well as the dust raised by the animals, could transmit anthrax, Q fever, brucellosis, and tuberculosis. Cultivation of the soil that requires breaking up the sod exposes people to chigger bites that carry the bacteria *Orientia tsutsugamushi* that causes scrub typhus [47]. Livingstone (1958) argues that slash-and-burn agriculture in West Africa exposed populations to *Anopheles gambiae*, a mosquito that is the vector for *Plasmodium falciparum*, which causes malaria. These agricultural practices created settled populations with an open environment with sun lit water that is required for *Anopheles* mosquito propagation.

The combination of disruptive environmental farming practices and the presence of domestic animals also increases human contact with arthropod vectors carrying yellow fever, trypanosomiasis, and filariasis, which developed a preference for human blood meals. Some vectors thrived in human habitats; the best example of which is *Aedes aegypti* (the vector for yellow fever and dengue), an artificial container breeder. Various agricultural practices such as irrigation and the use of human feces as fertilizer increased contact with non-vector parasites (schistosomal cercariae) [37].

The shift to agriculture changed the disease ecology created an environment for diseases not part of the foraging populations' disease-scape. The shift to agriculture resulted in a reduction of the dietary niche that predisposed them to dietary deficiencies. Food was stored in large quantities and widely distributed, probably resulting in outbreaks of food poisoning [48] and exposed them to bacteria that caused fermentation. Cohen and Armelagos [26] provide a number of case studies that show a decline in health following the Neolithic transformation. The increasing class inequalities, epidemic disease, and dietary insufficiencies added mental stress to the list of illnesses that plagues agriculturalist.

Urban development and disease

Urbanization is a recent development in human evolutionary history. In the Near East, large cities were established 6,700 years ago. In the New World, large urban settlements were in existence by 1,400 years ago. Urban centers at Memphis (Egypt) reached a population of 30,000 by 3100 BCE. The Babylonian center at Ur reached 65,000 inhabitants by 2030 BCE and reached 200,000 by 612 BCE [49]. Settlements of this size increased the already difficult problem of removing human wastes and delivering uncontaminated water. Cholera, which is transmitted in contaminated water, was a potential problem. Diseases such as typhus (carried by lice) and the plague bacillus (transmitted by fleas or by the respiratory route), could be spread from person to person. Viral diseases such as measles, mumps, chicken pox, and smallpox could be spread in a similar fashion. There were for the first time, during the period of urbanization, populations large enough to maintain disease in an endemic form. Cockburn [36] estimates that populations of one million would be necessary to maintain measles as an endemic disease. Others [50] suggest that a population of only 200,000 would be required to maintain the disease. Black and colleagues [50] argue that a population of only a 1,000 people needed to sustain chickenpox as an endemic disease. What was endemic disease in one population could be the source of a serious epidemic disease in another group. Cross-continental trade and travel resulted in intense epidemics [51, 52]. The Black Death took its toll in Europe in the 1300s in which epidemic eliminated 25 million people that represented a quarter of the European population (Laird, 1989).

The period of urban development can also be characterized by the exploration and expansion of populations into new areas, which resulted in the introduction of novel diseases to groups that had little resistance to them [51]. McNeill [53] describes the process in which urban populations 'digest' people they encounter in the course of exploration, as disease vectors cleared a path that allowed easy access for expansion. European population that carried smallpox and measles decimated Native Americans following contact [54–56].

The exchange of disease can be a two-way street with imbalanced travel between the worlds. Given the myriad of pathogens brought to the New World, only one was transmitted in return. The exploration of the New World has been the source of the treponemal infection that was transmitted to the Old World [57]. The treponemal infection in the New World was endemic and not sexually transmitted. When introduced into the Old World there was a different mode of disease transmission [58]. The sexual transmission of the treponeme created a different environment for the pathogen, and it resulted in a more severe and acute infection [59]. Furthermore, crowding in the urban centers created changes in sexual practices, such as prostitution, and an increase in sexual promiscuity may have been a factor in the new venereal transmission of the pathogen [60]. Assertions that pre-Columbian syphilis existed in Europe have been made in response to the claims of New World origin of the disease [61].

The process of industrialization, which began a little over 200 years ago, accelerated environmental degradation and social transformation. By 1800, London was the only world city with a million inhabitants. Inhabitants of cities were forced to contend with polluted water and air and industrial waste. Slums that rose in industrial cities would become the focal point for poverty and the spread of disease. Epidemics of smallpox, typhus, typhoid, diphtheria, measles, and yellow fever in urban settings are well documented [62]. Tuberculosis and respiratory diseases such as pneumonia and bronchitis are associated with harsh working situations and crowded living conditions. Urban population centers were population 'sinks'. The high rate of urban mortality made it difficult for cities to maintain their population base by reproductive capacity of those living in the city. Mortality outstripped fertility, requiring in-migration of rural populations to the city in order to maintain their numbers.

Recently much attention has been focused on the detrimental effects of industrialization on the international environment, including water, land, and atmosphere. Massive industrial production of commodities has caused pollution. Increasingly there is concern over the health implications of contaminated water supplies, overuse of pesticides in commercialized agriculture, atmospheric chemicals, and the future effects of depleted ozone on human health and food production. At no other time in human history have the changes in the environment been more rapid and so extreme. These environmental changes have been implicated in the increasing incidence of cancer among young people and the increase in respiratory disease.

In 2000, the United Nations Population Fund (2001 Briefing Paper) reported that nearly half (47% or 2.9 billion) of the world population is living in an urban setting increasing to 60% by 2030. The issue of health in urban settings has been addressed by WHO with their healthy cities initiative [63–65]. This program attempts to systematically address poverty, the vulnerability of segments of the populations, and the lack of access that these populations have to health care.

The second epidemiological transition: the rise of chronic and degenerative disease

Originally, 'epidemiological transition' refers to the shift from acute infectious diseases to chronic non-infectious, degenerative diseases [4, 5]. The increasing prevalence of chronic and degenerative diseases is related in part to an increase in longevity that resulted from the control of infectious diseases. These health advances result in a greater percentage of individuals reaching the oldest age segment of the population. In addition, the technological advances that characterize the second epidemiological transition for some populations also resulted in an increase in environmental degradation for others. An interesting characteristic of many of the chronic diseases is that they are particularly prevalent and 'epidemic'-like in transitional societies, or in those populations undergoing the shift from developing to developed modes of production. In developing countries, many of the chronic diseases associated with the epidemiological transition appear first in members of the upper socioeconomic strata [66], because of their access to Western products, public health, nutritional and medical practices. It is not uncommon to see segmented society where some are seeing the benefits of the second epidemiological transitions while others remain exposed to the ravages of infectious diseases. In India, for example, there are segments of the population experiencing the second and the third epidemiological transitions.

The source of the source of the second epidemiological transition remains controversial. Some have argued that it was an outgrowth of the developments in technology, medicine, and science in which the germ theory of disease causation was developed. The role that medicine played in the decline of some of the infectious diseases is one source of contention [67]. While there was a better understanding of the source of infectious disease and this admittedly resulted in increasing control over many contagious diseases it foreshadows other key elements in the decline. For example, the development of immunization resulted in the control of many infections and recently was the primary factor in the eradication of smallpox. In the developed nations, a number of other communicable diseases have diminished in importance. The decrease in infectious diseases and the subsequent reduction in infant mortality have resulted in greater life expectancy at birth. The increase in longevity for adults has resulted in an increase in chronic and degenerative diseases. Critics of McKeown have focused on his use of evidence for improved nutrition [68–70] and failure to consider improvements in public health practices [68–72].

Recently, Bremner and co-workers [73–75] have shown that it is not exposure to childhood infections that protects against later potential to contracting allergies. It is the exposure to saprophytic environmental bacteria and helminthic parasites in childhood that protects against the autoimmune disease and other chronic inflammatory disorders in adulthood [3, 15, 76–80]. While the exposure to helminthic parasites would be diminished in the second epidemiological transition, the diminished contact with saprophytic bacteria would be even more recent and did not occur until the introduction of advances in the delivery of a sanitary water supply and an improved sewage system.

The third epidemiological transition

Human populations are in the midst of the third epidemiological transition. There is a reemergence of infectious diseases that have multiple antibiotic resistances. Furthermore, the emergence of diseases has a potential for global impact. In a sense, the contemporary transition does not eliminate the possible co-existence of infectious diseases typical of the first epidemiological transition (some 10,000 years ago) in our own time; the World Health Organization reports that of the 50,000,000 deaths each year, 17,500,000 are the result of infectious and parasitic disease. WHO states that two billion people in the world are infected with hepatitis B virus [81]. Two billion of the world's population have tuberculosis (8 million cases contracted every year and 3 million die in that period). In the last 30 years, 40 million people have become infected with HIV and 3 million people have died during that period. To this list we can add the array of autoimmune diseases that arise with advances in sanitation and the lack of exposure to commensal bacteria that would challenge the immunological system [82, 83].

Humans have lived in urban centers for only 3.4% of human history. This may explain the paucity of evidence for the genetic response to specific disease. Catharina Svanborg-Eden and Bruce Levin [84, 42, 43] challenge the proposition that infectious disease is a major force in the selection and evolution of genetic variability in human populations. They argue that there are three constraints for infectious disease to act as an effec-tive agent of natural selection. First, most variation in the frequency of infectious disease is the result of environmental factors. Second, the array of host defenses is general in their actions and overlapping in their functions. Third, the specific immune defenses are adaptive at the somatic level and therefore there is less of a need for selection leading to germ-line evolution. Four decades ago, Lederberg (1963) suggested that diseases that had animal reservoirs could lead to the development of disease resistance in human populations. He argues that the persistence of 'small differentials' could lead to genetic immunity.

The reemergence of infectious diseases has been one of the most interesting evolutionary stories of the last decade and has captured the interest of scientists and the public. Satcher [85] and Lederberg [86] list almost 29 diseases that have emerged in the last 28 years. Humans encounter zoonotic diseases because of anthropogenic factors [87]. Lebarbenchon argues,

> *"Humanity continuously encounters new pathogens of animal origin because anthropogenic changes often increase the pathogen's opportunities to enter the human populations and to generate subsequent human-to-human transmission"* [87: p. 480].

The past decade has even seen the emergence and resurgence of many infectious diseases around the world with the majority being zoonoses [88]. HIV, Ebola, and new hanta viruses in humans are believed to have arisen through that route.

The Institute of Medicine (IOM) [89] reports that the emergence of disease is the result of an interaction of social, demographic, and environmental changes in a global ecology and in the adaptation and genetics of the microbe. Similarly, Morse [90] sees emerging disease as a result of demographic changes, international commerce and travel, technological change, breakdown of public health measures and microbial adaptation.

Among the ecological changes Morse describes are agricultural development projects, dams, deforestation, floods, droughts and climatic changes that resulted in the emergence of diseases such as Argentine hemorrhagic fever, Korean hemorrhagic fever (Hantaan) and Hantavirus pulmonary syndrome. Human demographic behavior has been a factor in the spread of dengue, the source for the introduction and spread of HIV and other sexually transmitted disease.

Conclusion

The hygiene hypothesis has argued that the disease burden in the Paleolithic exposed the populations to a vast array of pathogens that are missing from developed societies. I have shown that the Paleolithic would have been free of many of the infectious diseases that existed in the post Neolithic era, which experienced the first epidemiological transition. The last 10,000 years has seen the acceleration of infectious diseases and the rise of social inequality. The modification of the hygiene hypothesis framed by Rook [91] discounts the exposure to the childhood infections that arose relatively recently in the post-Neolithic era, and focuses on helminths, and saprophytic organisms such as mycobacteria and lactobacilli present in mud, untreated water and fermenting vegetable matter. These organisms were part of the human environment as far back as the Paleolithic, and diminished exposure to them might be an essential element in the contemporary increases in chronic inflammatory conditions. In the last 200 years, a number of developed nations have undergone a second epidemiological transition in which there was the control of infectious disease, the rise of chronic disease and a decline in exposure to these organisms that accompanied man from his earliest beginnings. This dramatic shift in the disease-scape is the basis of the hygiene hypothesis.

References

1 Strachan DP (1989) Hay fever, hygiene, and household size. *British Medical Journal* 299: 1259–1260

2 Strachan DP (2000) Family size, infection and atopy: the first decade of the "hygiene hypothesis". *Thorax* 55 (Suppl 1): S2–10

3 Rook GA, Brunet LR (2005) Old friends for breakfast. *Clinical & Experimental Allergy* 35: 841–842
4 Omran AR (1971) The epidemiologic transition: A theory of the epidemiology of population change. *Millbank Memorial Fund Quarterly* 49: 509–538
5 Omran AR (1977) A century of epidemiologic transition in the United States. *Preventive Medicine* 6: 30–51
6 Omran AR (1982) Epidemiological Transition. In: JA Ross (ed): *International Encyclopedia of Population.* The Free Press, London, 172–183
7 Omran AR (1983) The epidemiologic transition theory: A preliminary update. *Journal of Tropical Pediatrics* 29: 305–316
8 Barrett R, Kuzawa CW, McDade T, Armelagos GJ (1998) Emerging infectious disease and the third epidemiological transition. *Annual Review of Anthropology* 27: 247–271
9 Smith VS (2007) *Clean: a history of personal hygiene and purity.* Oxford University Press, Oxford ; New York
10 Ashenburg K (2007) *The dirt on clean: an unsanitized history.* North Point Press, New York
11 Logan WB (1995) *Dirt: the ecstatic skin of the earth.* Riverhead Books, New York
12 Obihara CC, Bardin PG (2008) Hygiene hypothesis, allergy and BCG: a dirty mix? *Clinical & Experimental Allergy* 38: 388–392
13 Koloski NA, Bret L, Radford-Smith G (2008) Hygiene hypothesis in inflammatory bowel disease: a critical review of the literature. *World Journal of Gastroenterology* 14: 165–173
14 Zavos C, Vini D, Kountouras J, Zavos N, Trivara E (2007) Hygiene hypothesis and protection against asthma in infants: spending time in the countryside encountering natural allergens may boost maternal immunity.[comment]. *Medical Hypotheses* 68: 914–915
15 Rook GA (2007) The hygiene hypothesis and the increasing prevalence of chronic inflammatory disorders. *Transactions of the Royal Society of Tropical Medicine & Hygiene* 101: 1072–1074
16 Schaub B, Lauener R, von Mutius E (2006) The many faces of the hygiene hypothesis. *Journal of Allergy & Clinical Immunology* 117: 969–977; quiz 978
17 von Mutius E (2007) Allergies, infections and the hygiene hypothesis – the epidemiological evidence. *Immunobiology* 212: 433–439
18 Armelagos GJ, Barnes KC, Lin J (1996) Disease in human evolution: The re-emergence of infectious disease in the third epidemiological transition. *AnthroNotes* 18: 1–7
19 Armelagos GJ (1998) The Viral Superhighway. *The Sciences* 38: 24–30
20 Armelagos GJ, Barnes K (1999) The evolution of human disease and the rise of allergy: Epidemiological transitions. *Medical Anthropology* 18: 187–213
21 Barnes KC, Armelagos GJ, Morreale SC (1999) Darwinian medicine and the emergence of allergy. In: W Trevethan, J McKenna, EO Smith (eds): *Evolutionary Medicine.* Oxford University Press, New York
22 Armelagos GJ, Harper. KN (2007) Disease globalization in the third epidemiological

transition. In: G Guest (ed): *Globalization, Health and the Environment: An Integrated Perspective*. AltaMira Press, Walnut Creek, CA, 27–33

23 Armelagos GJ (2004) Emerging disease in the third epidemiological transition. In: CGN Mascie-Taylor, J Peters, ST McGarvey (eds): *The Changing Face of Disease: Implications for Society*. CRC Press, Boca Raton, 7–22

24 Desowitz RS (1980) Epidemiological-ecological interactions in savanna environments. In: DR Harris (ed): *Human Ecology in Savanna Environments*. Academic Press, London, 457–477

25 Hobbes T, Crooke A, Nodin J, Oliver Wendell Holmes Library Collection (Library of Congress) (1651) *Leviathan, or, The matter, forme, & power of a common-wealth ecclesiasticall and civill*. Printed for Andrew Ckooke [i.e., Crooke], at the Green Dragon in St. Pauls Church-yard, London

26 Cohen MN, Armelagos GJ (eds) (1984) *Paleopathology at the Origins of Agriculture*. Academic Press, New York

27 Armelagos GJ, Harper KN (2005) Genomics at the origins of agriculture, part one. *Evolutionary Anthropology: Issues, News, and Reviews* 14: 68–77

28 Armelagos GJ, Harper KN (2005) Genomics at the origins of agriculture, part two. *Evolutionary Anthropology: Issues, News, and Reviews* 14: 109–121

29 Hoberg EP, Jones A, Rausch RL, Eom KS, Gardner SL (2000) A phylogenetic hypothesis for species of the genus Taenia (Eucestoda: Taeniidae) *Journal of Parasitology* 86: 89–98

30 Hoberg EP, Alkire NL, de Queiroz A, Jones A (2001) Out of Africa: origins of the Taenia tapeworms in humans. *Proceedings of the Royal Society of London – Series B: Biological Sciences* 268: 781–787

31 Bengmark S (2000) Bacteria for optimal health. *Nutrition* 16: 611–615

32 Sprent JFA (1969) Helminth "zoonoses": an analysis. *Helminthol Abstr* 38: 333–351

33 Sprent JFA (1969) Evolutionary aspects of immunity of zooparasitic infections. In: GJ Jackson (ed): *Immunity to Parasitic Animals*. Appleton, New York, 3–64

34 Laird M (1989) Vector-borne disease introduced into new areas due to human movements: A historical perspective. In: MW Service (ed): *Demography and Vector-Borne Diseases*. CRC Press, Inc., Boca Raton, Florida,., 17–33

35 Cockburn TA (1967) Infections of the order primates. In: TA Cockburn (ed): *Infectious Diseases: Their Evolution and Eradication*. C.C. Thomas, Springfield, Illinois

36 Cockburn TA (1967) The evolution of human infectious diseases. In: TA Cockburn (ed): *Infectious Diseases: Their Evolution and Eradication*. Charles C. Thomas, Springfield, IllinoisL, 84–107

37 Cockburn TA (1971) Infectious disease in ancient populations. *Current Anthropology* 12: 45–62

38 Livingstone FB (1958) Anthropological implication of sickle-cell gene distribution in West Africa. *American Anthropologist* 60: 533–552

39 Tishkoff SA, Varkonyi R, Cahinhinan N, Abbes S, Argyropoulos G, Destro-Bisol G, Drousiotou A, Dangerfield B, Lefranc G, Loiselet J et al (2001) Haplotype diversity

and linkage disequilibrium at human G6PD: recent origin of alleles that confer malarial resistance. *Science* 293: 455–462

40 Dicke BH (1932) The Tseste fly's influence on South African history. *South African Journal of Science* 29: 792

41 Lambrecht FL (1985) Trypanosomes and hominid evolution. *Bioscience* 35: 640–646

42 Lambrecht FL (1980) Paleoecology of tsetse flies and sleeping sickness in Africa. *Proceedings of the American Philosophical Society* 124: 367–385

43 Lambrecht FL (1967) Trypanosomiasis in prehistoric and later human populations: a tentative reconstruction. In: D Brothwell, AT Sandison (eds): *Diseases in Antiquity.* Thomas, Springfield, Illinois, 132–151

44 Lambrecht FL (1964) Aspects of evolution and ecology of Tsetse tsetse flies and Trypanosomaiasis in prehistoric African environments. *Journal of African History* 5: 1–24

45 Burnet FM (1962) *Natural History of Infectious Disease.* Cambridge University Press, Cambridge

46 Woolhouse MEJ, Gowtage-Sequeria S (2005) Host range and emerging and reemerging pathogens. *Emerging Infectious Disease* 11: 1842–1847

47 Audy JR (1961) The ecology of scrub typhus. In: JM May (ed): *Studies in Disease Ecology.* Hafner Publishing, New York, 389–432

48 Käferstein F (2005) Food Safety: A Pressing Public-Health and Economic Issue In: SWA Gunn, PB Mansourian, AM Davies, A Piel, BM Sayers (eds): *Understanding the Global Dimensions of Health.* Springer, New York, 199–212

49 Chandler T (1987) *Four Thousand Years of Urban Growth: An Historical Census.* Edward Mellen Press, Lewiston, New. York

50 Black FL, Hierholzer WJ, Pinheiro AS, Evans JP, Woodall EM, Opton JE, Emmons BS, West GE, Downs WG, Wallace GD (1974) Evidence for persistence of infectious agents in isolated human populations. *American Journal of Epidemiology* 100: 230–250

51 McNeill WH (1976) *Plagues and People.* Anchor/Doubleday, Garden City

52 Zinsser H (1935) *Rats, Lice and History.* Little, Brown and Company, Boston

53 McNeill WH (1978) Disease in History. *Social Science and Medicine* 12: 79–81

54 Ramenofsky A (1993) Diseases in the Americas, 1492–1700. In: K Kiple (ed): *The Cambridge World History of Human Disease.* Cambridge University Press, New York

55 Ramenofsky AF (1987) *Vectors of death: the archaeology of European contact.* University of New Mexico Press in association with the Center for Documentary Studies at Duke University, Albuquerque, NM

56 Dobyns H (1983) *Their Numbers Become Thinned: Native American Population Dynamics in Eastern United States.* University of TennesseeTennessee Press, Knoxville

57 Baker B, Armelagos GJ (1988) Origin and antiquity of syphilis: A dilemma in paleopathological diagnosis and interpretation. *Current Anthropology* 29: 703–737

58 Harper KN, Ocampo PS, Steiner BM, George RW, Silverman MS, Bolotin S, Pillay A, Saunders NJ, Armelagos GJ (2008) On the origin of the treponematoses: a phylogenetic approach. *PLoS Neglected Tropical Diseases* 2: e148

59 Harper K, Liu H, Ocampo P, Steiner B, Martin A, Levert K, Wang D, Sutton M, Armela-

gos G (2008) The sequence of the acidic repeat protein (arp) gene differentiates venereal from non-venereal T. pallidum subspecies, and the gene has evolved under strong positive selection in the subspecies that causes syphilis. *FEMS Immunology & Medical Microbiology* 53: 322–332
60 Hudson EH (1965) Treponematosis and man's social evolution. *American Anthropologist* 67: 885–901
61 Dutour O, Pálfi G, Bérato J, Brun J-P (1994) *L'origine de la syphilis en Europe – avant ou aprés 1493?* Errance, Paris
62 Polgar S (1964) Evolution and the ills of mankind. In: S Tax (ed): *Horizons of Anthropology*. Aldine Pub Comp, Chicago, 200–211
63 Goldstein G (2000) Healthy cities: overview of a WHO international program. *Reviews on Environmental Health* 15: 207–214
64 Kenzer M (2000) Healthy Cities: a guide to the literature. *Public Health Reports* 115: 279–289
65 Tsouros AD (2000) Why urban health cannot be ignored: the way forward. *Reviews on Environmental Health* 15: 267–271
66 Burkitt DP (1973) Some disease characteristics of modern western medicine. *British Medical Journal* 1: 274–278
67 McKeown T (1979) *The Role of Medicine: Dream, Mirage or Nemisis*. Princeton University Press, Princeton
68 Schofield R, Reher D (1991) The decline of mortality in Europe. In: R Schofield, D Reher, A Bideau (eds): *The Decline of Mortality in Europe*. Clarendon Press, Oxford, 1–17
69 Johansson SR (1991) The health transition: the cultural inflation of morbidity during the decline of mortality. *Health Transition Review* 1: 39–68
70 Johansson SR (1992) Measuring the cultural inflation of morbidity during the decline in mortality. *Health Transition Review* 2: 78–89
71 Kunitz SJ (1991) The personal physician and the decline of mortality. In: DR R. Schlofield, and A. Bideau (ed): *The Decline of Mortality in Europe*. Clarendon Press, Oxford, 248–262
72 Woods R (1990) The role of public health in the nineteenth-century mortality decline. In: J Caldwell, S Findley, P Caldwell, G Santow, W Cosford, J Braid, D Broers Freeman (eds): *What We Know About Health Transition: The Cultural, Social, and Behavioral Determinants of Health*. Health Transition Centre, Canberra, 110–115
73 Bremner SA, Carey IM, DeWilde S, Richards N, Maier WC, Hilton SR, Strachan DP, Cook DG (2008) Infections presenting for clinical care in early life and later risk of hay fever in two UK birth cohorts. *Allergy* 63: 274– 283
74 Benn CS, Melbye M, Wohlfahrt J, Bjorksten B, Aaby P (2004) Cohort study of sibling effect, infectious diseases, and risk of atopic dermatitis during first 18 months of life. *British Medical Journal* 328: 1223
75 Dunder T, Tapiainen T, Pokka T, Uhari M (2007) Infections in child day care centers and later development of asthma, allergic rhinitis, and atopic dermatitis: prospective follow-

up survey 12 years after controlled randomized hygiene intervention. *Arch Pediatr Adolesc Med* 161: 972–927

76 Rook GA, Adams V, Hunt J, Palmer R, Martinelli R, Brunet LR (2004) Mycobacteria and other environmental organisms as immunomodulators for immunoregulatory disorders. *Springer Seminars in Immunopathology* 25: 237–255

77 Rook GA, Brunet LR (2005) Microbes, immunoregulation, and the gut. *Gut* 54: 317–320

78 Rook GA, Hamelmann E, Brunet LR (2007) Mycobacteria and allergies. *Immunobiology* 212: 461–473

79 Yazdanbakhsh M, Kremsner PG, van Ree R (2002) Allergy, parasites, and the hygiene hypothesis. *Science* 296: 490–494

80 Yazdanbakhsh M, Matricardi PM (2004) Parasites and the hygiene hypothesis: regulating the immune system? *Clinical Reviews in Allergy & Immunology* 26: 15–24

81 WHO (1995) Executive Summary. The World Health Report: Bridging the Gaps. World Health Organization, Geneva

82 Bengmark S (1999) Immunonutrition – concluding remarks [editorial]. *Nutrition* 15: 57–61

83 Bengmark S (1999) Gut microenvironment and immune function. *Current Opinion in Clinical Nutrition & Metabolic Care* 2: 83–85

84 Svanborg-Eden C, Levin BR (1990) Infectious disease and natural selection in human populations. In: AC Swedlund, GJ Armelagos (eds): *Disease in Populations in Transition.* Bergfin and Garvey, New York, 31–48

85 Satcher D (1995) Emerging infections: Getting ahead of the curve. *Emerging Infectious Diseases* 1

86 Lederberg J (1997) Infectious disease as an evolutionary paradigm. *Emerging Infectious Diseases* 3

87 Lebarbenchon C, Brown SP, Poulin R, Gauthier-Clerc M, Thomas F (2008) Evolution of pathogens in a man-made world. *Molecular Ecology* 17: 475–484

88 Guégan JF, Prugnolle F, Thomas F (2007) Global spatial patterns of infectious diseases and human evolution. In: S Stearns, JC Koella (eds): *Evolution in Health and Disease.* Second Edition. Oxford University Press, New York, 19–29. Second Edition

89 Lederberg J, Shope RE, Oaks SC (eds) (1992) *Emerging Infection: Microbal Threats to Health in the United States.* Institute of Medicine, National Academy Press

90 Morse SS (1995) Factors in the emergence of infectious diseases. *Emerging Infectious Diseases* 1: 7–15

91 Rook GAW (2008) The broader implications of the hygiene hypothesis. *Immunology* 126: 3–11

Immunoregulation by microbes and parasites in the control of allergy and autoimmunity

Rick M. Maizels[1] *and Ursula Wiedermann*[2]

[1]Institute of Immunology and Infection Research, University of Edinburgh, Edinburgh, UK
[2]Department of Specific Prophylaxis and Tropical Medicine, Medical University of Vienna, Vienna, Austria

Abstract

Key changes in our microbial environment, encompassing mycobacteria, helminth worms and commensal microflora, may each have had a major impact on the development and reactivity of our immune system not only to infections, but also in the context of harmless antigens in autoimmunity and allergy. We discuss here recent advances in understanding host-microbe and host-parasite interactions, and their impact on the balance between immunoregulation and immunopathology in the mammalian immune system.

Introduction

The increased prevalence of immune-mediated allergic and autoimmune disorders in developed countries is of growing concern [1]. Both Th1-mediated autoimmune diseases such as multiple sclerosis [2] and Th2-associated allergic diseases such as asthma [3] have become markedly more common over recent decades. The inverse correlation between incidence of these pathologies and infectious diseases, whether measured on a national scale [2, 4] or according to individual case histories [5, 6] has reinforced the concept that infections may protect against immune overreactions to innocuous (self or environmental) antigens; this concept is commonly termed the hygiene hypothesis [7–10]. Intrinsic to the original hygiene hypothesis, is the proposition that early-life exposure to infectious agents exerts a lasting effect on the immune system of the individual [11, 12].

Since the original formulation of the hygiene hypothesis [13], our understanding of host-microbe interactions has been transformed in significant ways. The initial model envisaged Th1-promoting microbes counteracting Th2 responsiveness, and thereby protecting against Th2-dependent allergies. This formulation is not compatible with observations that developed countries have experienced an increase in Th1-associated autoimmune pathologies as well as Th2 allergies. Indeed, in some studies, individuals with Th1-autoimmune disorders such as Type 1 diabetes are actually more likely to suffer from asthma than non-autoimmune individuals [14, 15]. A recasting of the hygiene hypothesis must therefore provide an alternative model to a simple Th1/Th2 imbalance; as shown in Figure 1 one such alternative would be to invoke the activity of a suppressive regulatory T cell that can down-modulate both Th1 and Th2 [16]. In addition, the different specialisms of microbiology, virology and parasitology have become more integrated since the earliest days of the hygiene hypothesis, and we now take much greater account of the impact of the commensal microflora on immune status. Hence, we seek a more general explanation of how the microbial environment in all its forms determines the development of the immune response.

The hygiene hypothesis sparked extensive research into human cohorts, particularly children, as well as investigations across a range of animal models. In all cases, the link between infections and allergic (or auto-immune) outcomes were tested, as will be discussed below. At this stage, it would be useful to review several important general points. First, it is assumed that where an infectious agent can modify an allergen or auto-antigen specific immune response, this represents an antigen non-specific (bystander) effect, acting in the first instance through soluble cytokines such as IL-10, or by modulating key stimulatory cells such as the dendritic cell. Subsequently, allergen or auto-antigen specific T cells may well adopt a different phenotype as a result of changes to the initial stimulatory environment. We shall discuss such mechanisms later in this review.

Second, it is not necessary to postulate that active infection occurs, as it may be sufficient that certain microbes (or their products) are present in the environment. As will be detailed below, if bacterial endotoxins alone can modulate host immune status (through TLRs), then exposure rather than live infection may be the most important determining factor [17].

Third, as will be apparent in the discussion of the various microbial organisms below, the effect of each infective species is likely to be different [8], reflecting in part evolutionarily distinct host–pathogen relationships. Each species can develop very different infection dynamics over time, and/or peak intensity of infection. In this context, it is not surprising that the effect of infections on, say, allergy will depend entirely on the species in question. Hence, when data from multiple species (and from diverse prevalence environments) are aggregated, the protective effects associated with certain individual parasites are lost [18]. Moreover, the same species may exert opposite effects under different conditions; this has been well illustrated in murine infections with the helminth *Schistosoma mansoni*: acute infections are more pro-allergic, while a protective effect is seen in chronic infections [19]. An additional caveat to be considered, in the context of asthma, is that bacterial or viral infections in the airways are as likely to be exacerbating factors as they are to be protective [20], and again the exact nature of the infection itself is likely to be extremely important.

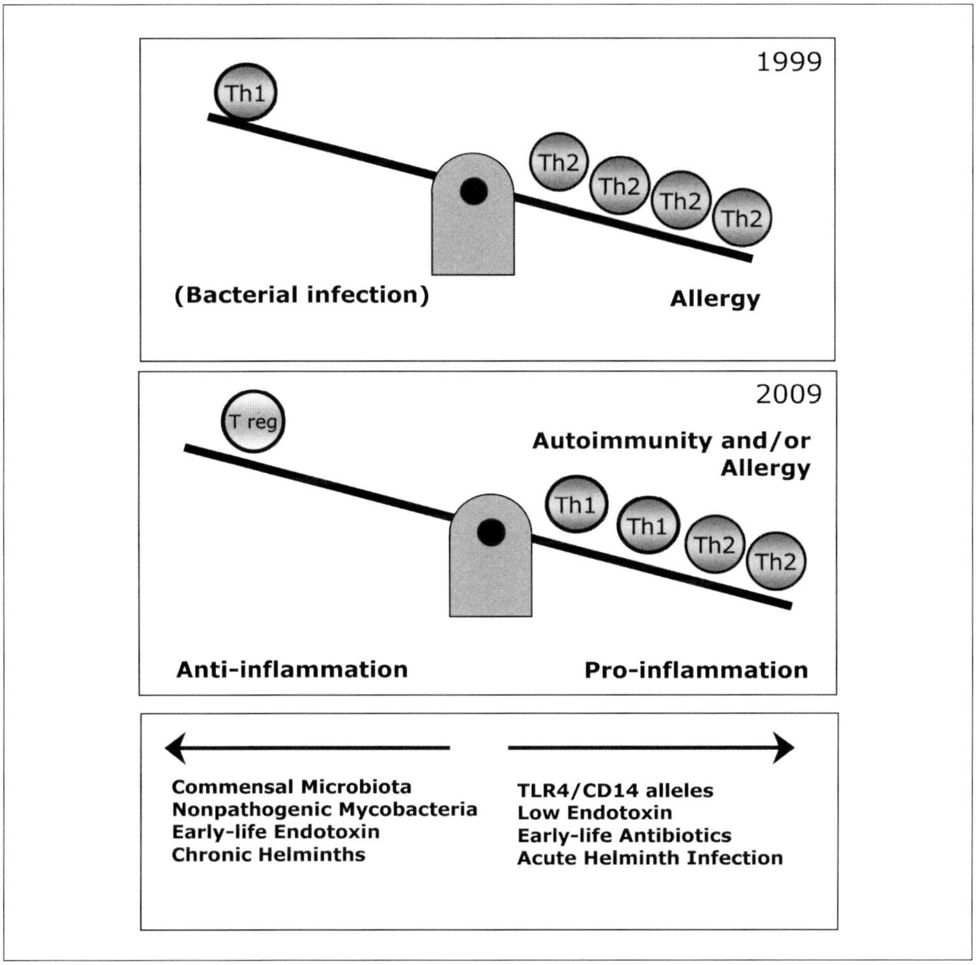

Figure 1
The hygiene hypothesis, ten years ago and today. Originally conceived as a balance between Th1 and Th2 immune responsiveness, the recognition of a Regulatory T cell subset (Treg) has recast thinking in this area, with Tregs balanced against both Th1- and Th2-mediated immunopathologies. Factors proposed to accentuate, or mitigate, allergic susceptibility are shown in the bottom panel.

A further prominent factor is the genetic status of the host. If susceptibility to allergy and autoimmunity is genetically-influenced, so too may be the propensity for infections to modulate the immunopathology of disease. Such interactions would result in particular infections being protective only in certain genotypes rather than in the population as a whole [21]. Finally, it is arguable that the identity of individual infections is less important than the longevity or frequency of infective episodes, particularly in the more formative stages of immune system development [22]. Hence, there may be some endogenous sequelae of infections which, over time, can drive the immune system to an alternative outcome that later is seen as predisposition (or resistance) to immunopathologies.

Evolution of host, pathogen – and environment

The infection-allergy/autoimmunity interaction is one part of a broader question, of how the host and its immune system have co-evolved with the panoply of associated microorganisms. This is evident firstly in the extraordinary polymorphism of the human immune system, almost certainly driven by the diversity of pathogens, and in another sense, by the exquisite host-specificity and complexity of the innate immune system which has evolved to recognise and respond to pathogens (perhaps the pathogens most prevalent many generations ago). In both senses, we carry the evolutionary imprint of microbes throughout the immune system. Moreover, our co-evolution is a highly dynamic and ongoing process; hence, while some interrelationships could be considered 'mature', perhaps with a higher degree of co-adaptation, others are evidently more recent and either or both host and pathogen may generate a more damaging outcome.

It is also clear that anthropogenic changes to our environment are impacting on allergies in particular; the marked contrast between economically developed Western countries and those in poorer economic circumstances have been well documented [4, 23]. One suggestion has been that antibiotic use may increase likelihood of allergies; two independent studies indicated so (reviewed by [24]), although other authors [25] have argued otherwise. As will be discussed, in a study of Swedish children, those from anthroposophic lifestyle families practising minimalist medical intervention and higher probiotic use, had lower indicators of allergy levels [26, 27]. In this setting, lower antibiotic use cannot be disentangled from exposure to probiotics, which may contribute to protection against allergy as discussed below. Interestingly, even within individual developed nations, children in rural environments suffer significantly lower levels of allergy and asthma than age-matched individuals in cities [28], spurring investigations of whether environmental products from bacteria may influence human immune status, as also described below.

In this review we set out firstly the system-level information which relates to the hygiene hypothesis and then review the specific molecular interactions, signalling pathways and immune cells which appear to be involved in the interaction between infection and protection from immunopathology. Infectious agents of all types, from viruses through to parasites have been reported to down-modulate allergic responses [29, 30], but for the purposes of this review we will focus on three groups of organisms. At first glance these appear poles apart, but they are each heavily implicated in influencing the host immune system in a favourable manner with regards to allergies and other immunopathologies. These are, firstly mycobacteria, particularly in the respiratory tissues; secondly the helminth (worm) parasites which are primarily in the gastrointestinal locale but often transit the lung; and thirdly commensal bacteria which dominate the intestinal tract.

Mycobacteria

A seminal paper in establishing the hygiene hypothesis reported that Japanese children who were strongly skin-test positive for *Mycobacterium tuberculosis* (MTB) had substantially lower allergic symptoms than observed in low or non-responders to tuberculin [31]. Those protected from allergy were considered to have been exposed to *M. tuberculosis*, as BCG (*M. bovis*) vaccination induces a lower level of tuberculin-positivity. A similar pattern could be discerned in comparing TB and allergy data from 23 countries worldwide [32]. Subsequently, numerous cohorts were studied with respect of either tuberculin reactivity or BCG vaccination, with mixed outcomes of allergic measures (reviewed by [33, 34]), indicating that an effect can occur depending on the environmental and genetic setting of the population. As well as an influence on allergies, there is epidemiological evidence to associate tuberculin-positivity with lower risk of multiple sclerosis (MS) [35].

In animal models, both heat-killed MTB and live BCG protect mice from the MS-like experimental autoimmune encephalomyelitis (EAE), which is considered a Th1-mediated disease [36]. BCG also exerts an anti-allergic effect when administered to animals shortly before sensitisation [37, 38]; these reports indicated that infection redirected immunity to the Th1 pathway, so it remained unclear how EAE is also prevented by the same organism.

A resolution of this question was offered by studies with *M. vaccae*. In this case, killed bacteria given 3 weeks before [39] or up to 5 weeks following [40] initial allergen sensitisation, reduced airway reactivity. In the study from the Rook laboratory, suppression occurred in an IFN-γ (and thus presumably a Th1-) independent fashion [39]. These authors additionally showed that *M. vaccae* immunisation generated a regulatory T cell population capable of transferring protection against allergy, dependent upon IL-10 and TGF-β, consistent with a model of regulatory T cell activity [41]. It is interesting to note that in this instance, no active infection occurs; it is sufficient to expose the immune system to the molecular stimulus represented by this mycobacterium.

An instructive report in view of comment above on gene–environment interactions, is that of Smit and colleagues; these authors showed that the protective effect of *M. vaccae* on allergy in mice is only observed in carriers of the susceptible allele of murine Slc11a1/Nramp1 [42]. Remarkably, the same relationship to allelic forms of human SLC11A1/NRAMP1 has been observed for reduced atopy in BCG-vaccinated children [43], offering one explanation for the variable strengths of effect observed in the other studies mentioned above.

M. vaccae is an environmental saprophytic mycobacterial species, which is widespread in the rural setting. According to Rook and colleagues, contact with such organisms may have been ubiquitous in the pre-industrial age, so that the immune system has evolved to recognise them in a non-inflammatory manner [44]. With declining exposure to these 'old friends' in urbanised societies, immune reactivity may be driven into a more proinflammatory mode with detrimental results for allergies and autoimmunity.

Whether considering tuberculoid, bovine or saprophytic mycobacteria, it is clear that a major part of the innate immune system is directed towards the detection of, and the response to, these organisms. For example, the CD1 receptor is an MHC-like molecule that binds to mycobacterial glycolipid ligands [45], and can stimulate a variety of T cells including NKT cells with an invariant (i.e., innately encoded) mycobacterial-specific TCR for this molecule. In addition the mycolic lipids can be anti-inflammatory for macrophages [46].

Helminths

Helminth parasites are still extraordinarily prevalent in developing countries, infecting over 25% of the global population [47]. Before modern sanitation, most humans would have been infected for significant periods of their lives, so that these parasites have acted as companions to the immune system throughout evolutionary history. Notably, many of these infections exact a relatively low pathological burden, although over time (and probably dependent on host factors) a significant proportion develop severe chronic disease such as hepatosplenic schistosomiasis or elephantiasis [48]. In a large number of cases, however, infection is essentially asymptomatic, although major immunological changes occur which result in loss of antigen-specific peripheral T cell responsiveness, together with features reminiscent of the 'modified Th2' with high IgG4 antibody levels [49]. Recently, it has been demonstrated in animal models at least, these parasites can stimulate regulatory T cells, which may protect the worms from expulsion, but also dampen overall immune responsiveness in the infected host [50–52].

Over recent years, evidence has accumulated that helminth infections in humans can be protective against immunopathological diseases of both allergic and autoimmune pathologies. In diverse geographical areas endemic for a variety of helminth parasites, a similar negative association between infection and allergy (measured as skin test reactivity or wheeze) has been reported [53–56]. Where asthma itself has been measured, it has been found to follow a less severe course in helminth-infected patients [57]. Moreover, when infected children are treated with anthelmintic drugs to remove gastrointestinal parasites, increased allergic reactivities can result [58].

Important insights have been gained at the mechanistic level from some of these studies, as in the demonstration by Van den Biggelaar et al., in Gabonese schoolchildren, of a significant reduction in skin allergy among Schistosome-infected individuals. Not only did this report establish that a Th2-associated infection can counteract a Th2 pathology, but by showing high parasite-specific IL-10 responses in the infected cohort, these authors indicated that the inhibition was due to a regulatory cytokine [59]. Consistent with earlier studies in Latin America [58], once the Gabonese children had been treated with anthelmintics, levels of skin-test reactivity rose [60]. Araujo et al. found that *S. mansoni*-infected patients showed reduced allergen reactivities, but that this increased after drug therapy while IL-10 responses declined. As with the Gabonese scenario, these data suggest that IL-10 may mediate the antigen non-specific suppression of bystander (i.e., allergen) responses [61].

Clearly not every species confers protection and, as with mycobacterial infections, a number of studies have found no significant association between helminth infections and allergies, or even reduced reactivity following chemotherapy [62]. It is to be expected that the circumstances in which some species may influence immunity will vary with intensity and transmission, as well as with host genotype. For example, in Ethiopia it was found that negative associations between helminth infections and allergic wheeze were strongly significant for *Ascaris*, weak but not statistically significant for hookworm, and without effect for *Trichuris* [56]. Another question surrounds exactly which measures of allergy are employed: in Ecuador it was found that skin test reactivity was markedly reduced in *Ascaris* and *Trichuris* patients [63]. However, while each symptomatic marker (wheeze, rash, rhinitis, etc.) was reduced in infection, no individual comparison achieved statistical significance. The important point is that the epidemiological data provides sufficient support for the hypothesis that helminth infection can materially alter the immune response in many settings. This conclusion forms the platform upon which clinical treatment is now being offered using live *Trichuris suis* ova to alleviate the chronic inflammatory diseases of Ulcerative Colitis and Crohn's Disease [64–66], and has stimulated similar studies with human hookworm as a potential therapy for allergies [67].

In parallel to these human studies, work in animal models has added compelling data which underpin and extend our concepts at the infection/immunopathology interface [68, 69]. In terms of allergy, various gastrointestinal nematodes, tissue-dwelling filarial nematodes and schistosomes have all been shown to inhibit allergies to food antigens [70], airway allergens [51, 71–73], and allergens introduced intravenously to induce anaphylaxis [74]. Because protection against allergy can be transferred by lymphocytes from infected mice to uninfected, sensitised animals, these anti-allergic effects cannot be ascribed to antigen competition, but appear to be caused by a suppressive down-regulatory influence on the allergen response itself.

Several investigators have studied the down-regulatory mechanisms in more detail. In one of the transfer systems, $CD4^+CD25^+$ regulatory T cells from *H. polygyrus* infected mice could protect recipients from airway allergy [51], but it was found that protection did not require IL-10. However, in a number of other systems, IL-10 does appear to play an important role: for example in the blocking of food allergy by the same parasite [70], and in protection against anaphylaxis [74].

Autoimmune reactivities are also profoundly affected by helminth infections [75]. Worldwide, human MS incidence is inversely correlated to intestinal parasitism [2], although many confounding socio-economic factors are likely to contribute to the differing prevalence of this disease. Because autoimmune diseases have a much lower prevalence in humans than allergies, most epidemiological studies of the effects of infection involve smaller cohort sizes (and longer timescales) with problems of achieving statistical power. However, it was recently reported that in a small cohort of MS patients in Argentina, 12 individuals who acquired helminth infections showed dramatically improved prognosis compared to age- and severity-matched uninfected patients [5]; importantly the same infected individuals were found to have a higher level of regulatory T cell activity (see the chapter by Jorge Correale in this volume).

Currently, most of our information on helminths and autoimmunity is derived from animal models, in particular Type 1 diabetes, colitis, and the EAE model for MS. In the first case, *S. mansoni* infection blocks the development of diabetes in the genetically-prone NOD mouse strain [76], although it has been noted that a range of infections common to animal husbandry units have the same effect [1]. More specifically, soluble extracts from *S. mansoni* can equally ablate diabetes development [77], implying a more exquisite level of interaction. TNBS-induced antigen-specific colitis can be inhibited by schistosome eggs [78, 79], while *S. mansoni* infection can also ablate DSS-colitis [80] and EAE [81, 82].

Despite the extensive body of literature demonstrating a protective effect of helminth infection on both allergies and autoimmunity, we have relatively little information on the molecular products elaborated by these parasites. A limited number of candidate molecules have been described: a cysteine protease inhibitor (cystatin) from a filarial parasite is able to diminish allergic airway inflammation [83], and a further filarial protein (ES-62) can block both allergic and autoimmune reactions [84, 85]. Mechanistically, the former acts through macrophages and promotion of IL-10, while the latter blocks mast cell activation and complexes with TLR4 to compromise its signalling. In schistosomes, the release of lyso-phosphatidylserine is responsible for TLR2-mediated stimulation of dendritic cells, which in turn induces responding T cells to adopt a regulatory, IL-10-producing phenotype [86]. While we are only just beginning to understand the molecular basis of helminth modulation of host immunity, the information available amply illustrate that each interaction is part of a complex cascade, and that receptors such as TLR2 and 4 may be targeted as an immunosuppressive strategy.

An interesting question is why, if helminths are so protective against allergies and autoimmunity, has the marked increase in pathologies in the Western World not occurred earlier? In other words, with common gut helminths eradicated in the middle part of the 20th century, several decades elapsed before the sharp increase in asthma in children. One possibility is that pinworm (*Enterobius vermicularis*), not targeted by any health measures, persisted in developed countries until much more recently. Gale has suggested that perhaps this once ubiquitous and near-harmless infection (the 'hidden helminth') provided the protective cover [87].

Intestinal commensal flora and homeostasis

The mucosal surfaces, which cover the largest areas of the body, function as important interface between the host and the outside environment, guaranteeing on the one hand permeability and exchange of important nutrients and on the other a barrier against environmental threats. These barrier functions are accomplished by the concerted action of a specified epithelium covered by a mucus layer, the commensal microbiota and the mucosal-associated immune system. How the discrimination between harmless antigens, including the commensal bacteria, and dangerous pathogens is actually achieved is still unclear. As documented in germfree animals, colonisation of the mucosal surfaces with commensals has a protective and stimulatory effect on postnatal development of immune responses as well as metabolic and nutritional processes. Evidence from clinical and experimental studies is increasing that the commensal bacterial not only play an important role in competing with foreign incoming microorganisms for space and resources but also are actively involved in the induction and maintenance of mucosal tolerance to prevent untoward immunological responses to innocuous antigens. In this respect the 'hygiene hypothesis', which postulates that a lack of early microbial stimulation results in aberrant immune responses to harmless antigens later in life, has been reconsidered as a 'microflora hypothesis': it suggests that perturbations in the gastrointestinal microbiota, mainly as a consequence of increased antibiotic use and dietary differences in westernised countries, can disturb the homeostasis and microbiota-associated mechanisms of mucosal tolerance, thereby leading to increased incidence of allergic and inflammatory diseases [88, 89].

The process of colonisation of the gut starts immediately after birth to establish a stable microbiota until adulthood [90]. The stable adult microbiota harbours an inconceivable number of bacteria, more than 10^{14}, thereby exceeding more than 10 times the number of somatic and germ cells in our body. In contrast to the high density of bacteria, their diversity is relatively low compared to other ecosystems [91]. Only eight of 55 known bacterial divisions have so far been identified of which five are relatively rare. Sequencing bacterial ribosomal RNA (16S rRNA) from colonic mucosa and stool samples has made it possible to cluster bacterial RNA into relatedness groups. Clustering of sequences gave rise up to 800 species with more than 7,000 different strains. However, it is estimated that only a minority of species (about 30–40) predominate in the gut microbiota. Some 97% of the bacterial species are anaerobes and only 3% are aerobes. Most of the anaerobes in the faeces and overlying the epithelium belong to the Cytophaga-Flavobacterium Bacteroides (CFB, e.g., *Bacteroides*) and the Firmicutes (e.g., *Clostridium* and *Eubacterium*). Only a small fraction contains aerobic Proteobacteria phylum, such as Gram-negative *Escherichia coli* and *Salmonella* spp and Gram-positive cocci (*Enterococcus*, *Staphylococcus*, *Streptococcus*). Additionally, aerobic fungal species (e.g., Candida) or certain protozoa are members of the normal microbiota [90, 92]. The reason for this particular selection of bacteria, namely few divisions represented by tight clusters of related bacteria, may be due to a strong host selection for bacteria that mainly exhibit beneficial functions for the host. They ensure on the one hand that nutrients are easily broken down, thereby providing the host with energetic substances; on the other hand they support tolerogenic processes which are important for maintenance of the intestinal immune homeostasis [93].

The following possible interactions of commensal bacteria with the intestinal epithelium and the innate and adaptive immune system might explain how the microbiota can counteract inflammatory and promote tolerising responses: (a) bacteria may prevent or alter inflammatory responses by escaping the interaction with certain Toll-like receptors on epithelia cells – e.g., pentacylic lipid A from *Bacteriodetes* cannot signal *via* TLR4 – or by actively downregulating epithelial proinflammatory signalling, such as decreasing the transcription of NF-κB genes [94] or suppressing degradation of the inhibitor of NF-κB [95]. On the other hand it has been demonstrated that commensal bacterial products stimulate TLRs on epithelial cells under steady state conditions leading to production of protective factors which are involved in cytoprotection, tissue repair and angiogenesis, thereby playing an essential role in resistance to epithelial injury and maintenance of epithelial homeostasis [96]. (b) Commensals may be controlled and kept separated from the epithelial surfaces by a thick mucus layer covering the epithelial cells as well as by the epithelial production of antimicrobial defensin peptides. Additionally, brush border enzymes, such as alkaline phosphatase, can detoxify luminal LPS by dephosphorylating lipid A [97]. (c) Inappropriate responses to innate signals from commensals may be due to the compartmentalisation of TLRs to the basolateral side or within epithelial cells, while the exposed surface of epithelial cells is barely equipped with TLRs. Upon contact with commensals the epithelial cells can contribute to tolerance by the production of thymic stromal lymphopoietin (TSLP), a substance able to direct T cell responses towards non-inflammatory responses [98]. Interaction of commensals with dendritic cells may result in production of IL-10 and TGF-β *via* regulatory T cells, exhibiting the most important cellular effectors of tolerance induction [98]. (d) Dendritic cells loaded with commensal bacteria can lead to local production of specific anti-commensal IgA. The commensal-loaded dendritic cells are restricted to the mucosal compartment by the mesenteric lymph nodes. Thereby mucosal immune responses without activation of systemic immunity are induced, contributing to the mucosal barrier to protect against bacterial penetration [99, 100].

A new pathway of maintaining intestinal homeostasis by commensals *via* production of IL-25 by epithelial cells has been recently described [101]. In germfree mice as well as in conventional mice treated with antibiotics it was shown that the frequency of proinflammatory Th-17 cells in the large intestine is elevated in the absence of commensal bacteria. The differentiation of these cells is promoted by IL-6 and TGF-β dependent expression of the transcription factor ROR-γt, as well as by IL-23, which controls the expansion and survival of these cells. On the other hand, IL-25, a member of the IL-17 family, whose expression by epithelial cells is dependent on the presence of the commensal bacteria, counteracts the expression of IL-23 and thereby limits the frequency of Th-17 cells in the large intestine. Thus, in the large intestine at least, homeostasis is regulated by the IL-25-IL-23-IL-17 axis [101]. Subsequently, Ivanov and colleagues have reported a conflicting result from the small intestine, in which no Th17 cells develop in mice lacking CFB microflora [102]. Whether the contrasting observations reflect a strong regional specialisation within the intestinal tract, or opposing effects from different specific commensal organisms in the different studies, remains to be determined.

Recently, immunomodulatory molecules produced by commensals have been described with beneficial influence on the development of gut immune responses. Mazmanian et al. [103, 104] described that *Bacteroides fragilis*, a Gram-negative obligate anaerobe, produces the zwitterionic polysaccharide A (PSA) that activates $CD4^+$ T cells to correct certain immune defects: in germfree mice deficiencies in splenic $CD4^+$ T cells and lymphoid follicle development were corrected after introduction of *B. fragilis* expressing PSA, and similar effects were seen after administration of purified PSA. Immune responses, which had been skewed towards Th2 responses under germfree conditions, were normalised through PSA stimulation of IFN-γ *via* STAT-4 dependent IL-12 production. TLR-2 on dendritic cells was identified as the PSA receptor which initiated innate cytokine production and primed the adaptive Th1 immune response to the polysaccharide [105]. In a model of experimental colitis induced by *Helicobacter hepaticus* it was further demonstrated that administration of purified PSA suppressed proinflammatory IL-17 production by intestinal immune cells *via* IL-10 producing $CD4^+$ T cells [106]. These findings indicate that a single bacterial molecule has the capacity to balance the immune system in response to different dysregulating stimuli, which may imply that only a few commensals – but the right ones – are actually needed to keep immunological intestinal homeostasis. The question why we carry billions of bacteria in the intestine, if only a few are important to balance immunological responses, might be explained by their important role in forming an ecosystem also responsible for various metabolic and nutritional processes and for anti-infectious resistance including competition with pathogenic bacteria [107]. The observation that changes in the composition of the intestinal flora are associated with a constant increase of allergic and autoimmune diseases may support the notion that we are increasingly suffering from a lack of such health-promoting commensals, of which *B. fragilis* is only one example.

Probiotic bacteria and allergy prevention

There is increasing evidence from clinical studies that atopic children have an altered intestinal flora compared to non-allergic children, with reduced numbers of lactic acid bacteria and increased numbers of *Clostridia* and *Staphylococci* [108–110]. Recent studies show that these differences in the composition of the gut microbiota precede the development of atopic sensitisation during infancy [111]. These findings are supported by studies, mentioned above, showing that children with an anthroposophic lifestyle exhibited a more diverse *Lactobacillus* microflora, and had a lower prevalence of atopy compared to children living by the modern westernised lifestyle [112]. Besides differences in diets, e.g., high intake of organically produced food items fermented by lactobacilli, antibiotic use, which appears to constitute a pressure on the microflora, is strongly restricted in anthroposophic children. Accordingly, it was shown that antibiotic treatment in the neonatal period is a risk factor for wheezing in early infancy [113].

Based on these findings there is increasing interest in supplementing human diet with lactic acid bacteria to prevent allergy development. Several double blind placebo controlled studies using different probiotics show a reduction in the development of eczema by the treatment [110, 114]. Probiotic supplementation with *Lactobacillus rhamnosus* strain GG (LGG) to mothers 4 weeks before and 6 months after delivery showed that the cumulative risk for developing eczema in their infants during the first 7 years of life is significantly reduced compared to the placebo group [115, 116]. However, the frequency of allergic sensitisation was not influenced by the probiotic treatment, suggesting that the preventive effects were not IgE-mediated. Based on the observation that pre- and postnatal treatment with LGG in another study did not have any effects on eczema development, Kopp et al. [117] raised the question how far intrinsic and environmental differences in study populations can influence the outcome of probiotic interventions. Differences in the clinical outcome of probiotic supplementation are definitely related to the respective strains and time point of intervention. Thus, in a study applying *Lactobacillus acidophilus* LAVRI-A1 exclusively in the postnatal period (first 6 months) of high risk infants, these children showed significantly increased frequency of atopic dermatitis and sensitisation to common allergens [118]. In a study with cow milk allergic infants supplementation of LGG for 4 weeks elicited beneficial effects in children with IgE associated atopic dermatitis, but not in those without IgE sensitisation. These protective effects were not seen when LGG was mixed with other probiotic strains, indicating that distinct intrinsic immunomodulatory properties of the different strains may interfere with each other [119].

The lessons learned from all these clinical studies are that prenatal treatment appears to be necessary for successful primary intervention, while the benefits of ongoing postnatal or only postnatal treatment are not yet that clear. Careful selection of probiotic strains, knowledge of their immunological properties *in vitro*, preclinical studies and dose finding studies are essential prior to application in prophylactic or therapeutic clinical settings.

In animal models, antibiotic-induced disruption of the intestinal microbiota can break airway tolerance to aeroallergens leading to allergic airway inflammation and goblet cell metaplasia [120]. In a mouse model of bronchial asthma oral administration of LGG or *Bifidobacterium lactis* led to suppression of all aspects of an asthmatic phenotype, such as allergen-specific IgE, airway inflammation and airway hyperresponsiveness, indicating that the respiratory and intestinal flora use common pathways for induction of tolerogenic responses. It has been proposed that antigen acquired by dendritic cells within a homeostatic intestinal environment preferentially stimulates the generation of Treg cells, which can be recruited to the airways. Along these lines, the immunosuppressive effects seen in that study were associated with TGF-β producing Treg cells [121]. Similar effects were seen in a mouse model of birch pollen allergy, where application of *Lactobacillus plantarum* or *Lactococcus lactis via* the inhaled or oral route had similar effects on allergy prevention [122–124].

The potential of probiotics to ameliorate ongoing inflammatory responses was demonstrated in a murine model of colitis [125]. Daily administration of a mixture of Bifidobacteria, Lactobacilli and *Streptococcus salivarius* reduced recurrent TNBS-induced colitis. The protective activity was mediated by lamina propria mononuclear cells *via* production of IL-10 and TGF-β, as LPMCs could transfer protection to naïve recipients and the activity of these cells was blocked by anti-IL10R and anti-TGF-β antibodies.

However, it was demonstrated that probiotics may present distinct strain-specific immunomodulatory capacities both *in vivo* and *in vitro*. Recent *in vitro* studies revealed that different *Lactobacillus* species exhibit different tolerising properties; two of three lactobacillus species, *L. reuteri* and *L. casei*, but not *L. plantarum*, primed monocyte derived dendritic cells to promote the development of Treg cells *via* the production of IL-10. Activation of the dendritic cells occurred *via* binding the C-type lectin DC-specific intracellular adhesion molecule 3-grabbing non integrin (DC-SIGN) and blockade of this receptor inhibited Treg cell development. This study suggested DC-SIGN activated Treg cell development as one possible explanation for the beneficial effects of certain probiotics, which might indicate that improved therapeutic effects could be achieved by prior screening of probiotics for binding to DC-sign [126]. Along these lines, the important role of dendritic cells in the functionality of different probiotic strains was also demonstrated *in vivo* using a model of colitis [127]. Dendritic cells pulsed with L. salivarius and *L. rhamnosus*, but not with *L. lactis* or *L. acidophilus*, exhibited a tolerogenic phenotype able to protect mice against TNBS-induced colitis. The preventive effect of probiotic-pulsed DCs was dependent on MyD-88-, TLR-2- and NOD-2-dependent signalling as well as the induction of $CD4^{+}CD25^{+}$ regulatory cells, but in an IL-10 independent matter. The authors suggested that TLR-2 signalling through NOD2 interaction could be initiated by certain cell wall components, such as peptidoglycan, but they could not exclude that also bacterial DNA *via* TLR-9 signalling could play an important role in the interaction of probiotics with DC function. A study on probiotic mediated prevention of allergic asthma in mice demonstrated that immunosuppression by *L. reuteri* (but not by *L. salivarius*) was induced *via* TLR-9 signalling and increased systemic activity of the tolerogenic enzyme indoleamine 2,3-dioxygenase (IDO). Importantly, the immunosuppressive effects of the bacteria required their viability, as killed bacteria did not suppress airway inflammation [128].

The question of how certain probiotics elicit their anti-inflammatory potential was also addressed in a mouse model of *Salmonella typhimurium* infection [129]: consumption of *Bifidobacterium infantis* for at least 3 weeks prior to infection with *Salmonella* resulted in a significant induction of $CD4^{+}CD25^{+}Foxp3^{+}$ Treg cells, which protected mice from excessive inflammation during infection. The Treg cells mediated protection *via* reduction of proinflammatory cytokine production, reduction of T cell proliferation, decrease in DC co-stimulatory molecule expression and attenuation of NF-κB activation. In adoptive cell transfer experiments the protective role of $CD4^{+}CD25^{+}$T cells *via* suppression of NF-κB was also confirmed in response to systemic LPS application.

With regard to infections it is, however, critical that a balance between regulatory and effector T cells is kept within the intestine to guarantee adequate immune responses against invading pathogens or to oral vaccination. Commensal gut flora (gf) DNA have been shown to be one factor dictating this balance through TLR-9 engagement [130]. In TLR-9 deficient mice the number of Treg cells was significantly increased resulting in decreased mucosal immune responses to oral vaccination with ovalbumin and decreased control of parasite infection with *Encephalitozoon cuniculi*. Treatment with anti-CD25 antibodies improved mucosal responses and parasite control. It was possible to demonstrate that gfDNA can inhibit Treg cell induction in a TLR-9 dependent manner and that gfDNA restored immune responses and parasite control in antibiotic treated mice, suggesting that DNA derived from commensals serves as natural adjuvant for priming intestinal immune responses. However, since it was also demonstrated that TLR-9 signalling at the apical side of epithelial cells can mediate anti-inflammatory effects [131], a compartmentalisation of gfDNA mediated TLR9 signalling in responses to different stimuli (inflammatory or steady state condition or exposure to pathogens) is suggested to control gut homeostasis.

From all these data it becomes obvious that the diverse gut flora commensals/ probiotics are not equal in their capacity to stimulate different TLRs or similar receptors and that a disturbance in the composition with over- or under-representation of certain strains can drive inflammatory responses and immunopathology. It is thus of major interest to further exploit how re-arrangement of the gut flora or manipulation of gut signalling can serve as new treatment approaches against inflammatory and allergic diseases.

Pathogen recognition receptors and microbial products

As mentioned above, a limited number of microbial products have been identified which can drive immune responsiveness away from the more proinflammatory and/or pro-allergic pathways. Generally, these specific pathogen-derived molecules interact with dedicated host receptors, such as PAMP receptors (Pathogen-Associated Molecular Pattern receptors), of which Toll-Like Receptors (TLRs) are the most important. From the discussion above it is also apparent that different ligands can bind to these receptors with differing outcomes in terms of pro- or anti-inflammatory effects. The importance of TLRs in regulating allergies is well-established at several levels: the effect of numerous polymorphisms in TLR signalling gene loci; the differential expression of the TLR proteins in humans [132]; and the degree of TLR ligand exposure in the human environment [133].

If TLR signalling significantly influences the control of allergies in humans, polymorphisms at TLR-related loci may be expected to influence the incidence of disease. There is indeed some support for this hypothesis. For example, a TLR4 polymorphism (Asp299Gly) was found in Swedish children to confer a large difference both in *in vitro* LPS responsiveness and in the levels of atopic asthma [134]. Further, a TLR2 polymorphism (T at –16934) has been associated with protection from asthma in farmers' children although not in the general population [135], indicating an important gene-environment interaction.

CD14 is a key co-receptor with TLR4 for the detection of endotoxin by the host innate immune system. An allelic variant (-159 T) within the 5' promoter region of CD14 was first associated with higher IgE responses [136], and subsequently with a more severe clinical asthma status [137]. Most recently, it has been reported that the complementary allele (–159C) is associated with milder symptoms only in environments where endotoxin levels are low; however, in high endotoxin circumstances, this allele actually confers a higher level of wheeze [138]. This relationship elegantly illustrates the complexity of gene-environment interactions, and raises the question of whether exposure, rather than active infection, may be sufficient to stimulate appropriate regulatory responses in the human immune system.

In a US study, asthmatic children were found to be reside in homes with less than half the endotoxin levels than did infants who were not sensitised [139]. In a similar study, children living in a high-endotoxin environment had lower incidence of asthma (and lower cellular immune responsiveness) than those in low-endotoxin environments [17]. Interestingly, as with helminth infections, the intensity of exposure may be a determining factor: in mice, while low level LPS administration amplified allergic reactions, high levels inhibited allergy with a resultant skew towards Th1 [140]. The complex, non-linear relationships between dose, subject age [141] and genotype have thus far prevented an integrated assessment of how (and when) TLR responsiveness regulates the allergic response, but there is little doubt that this interaction is an extremely important determinant of the immunological outcome.

Many other pattern recognition receptors exist in the mammalian immune system, such as the C-type lectin surface receptors on macrophages and dendritic cells (such as DC-SIGN), and proteins such as NOD2 which acts as a detector for intracellular pathogens. As our information, and consequently our conceptual framework, is heavily based on bacterial systems, it will be interesting to see whether a similar set of interactions occur in the recognition of helminth parasites, and to determine whether there is one common system for helminths and bacteria, or two intersecting pathways, or even two fully independent mechanisms operating entirely apart but in parallel. A summary of our current understanding of interactions in these pathways is shown in Figure 2.

Cellular pathways for immune modulation by microbes

As mentioned above with respect to individual organisms, different host cells have been implicated in the ability of infections to down-modulate immune pathologies. A pivotal role can be attributed to dendritic cells in the selection of appropriate responses to commensals and pathogens. For example, murine lamina propria $CD11c^{hi}$ DC can be separated into two subpopulations based on the expression of CD11b: $CD11c^{hi}CD11b^{+}$ DC cells induce proinflammatory Th17 cells, while $CD11c^{hi}CD11b^{-}$ DCs induce IL-10 producing cells [142]. The latter cells also express CD103, which is involved in *de novo* conversion of $Foxp3^{+}$ Treg cells [143]. The location of DCs within the lamina propria (upper or lower part of the villi) and their direct contact with epithelial cells are important factors for the development of tolerance. DCs located in the upper part of the villi interact closely with epithelial cells which condition tolerogenic DCs *via* steady state secretion of TSLP, retinoic acid or TGF-β. The latter two are required to drive the development of $CD103^{+}$ DC, while TSLP supports IgA class switching of B cells in the lamina propria [144]. On the other hand, protective immune responses necessary to fight invading pathogens, may require unconditioned DCs newly recruited from the basolateral membrane of the epithelial cell monolayer. The capacity of these DCs to release IL-12 and promote Th-1 responses depends on high local concentrations of TSLP produced by epithelial cells upon entrance of bacteria across the epithelial barrier [98]. The conditioning processes of DCs also depend on the interaction of bacteria with TLRs differentially expressed on DCs. DCs in close contact with the epithelial cells and

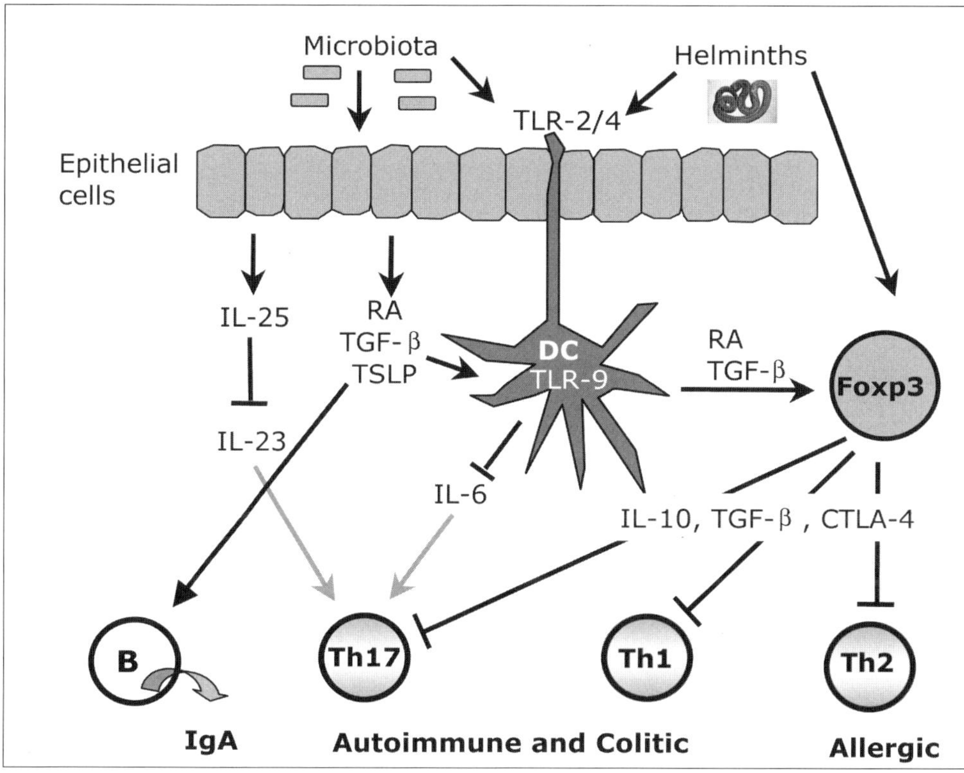

Figure 2
A simplified view of the mechanisms which may regulate propensity to allergy, autoimmunity or colitis. Commensal microbiota interact with epithelial cells and with dendritic cells (DCs), some of which can extend dendrites into the gut lumen. Epithelial cells produce immune mediators including IL-25 which inhibits IL-23 production, thereby reducing pro-inflammatory Th17 differentiation. TSLP, TGF-β and RA (retinoic acid) act on DCs, and according to the context can cause them to promote regulatory T cells (Tregs), dependent on further release of TGF-β and RA. In addition, TGF-β promotes IgA synthesis by B cells. DCs can also recognise bacterial DNA through intracellular TLR9. Tregs block all types of effector T cells including Th1, Th2 and Th17, through cytokines such as IL-10 and TGF-β, and by interfering with co-stimulation through CTLA-4. Some helminths are able to induce Tregs directly, bypassing DCs. Not shown on this schematic are γδ T cells, macrophages and NK cells, all of which play significant roles in immune regulation and responsiveness.

commensal bacteria do not respond because they do not express TLRs or respond to ligation in a non-inflammatory mode *via* IL-10 mediated mechanisms. DCs positioned distally from epithelial cells do express TLRs on their surface and may promptly respond to TLR ligation with upregulation of co-stimulatory molecules such as CD40 and CCR7 upon contact with the respective bacteria.

With respect to treatment approaches with probiotics, signalling *via* TLR-2 [127], TLR-4 [145], and TLR-9 [128] have been described to drive tolerogenic responses thereby inhibiting development of inflammation and allergy. Apart from TLRs, bacterial antigens can also be recognised *via* other receptors, such as nucleotide-binding oligomerisation domain receptor (NOD) or lectin receptors. *L. lactis* has been shown to activate cells, *via* NOD1 and NOD2 in addition to TLR-2 and TLR-4 [146]. *L. reuteri* and *L. casei* bind, as previously described, the C-type lectin DC-SIGN on dendritic cells to induce IL-10 producing Tregs [126]. The CD206 mannose receptor recognises a broad spectrum of molecules of different microbes, and plays a role in modulating inflammation. Recently it was demonstrated that oral application of the probiotic strain *L. casei* led to an increased number of $CD206^{+}$ cells as well as an increase of the receptor expression on cells, while non-probiotic strains, such as *E. coli*, did not [147].

Macrophages display similar properties and including selective expression of TLRs. Moreover, these cells can be unresponsive to TLR ligation in terms of proinflammatory cytokine production, while fully retaining their bactericidal activity. Tolerogenic macrophages might be involved in bacterial killing without inducing inflammation, they might locally support the maintenance of Treg cells in an environment that is continuously exposed to bacteria and bacterial products and they might influence DC function by inhibiting their potential to activate Th17 cells. Whether macrophages also migrate to mesenteric lymph nodes for tolerance induction is not yet clear [144].

One significant macrophage population with a broadly counter-inflammatory phenotype is the alternatively-activated macrophage. This type is markedly expanded in helminth infections, and is characterised by expression of arginase-1 (which countermands nitric oxide synthesis) and novel chitinase-like molecules [148]. Such macrophages are able to inhibit lymphocyte proliferation in a contact-dependent manner [149], and may be responsible for preventing inflammation in mucosal surfaces such as the lung. Recently a new colon-infiltrating macrophage population induced by *Schistosoma* infection was shown to prevent colitis in mice [80]. An increase of $CD11b^{+}CD11c^{-}$ macrophages in the lamina propria of infected mice was associated with limited intestinal inflammation in dextran sodium sulphate (DSS) induced colitis and transfer of these mononuclear cells also inhibited inflammation in non-infected recipients. As macrophages isolated from non-infected mice could not transfer protection from colitis, this indicated that schistosome infection modulates colon macrophages to become protective. In this model the protective properties of the cells were independent of Treg cells or regulatory cytokines, as infected mice depleted of $CD4^{+}CD25^{+}$ T-cells or treated with anti-IL-10 or TGF-β antibodies still did not develop inflammation upon DSS challenge. Similarly, in a mouse model of allergic airway inflammation it was demonstrated that prevention of allergic sensitisation and airway hyperresponsiveness was mediated by macrophages after application of filarial cystatin, a secreted protease inhibitor from filarial nematodes [83]. Also in this case the protective capacity was independent of Treg cells, but IL-10 proved to be a key element in cystatin induced immunomodulation, since IL10R-blockade reversed the protective effects. Transcription profiling of macrophages after exposure to filarial cystatin showed an increase in IL-10 expression but no changes in expression of other activation markers, which indicated that cystatin-activated macrophages resemble type-II activated macrophages, which differ from the alternatively activated macrophages described after schistosome infection. The receptor interaction of filarial cystatin on macrophages is still unknown, but the scavenger receptor CD36 is one potential candidate.

Beside Tregs and macrophages, B cells have also been shown to be a source of IL-10. In an experimental model of systemic anaphylaxis, *S. mansoni* infection led to prevention of anaphylaxis *via* B cell- and IL-10-dependent mechanisms [74]. Partial depletion of B cells in infected mice rendered them fully susceptible to anaphylaxis, while depletion of $CD4^{+}CD25^{+}$ cells had no effects on anaphylaxis. Prevention of anaphylaxis was not associated with a reduction in allergen-specific sensitisation but rather with a reduction of allergic responses during allergen challenge. The fact that inhibition of IL-10 led to development of lethal anaphylactic reactions upon allergen challenge, indicated that IL-10 producing B cells are necessary to maintain immunoregulation in the allergen-predisposed environment of schistosome infection. Along these lines, it was demonstrated that B cells from chronically, but not from acutely, schistosome infected mice can transfer inhibition of airway inflammation in an IL-10 dependent matter [19]. These data indicate that IL-10 producing B cells are induced during schistosome infection as part of the parasite's immune regulation of host immunity thereby also influencing unrelated responses to different antigens, including allergens but also vaccine antigens.

Regulatory T cells and a regulatory hygiene hypothesis

Perhaps the most consistent common thread which runs through studies on mycobacteria, helminths and commensal bacteria is the stimulation of an immunoregulatory network, which may be manifest in the activity of dendritic cells, macrophages, or regulatory T or B cells. There is no question that dendritic cells are essential to initiate the response of both the innate and adaptive immune systems, and there is good evidence that the phenotype of macrophages and B cells is dependent on the phenotype of the T cell population. Hence, in conclusion, it is attractive to suggest that the determining factor in whether an immune response reaches an inflammatory or counter-inflammatory conclusion is the level of activity within the regulatory T cell subset.

Because in humans there is good evidence that associates allergic and autoimmune conditions with a deficiency in regulatory T cell numbers and/or function [150, 151], a unifying model for how infections may down-modulate host immunopathology can plausibly be based on the expansion of regulatory T cell activity. Thus, in *M. vaccae*-immunized mice [41], in *H. polygyrus* infected mice [51], and in animals inoculated with commensal bacteria [129], similar populations of regulatory T cells can be found to be responsible for the suppression of inflammation.

With a new synthesis between previously disparate arms of immunology, between studies on allergy (a Th2 disease) and autoimmunity (Th1), and studies on bacteria and helminths, a reformulation of the original 'hygiene hypothesis' has taken place [7, 152, 153]. Tregs are a natural population of lymphocytes that are responsible for the control of immune pathologies, whether they be Th1-mediated autoimmune diseases or Th2-dependent allergies. We are dependent, in no small part, on our commensal bacteria for the appropriate stimulation and expansion of this Treg subset. However, many infectious agents can also enhance the activity of Tregs, in a manner which may depend on the precise genotype of the infected host. As the intensity and prevalence of these infections decline, the immune system is in danger of becoming over-reactive to innocuous substances such as allergens and autoantigens. The critical balance between subdued response and overt reactivity is therefore determined by a complex mix of environmental organisms, both pathogenic and non-pathogenic, in concert with the genetic status of the individual. As the field increasingly moves to a more mechanistic phase, to study molecular pathways and interactions which can explain the effect of infections on immunopathologies, the real test will be whether new therapies based upon this hypothesis will emerge which will be valuable in treating the spectrum of human immunopathological diseases.

References

1 Bach JF (2002) The effect of infections on susceptibility to autoimmune and allergic diseases. *N Engl J Med* 347: 911–920

2 Fleming JO, Cook TD (2006) Multiple sclerosis and the hygiene hypothesis. *Neurology* 67: 2085–2086

3 Eder W, Ege MJ, von Mutius E (2006) The asthma epidemic. *N Engl J Med* 355: 2226–2235

4 ISAAC (1998) Worldwide variation in prevalence of symptoms of asthma, allergic rhinoconjunctivitis, and atopic eczema: ISAAC. The International Study of Asthma and Allergies in Childhood (ISAAC) Steering Committee. *Lancet* 351: 1225–1232

5 Correale J, Farez M (2007) Association between parasite infection and immune responses in multiple sclerosis. *Annals of Neurology* 61: 97–108

6 Pelosi U, Porcedda G, Tiddia F, Tripodi S, Tozzi AE, Panetta V, Pintor C, Matricardi PM

(2005) The inverse association of salmonellosis in infancy with allergic rhinoconjunctivitis and asthma at school-age: a longitudinal study. *Allergy* 60: 626–630
7 Wills-Karp M, Santeliz J, Karp CL (2001) The germless theory of allergic disease: revisiting the hygiene hypothesis. *Nat Rev Immunol* 1: 69–75
8 Kamradt T, Göggel R, Erb KJ (2005) Induction, exacerbation and inhibition of allergic and autoimmune diseases by infection. *Trends Immunol* 26: 260–267
9 Schaub B, Lauener R, von Mutius E (2006) The many faces of the hygiene hypothesis. *J Allergy Clin Immunol* 117: 969–977; quiz 978
10 Fleming J, Fabry Z (2007) The hygiene hypothesis and multiple sclerosis. *Ann Neurol* 61: 85–89
11 Martinez FD, Holt PG (1999) Role of microbial burden in aetiology of allergy and asthma. *Lancet* 354 (Suppl 2): SII12–15
12 Holt PG, Upham JW, Sly PD (2005) Contemporaneous maturation of immunologic and respiratory functions during early childhood: implications for development of asthma prevention strategies. *J Allergy Clin Immunol* 116: 16–24
13 Strachan DP (1989) Hay fever, hygiene, and household size. *BMJ*mj 299: 1259–1260
14 Kero J, Gissler M, Hemminki E, Isolauri E (2001) Could TH1 and TH2 diseases coexist? Evaluation of asthma incidence in children with coeliac disease, type 1 diabetes, or rheumatoid arthritis: a register study. *J Allergy Clin Immunol* 108: 781–783
15 Simpson CR, Anderson WJ, Helms PJ, Taylor MW, Watson L, Prescott GJ, Godden DJ, Barker RN (2002) Coincidence of immune-mediated diseases driven by Th1 and Th2 subsets suggests a common aetiology. A population-based study using computerized general practice data. *Clin Exp Allergy* 32: 37–42
16 Sakaguchi S (2004) Naturally arising $CD4^+$ regulatory T cells for immunologic self-tolerance and negative control of immune responses. *Annu Rev Immunol* 22: 531–562
17 Braun-Fahrlander C, Riedler J, Herz U, Eder W, Waser M, Grize L, Maisch S, Carr D, Gerlach F, Bufe A et al (2002) Environmental exposure to endotoxin and its relation to asthma in school-age children. *N Engl J Med* 347: 869–877
18 Leonardi-Bee J, Pritchard D, Britton J (2006) Asthma and current intestinal parasite infection: systematic review and meta-analysis. *Am J Respir Crit Care Med* 174: 514–523
19 Smits HH, Hammad H, van Nimwegen M, Soullie T, Willart MA, Lievers E, Kadouch J, Kool M, Oosterhoud JK, Deelder AM et al (2007) Protective effect of *Schistosoma mansoni* infection on allergic asthma depends on intensity and chronicity of infection. *J Allergy Clin Immunol* 120: 932–940
20 Bisgaard H, Hermansen MN, Buchvald F, Loland L, Halkjaer LB, Bonnelykke K, Brasholt M, Heltberg A, Vissing NH, Thorsen SV et al (2007) Childhood asthma after bacterial colonization of the airway in neonates. *N Engl J Med* 357: 1487–1495
21 Smit JJ, Folkerts G, Nijkamp FP (2004) Ramp-ing up allergies: Nramp1 (Slc11a1), macrophages and the hygiene hypothesis. *Trends Immunol* 25: 342–347
22 Matricardi PM, Bonini S (2000) High microbial turnover rate preventing atopy: a

solution to inconsistencies impinging on the Hygiene hypothesis? *Clin Exp Allergy* 30: 1506–1510

23 von Mutius E, Fritzsch C, Weiland SK, Roll G, Magnussen H (1992) Prevalence of asthma and allergic disorders among children in united Germany: a descriptive comparison. *BMJ* 305: 1395–1399

24 Hopkin JM (1999) Early life receipt of antibiotics and atopic disorder. *Clin Exp Allergy* 29: 733–734

25 Foliaki S, Nielsen SK, Bjorksten B, Von Mutius E, Cheng S, Pearce N (2004) Antibiotic sales and the prevalence of symptoms of asthma, rhinitis, and eczema: The International Study of Asthma and Allergies in Childhood (ISAAC) *Int J Epidemiol* 33: 558–563

26 Alm JS, Swartz J, Lilja G, Scheynius A, Pershagen G (1999) Atopy in children of families with an anthroposophic lifestyle. *Lancet* 353: 1485–1488

27 Floistrup H, Swartz J, Bergstrom A, Alm JS, Scheynius A, van Hage M, Waser M, Braun-Fahrlander C, Schram-Bijkerk D, Huber M et al (2006) Allergic disease and sensitization in Steiner school children. *J Allergy Clin Immunol* 117: 59–66

28 Riedler J, Braun-Fahrlander C, Eder W, Schreuer M, Waser M, Maisch S, Carr D, Schierl R, Nowak D, von Mutius E (2001) Exposure to farming in early life and development of asthma and allergy: a cross-sectional survey. *Lancet* 358: 1129–1133

29 Matricardi PM, Rosmini F, Riondino S, Fortini M, Ferrigno L, Rapicetta M, Bonini S (2000) Exposure to foodborne and orofecal microbes versus airborne viruses in relation to atopy and allergic asthma: epidemiological study. *BMJ* 320: 412–417

30 Wagner A, Förster-Waldl E, Garner-Spitzer E, Schabussova I, Kundi M, Pollak A, Scheiner O, Joachim A, Wiedermann U (2009) Immunoregulation by Toxoplasma gondii infection prevents allergic immune responses in mice. *Int J Parasitol* 39: 465–472

31 Shirakawa T, Enomoto T, Shimazu S, Hopkin JM (1997) The inverse association between tuberculin responses and atopic disorder. *Science* 275: 77–79

32 von Mutius E, Pearce N, Beasley R, Cheng S, von Ehrenstein O, Bjorksten B, Weiland S (2000) International patterns of tuberculosis and the prevalence of symptoms of asthma, rhinitis, and eczema. *Thorax* 55: 449–453

33 Herz U, Lacy P, Renz H, Erb K (2000) The influence of infections on the development and severity of allergic disorders. *Curr Opin Immunol* 12: 632–640

34 Smit JJ, Folkerts G, Nijkamp FP (2004) Mycobacteria, genes and the 'hygiene hypothesis'. *Curr Opin Allergy Clin Immunol* 4: 57–62

35 Andersen E, Isager H, Hyllested K (1981) Risk factors in multiple sclerosis: tuberculin reactivity, age at measles infection, tonsillectomy and appendectomy. *Acta Neurol Scand* 63: 131–135

36 Sewell DL, Reinke EK, Hogan LH, Sandor M, Fabry Z (2002) Immunoregulation of CNS autoimmunity by helminth and mycobacterial infections. *Immunol Lett* 82: 101–110

37 Erb KJ, Holloway JW, Sobeck A, Moll H, Le Gros G (1998) Infection of mice with *Mycobacterium bovis*-Bacillus Calmette-Guerin (BCG) suppresses allergen-induced airway eosinophilia. *J Exp Med* 187: 561–569

38 Herz U, Gerhold K, Gruber C, Braun A, Wahn U, Renz H, Paul K (1998) BCG infection suppresses allergic sensitization and development of increased airway reactivity in an animal model. *J Allergy Clin Immunol* 102: 867–874
39 Zuany-Amorim C, Manlius C, Trifilieff A, Brunet LR, Rook G, Bowen G, Pay G, Walker C (2002) Long-term protective and antigen-specific effect of heat-killed *Mycobacterium vaccae* in a murine model of allergic pulmonary inflammation. *J Immunol* 169: 1492–1499
40 Smit JJ, Van Loveren H, Hoekstra MO, Schijf MA, Folkerts G, Nijkamp FP (2003) *Mycobacterium vaccae* administration during allergen sensitization or challenge suppresses asthmatic features. *Clin Exp Allergy* 33: 1083–1089
41 Zuany-Amorim C, Sawicka E, Manlius C, Le Moine A, Brunet LR, Kemeny DM, Bowen G, Rook G, Walker C (2002) Suppression of airway eosinophilia by killed *Mycobacterium vaccae*-induced allergen-specific regulatory T-cells. *Nat Med* 8: 625–629
42 Smit JJ, Van Loveren H, Hoekstra MO, Karimi K, Folkerts G, Nijkamp FP (2003) The Slc11a1 (Nramp1) gene controls efficacy of mycobacterial treatment of allergic asthma. *J Immunol* 171: 754–760
43 Alm JS, Sanjeevi CB, Miller EN, Dabadghao P, Lilja G, Pershagen G, Blackwell JM, Scheynius A (2002) Atopy in children in relation to BCG vaccination and genetic polymorphisms at SLC11A1 (formerly NRAMP1) and D2S1471. *Genes Immun* 3: 71–77
44 Rook GA, Adams V, Hunt J, Palmer R, Martinelli R, Brunet LR (2004) Mycobacteria and other environmental organisms as immunomodulators for immunoregulatory disorders. *Springer Semin Immunopathol* 25: 237–255
45 Sieling PA, Chatterjee D, Porcelli SA, Prigozy TI, Mazzaccaro RJ, Soriano T, Bloom BR, Brenner MB, Kronenberg M, Brennan PJ et al (1995) CD-1 restricted T cell recognition of microbial lipoglycan antigens. *Science* 269: 227–230
46 Korf JE, Pynaert G, Tournoy K, Boonefaes T, Van Oosterhout A, Ginneberge D, Haegeman A, Verschoor JA, De Baetselier P, Grooten J (2006) Macrophage reprogramming by mycolic acid promotes a tolerogenic response in experimental asthma. *Am J Respir Crit Care Med* 174: 152–160
47 Elliott DE, Summers RW, Weinstock JV (2007) Helminths as governors of immune-mediated inflammation. *Int J Parasitol* 37: 457–464
48 Muller R (2002) *Worms and Human Disease*. CABI Publishing, Wallingford
49 Maizels RM, Yazdanbakhsh M (2003) Regulation of the immune response by helminth parasites: cellular and molecular mechanisms. *Nat Rev Immunol* 3: 733–743
50 Taylor M, Le Goff L, Harris A, Malone E, Allen JE, Maizels RM (2005) Removal of regulatory T cell activity reverses hyporesponsiveness and leads to filarial parasite clearance *in vivo*. *J Immunol* 174: 4924–4933
51 Wilson MS, Taylor M, Balic A, Finney CAM, Lamb JR, Maizels RM (2005) Suppression of allergic airway inflammation by helminth-induced regulatory T cells. *J Exp Med* 202: 1199–1212
52 McSorley HJ, Harcus YM, Murray J, Taylor MD, Maizels RM (2008) Expansion of

Foxp3+ regulatory T cells in mice infected with the filarial parasite, *Brugia malayi*. *J Immunol* 181 : 6456–6466

53 Araujo MI, Lopes AA, Medeiros M, Cruz AA, Sousa-Atta L, Solé D, Carvalho EM (2000) Inverse association between skin response to aeroallergen and *Schistosoma mansoni* infection. *Int Arch Allergy Immunol* 123: 145–148

54 Nyan OA, Walraven GEL, Banya WAS, Milligan P, Van Der Sande M, Ceesay SM, Del Prete G, McAdam KPWJ (2001) Atopy, intestinal helminth infection and total serum IgE in rural and urban adult Gambian communities. *Clin Exp Allergy* 31: 1672–1678

55 Cooper PJ, Chico ME, Rodrigues LC, Ordonez M, Strachan D, Griffin GE, Nutman TB (2003) Reduced risk of atopy among school-age children infected with geohelminth parasites in a rural area of the tropics. *J Allergy Clin Immunol* 111: 995–1000

56 Dagoye D, Bekele Z, Woldemichael K, Nida H, Yimam M, Hall A, Venn AJ, Britton JR, Hubbard R, Lewis SA (2003) Wheezing, allergy, and parasite infection in children in urban and rural Ethiopia. *Am J Respir Crit Care Med* 167: 1369–1373

57 Medeiros M, Jr., Figueiredo JP, Almeida MC, Matos MA, Araujo MI, Cruz AA, Atta AM, Rego MA, de Jesus AR, Taketomi EA et al (2003) *Schistosoma mansoni* infection is associated with a reduced course of asthma. *J Allergy Clin Immunol* 111: 947–951

58 Lynch NR, Hagel I, Perez M, Di Prisco MC, Lopez R, Alvarez N (1993) Effect of anthelmintic treatment on the allergic reactivity of children in a tropical slum. *J Allergy Clin Immunol* 92: 404–411

59 van den Biggelaar A, van Ree R, Roderigues LC, Lell B, Deelder AM, Kremsner PG, Yazdanbakhsh M (2000) Decreased atopy in children infected with *Schistosoma haematobium*: a role for parasite-induced interleukin-10. *Lancet* 356: 1723–1727

60 van den Biggelaar AH, Rodrigues LC, van Ree R, van der Zee JS, Hoeksma-Kruize YC, Souverijn JH, Missinou MA, Borrmann S, Kremsner PG, Yazdanbakhsh M (2004) Long-term treatment of intestinal helminths increases mite skin-test reactivity in Gabonese schoolchildren. *J Infect Dis* 189: 892–900

61 Araujo MIAS, Hoppe B, Medeiros M Jr, Alcântara L, Almeida MC, Schriefer A, Oliveira RR, Kruschewsky R, Figueiredo JP, Cruz AA et al (2004) Impaired T helper 2 response to aeroallergen in helminth-infected patients with asthma. *J Infect Dis* 190: 1797–1803

62 Lynch NR, Palenque M, Hagel I, Di Prisco MC (1997) Clinical improvement of asthma after anthelmintic treatment in a tropical situation. *Am J Respir Crit Care Med* 156: 50–54

63 Cooper PJ, Chico ME, Bland M, Griffin GE, Nutman TB (2003) Allergic symptoms, atopy, and geohelminth infections in a rural area of Ecuador. *Am J Respir Crit Care Med* 168: 313–317

64 Summers RW, Elliott DE, Qadir K, Urban JF, Jr., Thompson R, Weinstock JV (2003) *Trichuris suis* seems to be safe and possibly effective in the treatment of inflammatory bowel disease. *Am J Gastroenterol* 98: 2034–2041

65 Summers RW, Elliott DE, Urban JF Jr, Thompson R, Weinstock JV (2005) *Trichuris suis* therapy in Crohn's disease. *Gut* 54: 87–90

66 Summers RW, Elliott DE, Urban JF, Jr., Thompson RA, Weinstock JV (2005) *Trichuris suis* therapy for active ulcerative colitis: a randomized controlled trial. *Gastroenterology* 128: 825–832
67 Mortimer K, Brown A, Feary J, Jagger C, Lewis S, Antoniak M, Pritchard D, Britton J (2006) Dose-ranging study for trials of therapeutic infection with *Necator americanus* in humans. *Am J Trop Med Hyg* 75: 914–920
68 Wilson MS, Maizels RM (2006) Regulatory T cells induced by parasites and the modulation of allergic responses. *Chem Immunol Allergy* 90: 176–195
69 Fallon PG, Mangan NE (2007) Suppression of Th2–type allergic reactions by helminth infection. *Nature Reviews Immunology* 7: 220–230
70 Bashir ME, Andersen P, Fuss IJ, Shi HN, Nagler-Anderson C (2002) An enteric helminth infection protects against an allergic response to dietary antigen. *J Immunol* 169: 3284–3292
71 Wang CC, Nolan TJ, Schad GA, Abraham D (2001) Infection of mice with the helminth *Strongyloides stercoralis* suppresses pulmonary allergic responses to ovalbumin. *Clin Exp Allergy* 31: 495–503
72 Wohlleben G, Trujillo C, Muller J, Ritze Y, Grunewald S, Tatsch U, Erb KJ (2004) Helminth infection modulates the development of allergen-induced airway inflammation. *Int Immunol* 16: 585–596
73 Dittrich AM, Erbacher A, Specht S, Diesner F, Krokowski M, Avagyan A, Stock P, Ahrens B, Hoffmann WH, Hoerauf A et al (2008) Helminth infection with *Litomosoides sigmodontis* induces regulatory T cells and inhibits allergic sensitization, airway inflammation, and hyperreactivity in a murine asthma model. *J Immunol* 180: 1792–1799
74 Mangan NE, Fallon RE, Smith P, van Rooijen N, McKenzie AN, Fallon PG (2004) Helminth infection protects mice from anaphylaxis *via* IL-10-producing B cells. *J Immunol* 173: 6346–6356
75 Dunne DW, Cooke A (2005) A worm's eye view of the immune system: consequences for evolution of human autoimmune disease. *Nat Rev Immunol* 5: 420–426
76 Cooke A, Tonks P, Jones FM, O'Shea H, Hutchings P, Fulford AJ, Dunne DW (1999) Infection with *Schistosoma mansoni* prevents insulin dependent diabetes mellitus in non-obese diabetic mice. *Parasite Immunology* 21: 169–176
77 Zaccone P, Fehervari Z, Jones FM, Sidobre S, Kronenberg M, Dunne DW, Cooke A (2003) *Schistosoma mansoni* antigens modulate the activity of the innate immune response and prevent onset of type 1 diabetes. *Eur J Immunol* 33: 1439–1449
78 Elliott DE, Li J, Blum A, Metwali A, Qadir K, Urban JF Jr, Weinstock JV (2003) Exposure to schistosome eggs protects mice from TNBS-induced colitis. *Am J Physiol Gastrointest Liver Physiol* 284: G385–391
79 Mo HM, Liu WQ, Lei JH, Cheng YL, Wang CZ, Li YL (2007) Schistosoma japonicum eggs modulate the activity of $CD4^+$ $CD25^+$ Tregs and prevent development of colitis in mice. *Exp Parasitol* 116: 385–389
80 Smith P, Mangan NE, Walsh CM, Fallon RE, McKenzie ANJ, van Rooijen N, Fallon

PG (2007) Infection with a helminth parasite prevents experimental colitis *via* a macrophage-mediated mechanism. *J Immunol* 178: 4557–4566
81 Sewell D, Qing Z, Reinke E, Elliot D, Weinstock J, Sandor M, Fabry Z (2003) Immunomodulation of experimental autoimmune encephalomyelitis by helminth ova immunization. *Int Immunol* 15: 59–69
82 La Flamme AC, Ruddenklau K, Backstrom BT (2003) Schistosomiasis decreases central nervous system inflammation and alters the progression of experimental autoimmune encephalomyelitis. *Infect Immun* 71: 4996–5004
83 Schnoeller C, Rausch S, Pillai S, Avagyan A, Wittig BM, Loddenkemper C, Hamann A, Hamelmann E, Lucius R, Hartmann S (2008) A helminth immunomodulator reduces allergic and inflammatory responses by induction of IL-10-producing macrophages. *J Immunol* 180: 4265–4272
84 Melendez AJ, Harnett MM, Pushparaj PN, Wong WSF, Tay HK, McSharry CP, Harnett W (2007) Inhibition of FceRI-mediated mast cell responses by ES-62, a product of parasitic filarial nematodes. *Nat Med* 13: 1375–1381
85 Harnett W, Harnett MM (2008) Therapeutic immunomodulators from nematode parasites. *Expert Rev Mol Med* 10: e18
86 van der Kleij D, Latz E, Brouwers JFHM, Kruize YCM, Schmitz M, Kurt-Jones EA, Espevik T, de Jong EC, Kapsenberg ML, Golenbock DT et al (2002) A novel host – parasite lipid cross talk: schistosomal lysophosphatidylserine activates Toll-like receptor 2 and affects immune polarization. *J Biol Chem* 277: 48122–48129
87 Gale EA (2002) A missing link in the hygiene hypothesis? *Diabetologia* 45: 588–594
88 Noverr MC, Huffnagle GB (2005) The 'microflora hypothesis' of allergic diseases. *Clin Exp Allergy* 35: 1511–1520
89 Macpherson AJ, Harris NL (2004) Interactions between commensal intestinal bacteria and the immune system. *Nat Rev Immunol* 4: 478–485
90 Noverr MC, Huffnagle GB (2004) Does the microbiota regulate immune responses outside the gut? *Trends Microbiol* 12: 562–568
91 Backhed F, Ley RE, Sonnenburg JL, Peterson DA, Gordon JI (2005) Host-bacterial mutualism in the human intestine. *Science* 307: 1915–1920
92 Pamer EG (2007) Immune responses to commensal and environmental microbes. *Nat Immunol* 8: 1173–1178
93 Pédron T, Sansonetti P (2008) Commensals, bacterial pathogens and intestinal inflammation: an intriguing menage a trois. *Cell Host Microbe* 3: 344–347
94 Kelly D, Campbell JI, King TP, Grant G, Jansson EA, Coutts AG, Pettersson S, Conway S (2004) Commensal anaerobic gut bacteria attenuate inflammation by regulating nuclear-cytoplasmic shuttling of PPAR-gamma and RelA. *Nat Immunol* 5: 104–112
95 Tien MT, Girardin SE, Regnault B, Le Bourhis L, Dillies MA, Coppee JY, Bourdet-Sicard R, Sansonetti PJ, Pedron T (2006) Anti-inflammatory effect of *Lactobacillus casei* on Shigella-infected human intestinal epithelial cells. *J Immunol* 176: 1228–1237
96 Rakoff-Nahoum S, Paglino J, Eslami-Varzaneh F, Edberg S, Medzhitov R (2004) Recog-

nition of commensal microflora by toll-like receptors is required for intestinal homeostasis. *Cell* 118: 229–241
97 Lievin-Le Moal V, Servin AL (2006) The front line of enteric host defense against unwelcome intrusion of harmful microorganisms: mucins, antimicrobial peptides, and microbiota. *Clin Microbiol Rev* 19: 315–337
98 Rimoldi M, Chieppa M, Salucci V, Avogadri F, Sonzogni A, Sampietro GM, Nespoli A, Viale G, Allavena P, Rescigno M (2005) Intestinal immune homeostasis is regulated by the crosstalk between epithelial cells and dendritic cells. *Nat Immunol* 6: 507–514
99 MacPherson AJ, Uhr T (2004) Induction of protective IgA by intestinal dendritic cells carrying commensal bacteria. *Science* 303: 1662–1665
100 Peterson DA, McNulty NP, Guruge JL, Gordon JI (2007) IgA response to symbiotic bacteria as a mediator of gut homeostasis. *Cell Host Microbe* 2: 328–339
101 Zaph C, Du Y, Saenz SA, Nair MG, Perrigoue JG, Taylor BC, Troy AE, Kobuley DE, Kastelein RA, Cua DJ et al (2008) Commensal-dependent expression of IL-25 regulates the IL-23–IL-17 axis in the intestine. *J Exp Med* 205: 2191–2198
102 Ivanov, II, Frutos Rde L, Manel N, Yoshinaga K, Rifkin DB, Sartor RB, Finlay BB, Littman DR (2008) Specific microbiota direct the differentiation of IL-17-producing T-helper cells in the mucosa of the small intestine. *Cell Host Microbe* 4: 337–349
103 Mazmanian SK, Kasper DL (2006) The love-hate relationship between bacterial polysaccharides and the host immune system. *Nat Rev Immunol* 6: 849–858
104 Mazmanian SK, Liu CH, Tzianabos AO, Kasper DL (2005) An immunomodulatory molecule of symbiotic bacteria directs maturation of the host immune system. *Cell* 122: 107–118
105 Wang Q, McLoughlin RM, Cobb BA, Charrel-Dennis M, Zaleski KJ, Golenbock D, Tzianabos AO, Kasper DL (2006) A bacterial carbohydrate links innate and adaptive responses through Toll-like receptor 2. *J Exp Med* 203: 2853–2863
106 Mazmanian SK, Round JL, Kasper DL (2008) A microbial symbiosis factor prevents intestinal inflammatory disease. *Nature* 453: 620–625
107 Fiocchi C (2005) One commensal bacterial molecule – all we need for health? *N Engl J Med* 353: 2078–2080
108 Bjorksten B, Naaber P, Sepp E, Mikelsaar M (1999) The intestinal microflora in allergic Estonian and Swedish 2-year-old children. *Clin Exp Allergy* 29: 342–346
109 Watanabe S, Narisawa Y, Arase S, Okamatsu H, Ikenaga T, Tajiri Y, Kumemura M (2003) Differences in fecal microflora between patients with atopic dermatitis and healthy control subjects. *J Allergy Clin Immunol* 111: 587–591
110 Kalliomaki M, Salminen S, Arvilommi H, Kero P, Koskinen P, Isolauri E (2001) Probiotics in primary prevention of atopic disease: a randomised placebo-controlled trial. *Lancet* 357: 1076–1079
111 Penders J, Thijs C, van den Brandt PA, Kummeling I, Snijders B, Stelma F, Adams H, van Ree R, Stobberingh EE (2007) Gut microbiota composition and development of atopic manifestations in infancy: the KOALA Birth Cohort Study. *Gut* 56: 661–667
112 Alm JS, Swartz J, Bjorksten B, Engstrand L, Engstrom J, Kuhn I, Lilja G, Mollby R,

Norin E, Pershagen G et al (2002) An anthroposophic lifestyle and intestinal microflora in infancy. *Pediatr Allergy Immunol* 13: 402–411

113 Alm B, Erdes L, Mollborg P, Pettersson R, Norvenius SG, Aberg N, Wennergren G (2008) Neonatal antibiotic treatment is a risk factor for early wheezing. *Pediatrics* 121: 697–702

114 Wickens K, Black PN, Stanley TV, Mitchell E, Fitzharris P, Tannock GW, Purdie G, Crane J (2008) A differential effect of 2 probiotics in the prevention of eczema and atopy: a double-blind, randomized, placebo-controlled trial. *J Allergy Clin Immunol* 122: 788–794

115 Kalliomaki M, Salminen S, Poussa T, Arvilommi H, Isolauri E (2003) Probiotics and prevention of atopic disease: 4–year follow-up of a randomised placebo-controlled trial. *Lancet* 361: 1869–1871

116 Kalliomaki M, Salminen S, Poussa T, Isolauri E (2007) Probiotics during the first 7 years of life: a cumulative risk reduction of eczema in a randomized, placebo-controlled trial. *J Allergy Clin Immunol* 119: 1019–1021

117 Kopp MV, Hennemuth I, Heinzmann A, Urbanek R (2008) Randomized, double-blind, placebo-controlled trial of probiotics for primary prevention: no clinical effects of Lactobacillus GG supplementation. *Pediatrics* 121: e850–856

118 Taylor AL, Hale J, Hales BJ, Dunstan JA, Thomas WR, Prescott SL (2007) FOXP3 mRNA expression at 6 months of age is higher in infants who develop atopic dermatitis, but is not affected by giving probiotics from birth. *Pediatr Allergy Immunol* 18: 10–19

119 Viljanen M, Savilahti E, Haahtela T, Juntunen-Backman K, Korpela R, Poussa T, Tuure T, Kuitunen M (2005) Probiotics in the treatment of atopic eczema/dermatitis syndrome in infants: a double-blind placebo-controlled trial. *Allergy* 60: 494–500

120 Noverr MC, Falkowski NR, McDonald RA, McKenzie AN, Huffnagle GB (2005) Development of allergic airway disease in mice following antibiotic therapy and fungal microbiota increase: role of host genetics, antigen, and interleukin-13. *Infect Immun* 73: 30–38

121 Feleszko W, Jaworska J, Rha RD, Steinhausen S, Avagyan A, Jaudszus A, Ahrens B, Groneberg DA, Wahn U, Hamelmann E (2007) Probiotic-induced suppression of allergic sensitization and airway inflammation is associated with an increase of T regulatory-dependent mechanisms in a murine model of asthma. *Clin Exp Allergy* 37: 498–505

122 Repa A, Grangette C, Daniel C, Hochreiter R, Hoffmann-Sommergruber K, Thalhamer J, Kraft D, Breiteneder H, Mercenier A, Wiedermann U (2003) Mucosal co-application of lactic acid bacteria and allergen induces counter-regulatory immune responses in a murine model of birch pollen allergy. *Vaccine* 22: 87–95

123 Daniel C, Repa A, Mercenier A, Wiedermann U, Wells J (2007) The European LAB-DEL project and its relevance to the prevention and treatment of allergies. *Allergy* 62: 1237–1242

124 Schabussova I, Wiedermann U (2008) Lactic acid bacteria as novel adjuvant systems for prevention and treatment of atopic diseases. *Curr Opin Allergy Clin Immunol* 8: 557–564

125 Di Giacinto C, Marinaro M, Sanchez M, Strober W, Boirivant M (2005) Probiotics ameliorate recurrent Th1–mediated murine colitis by inducing IL-10 and IL-10-dependent TGF-β-bearing regulatory cells. *J Immunol* 174: 3237–3246
126 Smits HH, Engering A, van der Kleij D, de Jong EC, Schipper K, van Capel TM, Zaat BA, Yazdanbakhsh M, Wierenga EA, van Kooyk Y et al (2005) Selective probiotic bacteria induce IL-10-producing regulatory T cells *in vitro* by modulating dendritic cell function through dendritic cell-specific intercellular adhesion molecule 3–grabbing nonintegrin. *J Allergy Clin Immunol* 115: 1260–1267
127 Foligne B, Zoumpopoulou G, Dewulf J, Ben Younes A, Chareyre F, Sirard JC, Pot B, Grangette C (2007) A key role of dendritic cells in probiotic functionality. *PLoS ONE* 2: e313
128 Forsythe P, Inman MD, Bienenstock J (2007) Oral treatment with live Lactobacillus reuteri inhibits the allergic airway response in mice. *Am J Respir Crit Care Med* 175: 561–569
129 O'Mahony C, Scully P, O'Mahony D, Murphy S, O'Brien F, Lyons A, Sherlock G, MacSharry J, Kiely B, Shanahan F et al (2008) Commensal-induced regulatory T cells mediate protection against pathogen-stimulated NF-kappaB activation. *PLoS Pathog* 4: e1000112
130 Hall JA, Bouladoux N, Sun CM, Wohlfert EA, Blank RB, Zhu Q, Grigg ME, Berzofsky JA, Belkaid Y (2008) Commensal DNA limits regulatory T cell conversion and is a natural adjuvant of intestinal immune responses. *Immunity* 29: 637–649
131 Lee J, Mo JH, Katakura K, Alkalay I, Rucker AN, Liu YT, Lee HK, Shen C, Cojocaru G, Shenouda S et al (2006) Maintenance of colonic homeostasis by distinctive apical TLR9 signalling in intestinal epithelial cells. *Nat Cell Biol* 8: 1327–1336
132 Lauener RP, Birchler T, Adamski J, Braun-Fahrlander C, Bufe A, Herz U, von Mutius E, Nowak D, Riedler J, Waser M et al (2002) Expression of CD14 and Toll-like receptor 2 in farmers' and non-farmers' children. *Lancet* 360: 465–466
133 Horner AA, Raz E (2003) Do microbes influence the pathogenesis of allergic diseases? Building the case for Toll-like receptor ligands. *Curr Opin Immunol* 15: 614–619
134 Fageras Bottcher M, Hmani-Aifa M, Lindstrom A, Jenmalm MC, Mai XM, Nilsson L, Zdolsek HA, Bjorksten B, Soderkvist P, Vaarala O (2004) A TLR4 polymorphism is associated with asthma and reduced lipopolysaccharide-induced interleukin-12(p70) responses in Swedish children. *J Allergy Clin Immunol* 114: 561–567
135 Eder W, Klimecki W, Yu L, von Mutius E, Riedler J, Braun-Fahrlander C, Nowak D, Martinez FD (2004) Toll-like receptor 2 as a major gene for asthma in children of European farmers. *J Allergy Clin Immunol* 113: 482–488
136 Baldini M, Lohman IC, Halonen M, Erickson RP, Holt PG, Martinez FD (1999) A Polymorphism in the 5' flanking region of the CD14 gene is associated with circulating soluble CD14 levels and with total serum immunoglobulin E. *Am J Respir Cell Mol Biol* 20: 976–983
137 Koppelman GH, Reijmerink NE, Colin Stine O, Howard TD, Whittaker PA, Meyers

DA, Postma DS, Bleecker ER (2001) Association of a promoter polymorphism of the CD14 gene and atopy. *Am J Respir Crit Care Med* 163: 965–969

138 Simpson A, John SL, Jury F, Niven R, Woodcock A, Ollier WE, Custovic A (2006) Endotoxin exposure, CD14, and allergic disease: an interaction between genes and the environment. *Am J Respir Crit Care Med* 174: 386–392

139 Gereda JE, Leung DY, Thatayatikom A, Streib JE, Price MR, Klinnert MD, Liu AH (2000) Relation between house-dust endotoxin exposure, type 1 T-cell development, and allergen sensitisation in infants at high risk of asthma. *Lancet* 355: 1680–1683

140 Eisenbarth SC, Piggott DA, Huleatt JW, Visintin I, Herrick CA, Bottomly K (2002) Lipopolysaccharide-enhanced, toll-like receptor 4-dependent T helper cell type 2 responses to inhaled antigen. *J Exp Med* 196: 1645–1651

141 Tulic MK, Fiset PO, Manoukian JJ, Frenkiel S, Lavigne F, Eidelman DH, Hamid Q (2004) Role of toll-like receptor 4 in protection by bacterial lipopolysaccharide in the nasal mucosa of atopic children but not adults. *Lancet* 363: 1689–1697

142 Denning TL, Wang YC, Patel SR, Williams IR, Pulendran B (2007) Lamina propria macrophages and dendritic cells differentially induce regulatory and interleukin 17-producing T cell responses. *Nat Immunol* 8: 1086–1094

143 Sun CM, Hall JA, Blank RB, Bouladoux N, Oukka M, Mora JR, Belkaid Y (2007) Small intestine lamina propria dendritic cells promote *de novo* generation of Foxp3 T reg cells *via* retinoic acid. *J Exp Med* 204: 1775–1785

144 Rescigno M, Matteoli G (2008) Lamina propria dendritic cells: for whom the bell TOLLs? *Eur J Immunol* 38: 1483–1486

145 Bashir ME, Louie S, Shi HN, Nagler-Anderson C (2004) Toll-like receptor 4 signaling by intestinal microbes influences susceptibility to food allergy. *J Immunol* 172: 6978–6987

146 Debarry J, Garn H, Hanuszkiewicz A, Dickgreber N, Blumer N, von Mutius E, Bufe A, Gatermann S, Renz H, Holst O et al (2007) Acinetobacter lwoffii and Lactococcus lactis strains isolated from farm cowsheds possess strong allergy-protective properties. *J Allergy Clin Immunol* 119: 1514–1521

147 Dogi CA, Galdeano CM, Perdigon G (2008) Gut immune stimulation by non pathogenic Gram(+) and Gram(–) bacteria. Comparison with a probiotic strain. *Cytokine* 41: 223–231

148 Nair MG, Gallagher IJ, Taylor MD, Loke P, Coulson PS, Wilson RA, Maizels RM, Allen JE (2005) Chitinase and Fizz family members are a generalized feature of nematode infection with selective upregulation of Ym1 and Fizz1 by antigen-presenting cells. *Infection and Immunity* 73: 385–394

149 Loke P, MacDonald AS, Robb A, Maizels RM, Allen JE (2000) Alternatively activated macrophages induced by nematode infection inhibit proliferation *via* cell to cell contact. *Eur J Immunol* 30: 2669–2678

150 Akdis M, Verhagen J, Taylor A, Karamloo F, Karagiannidis C, Crameri R, Thunberg S, Deniz G, Valenta R, Fiebig H et al (2004) Immune responses in healthy and allergic

individuals are characterized by a fine balance between allergen-specific T regulatory 1 and T helper 2 Cells. *J Exp Med* 199: 1567–1575

151 Verhagen J, Akdis M, Traidl-Hoffmann C, Schmid-Grendelmeier P, Hijnen D, Knol EF, Behrendt H, Blaser K, Akdis CA (2006) Absence of T-regulatory cell expression and function in atopic dermatitis skin. *J Allergy Clin Immunol* 117: 176–183

152 Yazdanbakhsh M, Kremsner PG, van Ree R (2002) Allergy, parasites, and the hygiene hypothesis. *Science* 296: 490–494

153 Maizels RM (2005) Infections and allergy – helminths, hygiene and host immune regulation. *Curr Opin Immunol* 17: 656–661

Hepatitis A virus, TIM-1 and allergy

Dale T. Umetsu and Rosemarie H. DeKruyff

Harvard Medical School, Children's Hospital Boston, One Blackfan Circle, Boston, MA 02115, USA

Abstract

In 1989, Strachan first proposed the hygiene hypothesis to explain the rapid increase in the prevalence of allergy and asthma. While the significant rise in the prevalence of asthma and allergy remains unexplained, epidemiological data suggest that infection with the Hepatitis A virus (HAV) might protect against asthma and allergy, and genetic studies identifying the HAV receptor, TIM-1, as an important atopy susceptibility gene, support this idea. In this chapter, we review the genetics and immunobiology of TIM-1 and TIM gene family members, and the possibility that HAV and TIM-1 may regulate the development of asthma and allergy.

Introduction

Atopic diseases, which include asthma, hay fever (allergic rhinitis), atopic dermatitis (eczema) and IgE mediated food allergy, affect 20–40% of individuals in industrialized countries. In the US, asthma affects over 21 million individuals, atopic dermatitis affects 10–15% of individuals, and food allergy now affects 4% of all individuals. What is most alarming about the atopic diseases is that the prevalence of these problems has increased dramatically worldwide, particularly over the last two decades, for reasons that are not fully understood. This is clearly a puzzle that no one fully understands, but is due most likely to the dramatic changes that have occurred in our environment and not to changes in the genetic composition of the population (genetic changes would have to occur over several generations to explain this).

The precise environmental factors however, that are responsible for the increase in the prevalence of atopy are not fully understood, and multiple causes have been suggested. For example, the increase in atopy and asthma has been attributed to increases in air pollution [1], to an increase in aeroallergen exposure, in part due to global warming [2], or to an increase in the prevalence of obesity [3]. Epidemiological studies have also linked the increased prevalence of atopic diseases to an

The Hygiene Hypothesis and Darwinian Medicine, edited by Graham A.W. Rook

increased use of acetaminophen [4] or to vitamin D deficiency [5]. However, the hypothesis that has received the most attention is the hygiene hypothesis, the idea that "*the increase in allergy could be explained if allergic diseases were prevented by infection in early childhood, transmitted by unhygienic contact with older siblings, or acquired prenatally*" [6].

The hygiene hypothesis is based on epidemiological observations indicating that exposure to older siblings reduced the risk of developing atopy, and that the incidence of many infections have decreased over the past three decades, due to improved public health measures, increased use of vaccines and antibiotics, while the incidence of atopy has increased. For example, the incidence of measles, mumps, tuberculosis, rheumatic fever, as well as hepatitis A, has decreased quite dramatically over the past several decades [7]. However, the specific microorganisms that, and the mechanisms by which these infections, might protect against asthma and allergy are not fully understood. It is has been suggested that specific single infectious organisms, e.g., measles or tuberculosis, (either overt or unapparent infection) or possibly that specific commensal organisms (e.g., in the intestinal tract) might alter the immune system and protect against asthma and allergy. Alternatively, multiple infectious organisms in aggregate might affect the immune system and enhance the maturation or responsiveness of immunity. In this review, we will examine one possible infectious microorganism, the hepatitis A virus (HAV), infection with which has been proposed to have important protective effects against the development of asthma and allergy.

Hepatitis A protects against atopy

Epidemiological studies performed first in 1997 but repeated in different populations in subsequent years showed that infection with the HAV was associated with a reduced incidence of allergy and asthma. In these studies, the prevalence of allergy and asthma, as well as peanut food allergy, was significantly lower in HAV seropositive individuals compared to that in HAV seronegative individuals [8–10]. However, because HAV is not a respiratory virus, and because HAV is transmitted through fecal-oral routes, HAV infection was assumed to be merely a marker of poor hygiene and that poor hygiene was responsible for the protective effects against atopy associated with HAV infection. Although some studies in other populations have not found this relationship [11], the observed association of HAV and protection against atopy is quite remarkable, since antibody titers against very few other specific microorganisms have been associated with protection against atopy. For example, antibody titers against a few other gastrointestinal infectious organisms, *Salmonella*, *Helicobacter pylori* and *Toxoplasma gondii*, and herpes simplex virus 1, have been inversely associated with atopy [8, 10, 12, 13], but titers to other infectious agents, e.g., microorganisms affecting the respiratory tract, have not been associated with protection against atopy. Infection with mycobacteria or BCG vaccination has been reported to protect against asthma and allergy [14], but this has been controversial.

The relationship between atopy and HAV is also remarkable, because three decades ago childhood infection with HAV occurred in nearly all children [7]. Thus, prior to 1970 the incidence of HAV infection approached 100%, but is now < 10% in most industrialized countries, a rapid shift which could account for the rise in atopy over the past two decades, if infection with HAV indeed affected the immune system and protected children from the development of asthma and allergy. In the past, childhood infection with HAV was for the most part mild and clinically unapparent, suggesting a relationship between HAV and the host not unlike that of commensal microorganisms. The great reduction in the prevalence of HAV infection is due to improvements in public health and sanitation, and currently in the United States, infection with HAV occurs primarily during food-borne epidemics or in daycare settings [15]. Since protection against the development of atopy is also associated with early entrance into daycare [16], it is possible that in daycare settings HAV may be the important microbe that protects against atopy, although other infectious agents may also contribute to protection.

Importantly, the possibility that infection with HAV can indeed protect against atopy has been greatly strengthened by genetic studies demonstrating that certain polymorphisms in the receptor for the HAV, i.e., TIM-1/HAVCR1, are associated with the development of atopy [17] (see below). These observations strongly suggest that HAV, by acting through TIM-1/HAVCR1 has direct effects on the host immune system, which then protects against asthma and allergy. Together, the genetic and epidemiological observations demonstrating a protective effect of HAV against the development of atopy provide a compelling story that helps to explain the hygiene hypothesis, as discussed below.

Cloning of the TIM gene family

The identification of specific susceptibility genes for allergy and asthma, which are very complex genetic traits, has been extremely difficult because multiple atopy specific susceptibility genes interact with the environment in nonadditive ways and because each of these genes segregate independently. Geneticists have tried to address the complexity of the problem by performing linkage studies using larger and larger populations to increase statistical power. An alternative approach to address this problem has been to employ animal models of asthma and allergy. With such an approach, using unique congenic mice, the TIM gene family was identified [18]. The congenic strains were generated by genetically moving discrete chromosomal segments from one mouse strain, DBA/2 (asthma resistant), into another strain, BALB/c (asthma susceptible), by repeated backcrossing [19]. One congenic strain, called C.D2/Es-3/Hba, exhibited the DBA/2 phenotype (asthma resistance) and contained a chromosome 11 segment that was syntenic to human chromosome 5g23-35, a region that had been repeated linked to asthma and allergy in humans [20]. Using the C.D2/Es-3/Hba congenic mice, BALB/c mice and their offspring, a novel atopy susceptibility gene locus called *Tapr* (T cell and airway phenotype regulator) was identified, and within this locus the *TIM* (T cell, Ig domain and mucin domain) gene family was cloned.

The TIM gene family

The TIM gene family consists of eight members (TIM-1–8) on mouse chromosome 11B1.1, and three members, TIM-1, TIM-3 and TIM-4) on human chromosome 5q33.2 [21]. The TIM genes encode type 1 membrane proteins, consisting of an N-terminal Cys-rich IgV-like domain, a mucin-like domain, a transmembrane domain, and an intracellular tail. The intracellular tails of TIM-1, TIM-2 and TIM-3, but not TIM-4, contain predicted tyrosine phosphorylation motifs, suggesting that these TIMs are involved in transmembrane signaling. Within the mucin domain, TIM-1 contains 60 O-linked glycosylation sites located within the mucin-like domain, whereas TIM-3 has only three predicted glycosylation sites. The N-terminal Cys-rich regions of the TIM homologs have 40% sequence identity, whereas sequence identity between the mouse and human orthologs is approximately 60% [22]. The structural similarities between all of the TIMs suggest that they arose from an ancestral gene by successive gene duplication events.

Human TIM-1, an important atopy susceptibility gene

The first member of the TIM gene family, TIM-1, had been previously identified by virologists studying the biological functions of HAV, as the receptor for the HAV (*HAVCR1*), and by nephrologists as a kidney injury molecule, KIM-1. However, HAVCR1/TIM-1 is also a member of the TIM family of genes, which have been shown by more recent studies to regulate immune responses. TIM-1 is highly polymorphic in monkeys and in humans, as it is in mice, with single nucleotide polymorphisms (SNPs) as well as insertion/deletion variants in the mucin domain. The significant degree of polymorphisms is thought to be driven by infection with HAV, although data supporting this concept are not yet available.

TIM-1 resides on chromosome 5q23-35, a region that has been repeatedly linked with asthma, and an association between polymorphic variants of TIM-1 and protection against atopic diseases was identified, which was strongest in individuals who had prior infection with HAV [17]. The studies demonstrating an association between TIM-1 and atopy and that TIM-1 is an atopy susceptibility gene have been reproduced in a number of other populations, including one in African-American asthmatics [23], in children with atopic dermatitis in Arizona [24] and in Australia [25], in Koreans with asthma and atopic dermatitis [26], but not in Japanese children with asthma [27]. The lack of association of TIM-1 in Japanese children with asthma may be due to a very reduced incidence of HAV infection in Japan, now close to zero in young Japanese children. However, the precise immunological mechanisms by which HAV infection and TIM-1 alter the immune system to protect against atopy are not yet clear. The immunology of TIM-1 is only beginning to be understood (see below), and the results so far indicate that TIM-1 potently regulates immune responses through novel mechanisms. These results support the possibility that infection with HAV, by stimulating the immune system through TIM-1, can prevent the development of atopy, at least in children with a particular variant of TIM-1.

Human TIM-1 in autoimmune disease

TIM-1 has been associated not only with atopic diseases, but also with several autoimmune diseases, suggesting that TIM-1 regulates the immune system globally. For example, in rheumatoid arthritis polymorphisms in TIM-1 (5509_5511delCAA in exon 4) was associated disease [26], while polymorphisms in the promoter region of TIM-1 was associated with increased C-reactive protein or rheumatoid factor levels [28]. How TIM-1 regulates autoimmune disease, or whether HAV infection is associated with protection from autoimmunity is not yet known. However, TIM-1 mRNA is expressed in the cerebrospinal fluid mononuclear cells of patients with multiple sclerosis (MS), primarily in patients in remission rather than in patients in relapse, suggesting that TIM-1 regulates the development of MS, perhaps regulating the development of tolerance to autoantigens [29].

The relationship between TIM-1 and the hygiene hypothesis

The identification of the relationship between TIM-1, atopy and HAV infection was among the first demonstrating a link between environmental factors (HAV infection) and an important susceptibility gene (TIM-1). This concept linking the environment and genes is now commonly applied in most genetic studies of complex traits, for example in explaining the relationship between polymorphisms in CD14 (a component of the endotoxin receptor) and exposure to bacterial endotoxin in the development of atopy [30], in which the effects of CD14 polymorphisms develop only on exposure to endotoxin. The idea has been extended, and it has now been proposed that genetic polymorphisms modulate the effects of environmental exposures rather than directly influence the risk for asthma and allergy [31]. In any case, gene-environment interactions must always be assessed in complex diseases, and in the case of TIM-1, the receptor for HAV, infection with HAV appears to affect the host immune system and the development of atopy, through TIM-1.

Biological role of TIM-1 in the regulation of immunity

The relationship between TIM-1, atopy and HAV infection and the discovery of TIM-1 as the receptor for HAV implies a very specific molecular mechanism by which HAV infection might protect against atopy. Such a mechanism would help verify the hygiene hypothesis, which up to now is primarily based on epidemiological relationships, with only minimal data on specific infectious microorganisms that might protect against atopy. Since the discovery of the TIM gene family seven years ago, the normal biological functions of the TIM molecules have been intensively studied [32, 33]. Although the precise function of the TIM family of genes is not yet fully understood, the results of such studies of TIM-1 are likely to provide important insight into how HAV infection protects against atopy.

The *in vivo* function and effects of TIM-1 is determined by its expression: TIM-1 is expressed primarily by T cells, as well as by a small subset of dendritic cells (DCs). Activated but not naïve T cells express TIM-1. As T cells differentiate into Th2 cells, TIM-1 continues to be expressed by differentiated Th2 cells, but TIM-1 expression is lost as T cells differentiate into Th1 cells [34]. Cross-linking of TIM-1 on T cells with an agonist mAb provided a very potent costimulatory signal to CD4$^+$ T cells that increased T cell proliferation and cytokine production (IL-4, IFN-γ and IL-10). The costimulatory effect on resting T cells required simultaneous TCR signaling, and could not be observed with monomeric Fab fragments of the anti-TIM-1 mAb (which cannot cross-link TIM-1). *In vivo* administration of the agonist anti-TIM-1 mAb along with antigen provided a very potent adjuvant effect, resulting in greatly increased antigen-specific T cell proliferation and cytokine production. The adjuvant effect of anti-TIM-1 mAb prevented the development of respiratory tolerance [34], consistent with the idea that TIM-1 costimulation potently activates T cells. Normally, respiratory exposure to antigen induces T cell unresponsiveness, and is associated with the development of antigen-specific regulatory T (T_{Reg}) cells expressing Foxp3 [35, 36], but treatment with the agonist anti-TIM-1 mAb prevented this tolerance induction [34], possibly by enhancing Th cell development and hindering T_{Reg} cell development [37].

The function of TIM-1 may depend on how TIM-1 interacts with its ligands, since mAbs recognizing distinct epitopes of TIM-1, or possibly interacting with TIM-1 with distinct affinities, may have different effects on the TIM-1 function. For example, anti-TIM-1 mAbs recognizing exon 4 of the mucin/stalk domain greatly exacerbated airway inflammation and Th2 cytokine production (e.g., having agonistic effects), but another mAb blocked inflammation in a mouse model of asthma (e.g., having blocking effects) [38]. Alternatively, it is possible that the strength of the signal provided by TIM-1, for example by distinct mAbs, may induce different outcomes. Thus, a high affinity anti-TIM-1 mAb, 3B3, strongly costimulates T cells [34], whereas a mAb with much lower affinity, RMT1-10, appears to enhance tolerance induction and inhibit experimental autoimmune encephalomyelitis [39]. However, the precise events that regulate the various outcomes of TIM-1 signaling by distinct mAbs are not yet known, but could be related to the differential effects of HAV infection in patients with different TIM-1 alleles [17].

TIM-1 signal transduction

The molecular signal transduction mechanisms by which TIM-1 costimulates T cell activation are only beginning to be elucidated. For example, it is known that overexpression of TIM-1 in T cells results in an increase in production of IL-4 but not IFN-γ [40]. Furthermore, transfection of the D10 T cell line with TIM-1 results in increased transcription from the IL-4 promoter and activation of NFAT/AP1 elements [40], suggesting that TIM-1 preferentially enhances Th2 cytokine production, consistent with the preferential expression of TIM-1 on Th2 cells. Activation of T cells appears to result in the phosphorylation of a conserved tyrosine in the cytoplasmic tail (Y276) of TIM-1. In studies of TIM-1 utilizing overexpression of TIM-1 in Jurkat T cells, which normally do not express TIM-1 proteins, investigators have found that TIM-1 colocalizes on the T cell surface with CD3 [41]. TIM-1 coimmunoprecipitates with the TCR complex upon TCR cross-linking and T cell activation, and TCR signaling increases upon TIM-1 cross-linking. Furthermore, TIM-1 cross-linking caused rapid tyrosine phosphorylation of TIM-1, as well as phosphorylation of Zap70 and ITK. These results indicate that TIM-1 costimulation enhances TCR signaling and T cell activation, although other pathways may also be activated by TIM-1 cross-linking.

TIM-1 ligands

Several approaches have been taken to determine the natural ligands of TIM-1 (in addition to HAV), which may provide further insight into the natural function of TIM-1. These approaches have identified several ligands, including TIM-1 itself, TIM-4, IgAλ and phosphatidylserine (PtdSer) as molecules that can bind to TIM-1 [42–44]. The structure of TIM-1 includes a glycosylated mucin domain, which imparts a degree of promiscuity in terms of what binds to the TIM-1 molecule. This has made the identification of 'specific' TIM-1 ligands difficult, and may explain the identification of multiple ligands. *In vitro* studies demonstrated that a TIM-1-Ig fusion proteins (consisting of the TIM-1 IgV domain with or without the mucin domain coupled to the Fc portion of IgG) bound to cells expressing TIM-4 or to cells expressing TIM-1. This interaction could be specifically inhibited by anti-TIM-1 mAb, strongly suggesting that TIM-4 or TIM-1 were ligands of TIM-1 [44, 45]. Furthermore, *in vivo* administration of TIM-1-Ig fusion protein or TIM-4-Ig fusion protein along with antigen increased subsequent *in vitro* antigen-specific T cell proliferation and production of cytokines. Endogenous TIM-4 did not appear to participate in this response to exogenous fusion proteins, since TIM-4 expression is restricted to a few cells, including $CD11b^+$ and $CD11c^+$ cells, including macrophages and DCs, particularly on lymphoid $CD8^+$ DCs in the splenic marginal zone [46]. The initial interpretation of these results was that TIM-1-Ig or TIM-4-Ig bound to TIM-1 on T cells resulting in T cell activation [44]. However, the interpretation of these results was revised when the crystal structure of TIM-1 was solved in 2007 [22], demonstrating that PtdSer was an important ligand of TIM-1 and TIM-4, and suggesting that the TIM-1–TIM-1 interaction and the TIM-1–TIM-4 interaction might be mediated through a bridge created by PtdSer containing exosomes, which are micro membrane vesicles released from leukocytes and epithelial cells [47].

TIM-1 crystal structure and phosphatidylserine

Although TIM-4 can bind to TIM-1, TIM-1 (and TIM-4) also appears to very specifically bind PtdSer, a membrane phospholipid expressed by apoptotic cells. This binding has been confirmed both by biochemical and crystallographic studies [43, 48, 49]. Crystal structure analysis of TIM family members indicated that the IgV domain of all the TIMs has two anti-parallel β-sheets (Fig. 1), bridged by the first and last of six Cys residues in the IgV domain, similar to the structure of other Ig superfamily members. Four additional Cys residues link two loops, the FG loop and the CC' loop, forming a cleft (called MILIBS, for Metal Ion Ligand Binding Site). The six conserved Cys residues in the IgV domains of all of the TIM molecules appear to provide a distinctive conserved structural feature, resulting in a MILIBS cleft in TIM-1, TIM-4 and TIM-3, but not TIM-2. Importantly, *PtdSer fits snugly in this cleft of TIM-1 and TIM-4*, and the HAV also binds to human TIM-1 at this cleft region [48], with Ser37 in the CC' loop possibly the critical virus-binding residue [22]. The crystal structure of TIMs suggests most likely that PtdSer is the major natural ligand of TIM-1 and TIM-4. Moreover, studies suggest that the TIM-1 to TIM-1 and TIM-1 to TIM-4 binding may be mediated by exosomes from apoptotic cells, such that the exosomes expressing PtdSer, bridge the TIM molecules, as has been shown by electron microscopy [49].

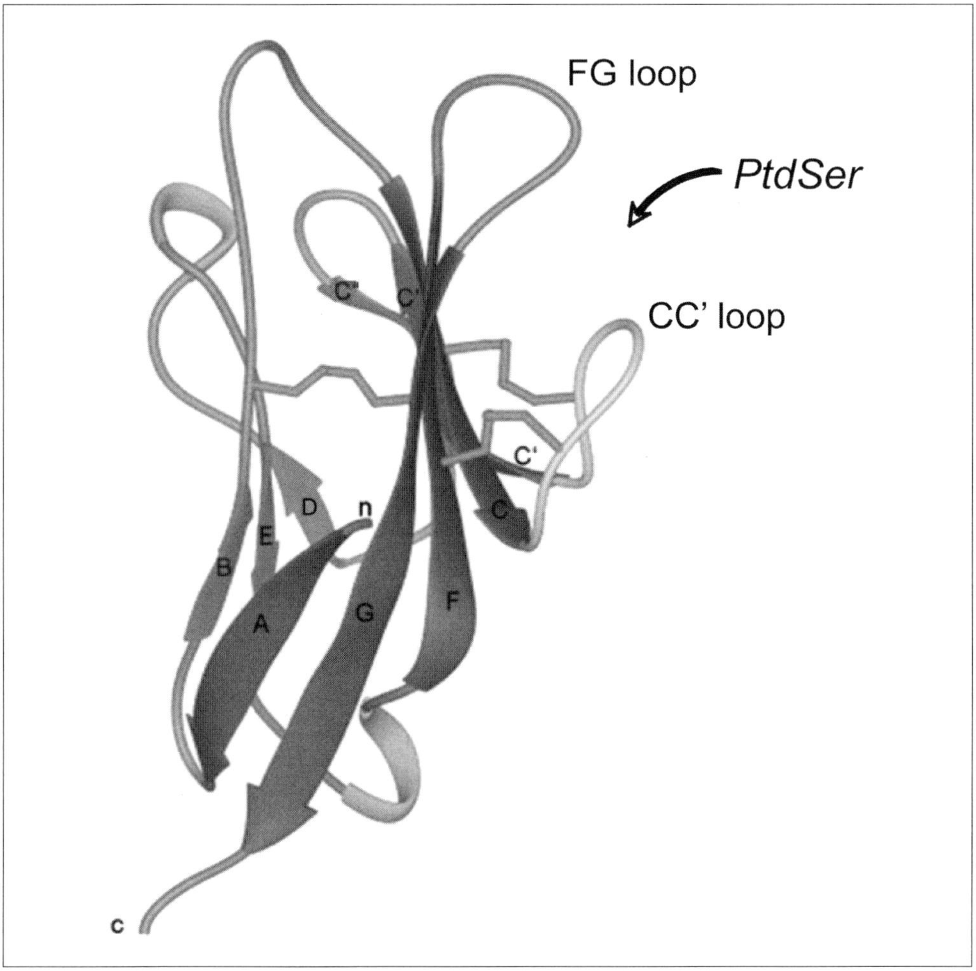

Figure 1
Ribbon diagram of the mTIM-1 Ig domain. β-sheets are labelled AGF and BED. The sticks represent the 3 Cys disulphide bonds. The FG and CC' loops form a conserved cleft, into which PtdSer snugly fits. Adapted from [22].

TIM-4 and TIM-3 regulate apoptosis and immune tolerance

Apoptotic cell death is a critical and evolutionally conserved process for elimination of unnecessary cells [50, 51], and is essential for maintenance of tissue homeostasis and self-tolerance [52]. Clearance of apoptotic cells by phagocytosis can result in powerful anti-inflammatory and immunosuppressive effects [53]. Defective clearance of apoptotic cells has been linked closely to autoimmune and inflammatory diseases [54–56]. Phagocytes such as macrophages or DCs engulf both apoptotic and necrotic cells and the biological response of the phagocyte differs depending on the balance. Necrotic cells induce release of proinflammatory cytokines and stimulate upregulation of costimulatory molecules on the dendritic cell surface, which in turn induces T cell activation and an inflammatory response. In contrast, engulfment of apoptotic cells does not trigger expression of costimulatory molecules, and is associated with production of anti-inflammatory cytokines including TGF-β and IL-10, thus preventing inappropriate inflammatory responses to self-proteins. Under some circumstances however, the ingestion of apoptotic cells by DCs can induce immune responses and inflammation [57].

Several pathways facilitating phagocyte recognition and uptake of apoptotic cells have been identified. However, the TIM-4 pathway is very different from these in several respects and is unlikely to be simply a redundant alternative pathway. Other molecules that facilitate phagocyte recognition of PtdSer on apoptotic cells that have been described include the Mer receptor tyrosine kinase [58], Milk fat globule-EGF-factor 8 (MFG-E8) which is secreted by macrophages and binds PtdSer [59], and Growth arrest-specific gene 6 (GAS6) [58]. These are component systems that act in concert, and consist of a soluble molecule such as GAS-6 and a cell bound tether molecule such as Mer. While MFG-E8 and GAS6 are widely expressed in somatic cells [60], TIM-4 expression in contrast, is tightly restricted to DCs and macrophages [43], particularly to APCs in the marginal zone of the spleen [46], where TIM-4 expressing cells may select out circulating apoptotic immune cells. TIM-4 recognition of apoptotic cells is highly specific for PtdSer [43], while other PtdSer binding molecules such as MFG-E8 and GAS-6 bind PtdSer as well as other phospholipids [59, 61]. Moreover, the maturation state under which a phagocyte expresses TIM-4 appears to differ from that when it expresses MFG-E8. While normal resident peritoneal macrophages express TIM-4 but not MFG-E8, stimulation with thioglycollate induces a peritoneal macrophage population which expresses MFG-E8 but little or no TIM-4 [49]. Thus TIM-4-mediated engulfment of apoptotic cells may have an important role in immune responses distinct from that of other PtdSer receptors.

Other ligands of TIM-1

Although TIM-1 specifically binds PtdSer, several other ligands for TIM-1 have been identified using other approaches. These include TIM-1 itself [22], TIM-4 [44], and IgA [42]. The structure of TIM-1 includes a glycosylated mucin domain, which imparts a degree of promiscuity to the TIM-1 molecule. This may explain the identification of multiple ligands, and has made it difficult to determine which if any of the already identified ligands is the primary ligand of TIM-1. TIM-1-Ig and

TIM-1 tetramer bound to CHO cells transfected with TIM-4 [44], and this interaction could be specifically inhibited by anti-TIM-1 mAb. In addition, TIM-1-Ig and TIM-1 tetramers bound to cells expressing TIM-1, and this binding was dependent on the presence of the glycosylated mucin stalk, although the mucin stalk alone was not sufficient for TIM binding, indicating that homotypic TIM-1-TIM-1 binding also occurred [45].

Administration of TIM-4-Ig *in vivo* resulted in hyperproliferation of T cells and enhancement of T cell cytokine production [44]. The interpretation was that TIM-4-Ig bound to TIM-1 on T cells resulting in T cell activation. Administration of a TIM-1-Ig fusion protein, however, produced similar results, which was initially difficult to understand. In this instance, it is possible that TIM-1 binds to T cells expressing TIM-1, which is suggested by the crystal structure of TIM-1, which was solved in 2007 [22]. However, there is evidence that some apparent TIM-1–TIM-1 or TIM-1–TIM-4 interactions could occur *via* a 'bridge' with two TIM proteins binding to a membrane fragment or exosome expressing exposed PtdSer. Observation by electron microscopy of TIM-1 and TIM-4 transfected cell lines indicated the presence of exosome-like vesicles bound to the cells [49].

Conclusions

The significant rise in the prevalence of asthma and allergy over the past two decades remains unexplained, but is likely to be related at least in part to reductions in the incidence of some infections, as articulated by the hygiene hypothesis. Epidemiological data suggest that infection with HAV might protect against asthma and allergy, and genetic studies identifying the HAV receptor, TIM-1, as an important atopy susceptibility gene, support this idea. Recent immunological and functional studies of TIM-1 and of another member of the TIM gene family, TIM-4, indicate that TIM-1 and TIM-4 have novel, unique and important roles in regulating immune responses. TIM-1 has been shown to be a potent costimulatory molecule for T cells, and both TIM-1 and TIM-4 are receptors for phosphatidyserine (PtdSer), a key molecule expressed by apoptotic cells. Both TIM-1 and TIM-4, as receptors for PtdSer, mediate the recognition and engulfment of apoptotic cells, an important event that regulates immune responses and peripheral tolerance. Since defective clearance of apoptotic cells results in increased immune responses, including autoimmunity, we suggest that TIM-1 and TIM-4 expressing cells, by controlling the recognition and clearance of apoptotic cells, regulate immune responsiveness, tolerance and the development of atopy. Moreover, infection with HAV, by activating TIM-1 expressing cells, may alter this recognition and clearance of apoptotic cells, resulting in protection against atopy in individuals expressing the relevant polymorphic variant of TIM-1.

The precise molecular pathways by which TIM-1 and HAV control immunity require further elucidation. We suggest that HAV by binding its receptor TIM-1 has broad effects on innate and adaptive immunity that somehow enhance protection against allergic responses. Because HAV in the past infected nearly 100% of individuals, it is likely that HAV evolved with its host, and drove the development of the striking degree of TIM-1 polymorphisms. Further, recent studies of the TIM gene family suggest that the TIMs evolved to play an important role in regulating immunity involving cell survival, death, immune activation and immune tolerance. Together these observations strongly suggest the HAV, interacting with TIM-1 expressing cells, greatly alters immunity, resulting in protection against atopy. Therefore, the study of the mechanisms by which TIM-1 and HAV affect immunity should provide important insight into the regulation of immune responses, and help to elucidate novel features of immunity and tolerance. We believe that this understanding of how HAV and TIM-1 interact may allow the protective effects of HAV infection to be replicated as a novel therapy without the detrimental effects of infection, that could prevent the development of atopic diseases, including asthma, eczema and food allergy.

References

1 Peden, D.B. 2005. The epidemiology and genetics of asthma risk associated with air pollution. *J Allergy Clin Immunol* 115: 213–219; quiz 220

2 D'Amato, G., and L. Cecchi. 2008. Effects of climate change on environmental factors in respiratory allergic diseases. *Clin Exp Allergy* 38: 1264–1274

3 Camargo, C.A., Jr., S.T. Weiss, S. Zhang, W.C. Willett, and F.E. Speizer. 1999. Prospective study of body mass index, weight change, and risk of adult-onset asthma in women. Arch Intern Med 159: 2582–2588

4 Beasley, R., T. Clayton, J. Crane, E. von Mutius, C.K. Lai, S. Montefort, and A. Stewart. 2008. Association between paracetamol use in infancy and childhood, and risk of asthma, rhinoconjunctivitis, and eczema in children aged 6–7 years: analysis from Phase Three of the ISAAC programme. *Lancet* 372: 1039–1048

5 Camargo, C.A., Jr., S.L. Rifas-Shiman, A.A. Litonjua, J.W. Rich-Edwards, S.T. Weiss, D.R. Gold, K. Kleinman, and M.W. Gillman. 2007. Maternal intake of vitamin D during pregnancy and risk of recurrent wheeze in children at 3 y of age. *Am J Clin Nutr* 85: 788–795

6 Strachan, D.P. 1989. Hay fever, hygiene, and household size. *BMJ* 299: 1259–1260

7 Bach, J. 2002. The effect of infections on susceptibility to autoimmune and allergic diseases. *N Engl J Med* 347: 911–920

8 Matricardi, P., F. Rosmini, S. Riondino, M. Fortini, L. Ferrigno, M. Rapicetta, and S. Bonini. 2000. Exposure to foodborne and orofecal microbes versus airborne viruses in relation to atopy and allergic asthma: epidemiological study. *BMJ* 320: 412–417

9 Matricardi, P.M., F. Rosmini, L. Ferrigno, R. Nisini, M. Rapicetta, P. Chionne, T. Stroffolini, P. Pasquini, and R. D'Amelio. 1997. Cross sectional retrospective study of

prevalence of atopy among Italian military students with antibodies against hepatitis A virus. *BMJ* 314: 999–1003

10 Linneberg, A., C. Ostergaard, M. Tvede, L.P. Andersen, N.H. Nielsen, F. Madsen, L. Frolund, A. Dirksen, and T. Jorgensen. 2003. IgG antibodies against microorganisms and atopic disease in Danish adults: the Copenhagen Allergy Study. *J Allergy Clin Immunol* 111: 847–853

11 Jarvis, D., C. Luczynska, S. Chinn, and P. Burney. 2004. The association of hepatitis A and *Helicobacter pylori* with sensitization to common allergens, asthma and hay fever in a population of young British adults. *Allergy* 59: 1063–1067

12 Kosunen, T.U., J. Hook-Nikanne, A. Salomaa, S. Sarna, A. Aromaa, and T. Haahtela. 2002. Increase of allergen-specific immunoglobulin E antibodies from 1973 to 1994 in a Finnish population and a possible relationship to *Helicobacter pylori* infections. *Clin Exp Allergy* 32: 373–378

13 Illi, S., E. von Mutius, S. Lau, R. Bergmann, B. Niggemann, C. Sommerfeld, and U. Wahn. 2001. Early childhood infectious diseases and the development of asthma up to school age: a birth cohort study. *BMJ* 322: 390–395

14 Shirakawa, T., T. Enomoto, S. Shimazu, and J.M. Hopkin. 1997. The inverse association between tuberculin responses and atopic disorder. *Science* 275: 77–79

15 Haaheim, L., J. Pattison, and R. Whitley. 2002. *A practical guide to clinical virology*. John Wiley & Sons.,

16 Ball, T.M., J.A. Castro-Rodriguez, K.A. Griffith, C.J. Holberg, F.D. Martinez, and A.L. Wright. 2000. Siblings, day-care attendance, and the risk of asthma and wheezing during childhood. *N Engl J Med* 343: 538–543

17 McIntire, J., S. Umetsu, C. Macaubas, E. Hoyte, C. Cinnioglu, L. Cavalli-Sforza, G. Barsh, J. Hallmayer, P. Underhill, N. Risch, G. Freeman, R. DeKruyff, and D. Umetsu. 2003. *Immunology*: hepatitis A virus link to atopic disease. *Nature* 425: 576

18 McIntire, J.J., S.E. Umetsu, O. Akbari, M. Potter, V.K. Kuchroo, G.S. Barsh, G.J. Freeman, D.T. Umetsu, and R.H. DeKruyff. 2001. Identification of Tapr (an airway hyperreactivity regulatory locus) and the linked Tim gene family. *Nat Immunol* 2: 1109–1116

19 Ruscetti, S., R. Matthai, and M. Potter. 1985. Susceptibility of BALB/c mice carrying various DBA/2 genes to development of Friend murine leukemia virus-induced erythroleukemia. *J Exp Med* 162: 1579–1587

20 Marsh, D.G., J.D. Neely, D.R. Breazeale, B. Ghosh, L.R. Freidhoff, E. Ehrlich-Kautzky, C. Schou, G. Krishnaswamy, and T.H. Beaty. 1994. Linkage analysis of IL4 and other chromosome 5q31.1 markers and total serum immunoglobulin E concentrations. *Science* 264: 1152–1156

21 McIntire, J.J., D.T. Umetsu, and R.H. DeKruyff. 2004. TIM-1, a novel allergy and asthma susceptibility gene. *Springer Semin Immunopathol* 25: 335–348

22 Santiago, C., A. Ballesteros, C. Tami, L. Martinez-Munoz, G.G. Kaplan, and J.M. Casasnovas. 2007. Structures of T Cell immunoglobulin mucin receptors 1 and 2 reveal mechanisms for regulation of immune responses by the TIM receptor family. *Immunity* 26: 299–310

23 Gao, P.S., R.A. Mathias, B. Plunkett, A. Togias, K.C. Barnes, T.H. Beaty, and S.K. Huang. 2005. Genetic variants of the T-cell immunoglobulin mucin 1 but not the T-cell immunoglobulin mucin 3 gene are associated with asthma in an African American population. *J Allergy Clin Immunol* 115: 982–988
24 Graves, P.E., V. Siroux, S. Guerra, W.T. Klimecki, and F.D. Martinez. 2005. Association of atopy and eczema with polymorphisms in T-cell immunoglobulin domain and mucin domain-IL-2–inducible T-cell kinase gene cluster in chromosome 5 q 33. *J Allergy Clin Immunol* 116: 650–656
25 Page, N.S., G. Jones, and G.J. Stewart. 2006. Genetic association studies between the T cell immunoglobulin mucin (TIM) gene locus and childhood atopic dermatitis. *Int Arch Allergy Immunol* 141: 331–336
26 Chae, S., J. Song, Y. Lee, J. Kim, and H. Chung. 2003. The association of the exon 4 variations of Tim-1 gene with allergic diseases in a Korean population. *Biochem Biophys Res Commun* 12: 346–350
27 Noguchi, E., J. Nakayama, M. Kamioka, K. Ichikawa, M. Shibasaki, and T. Arinami. 2003. Insertion/deletion coding polymorphisms in hHAVcr-1 are not associated with atopic asthma in the Japanese population. *Genes Immun* 4: 170–173
28 Chae, S.C., Y.R. Park, J.H. Song, S.C. Shim, K.S. Yoon, and H.T. Chung. 2005. The polymorphisms of Tim-1 promoter region are associated with rheumatoid arthritis in a Korean population. *Immunogenetics* 56: 696–701
29 Khademi, M., Z. Illes, A.W. Gielen, M. Marta, N. Takazawa, C. Baecher-Allan, L. Brundin, J. Hannerz, C. Martin, R.A. Harris, D.A. Hafler, V.K. Kuchroo, T. Olsson, F. Piehl, and E. Wallstrom. 2004. T Cell Ig- and mucin-domain-containing molecule-3 (TIM-3) and TIM-1 molecules are differentially expressed on human Th1 and Th2 cells and in cerebrospinal fluid-derived mononuclear cells in multiple sclerosis. *J Immunol* 172: 7169–7176
30 Martinez, F.D. 2007. CD14, endotoxin, and asthma risk: actions and interactions. *Proc Am Thorac Soc* 4: 221–225
31 Martinez, F.D. 2008. Gene-environment interaction in complex diseases: asthma as an illustrative case. *Novartis Found Symp* 293: 184–192; discussion 192–187
32 Umetsu, D.T., S.E. Umetsu, G.J. Freeman, and R.H. DeKruyff. 2008. TIM gene family and their role in atopic diseases. *Curr Top Microbiol Immunol* 321: 201–215
33 Kuchroo, V.K., V. Dardalhon, S. Xiao, and A.C. Anderson. 2008. New roles for TIM family members in immune regulation. *Nat Rev Immunol* 8: 577–580
34 Umetsu, S., W. Lee, J. McIntire, L. Downey, B. Sanjanwala, O. Akbari, G. Berry, H. Nagumo, G. Freeman, D. Umetsu, and R. DeKruyff. 2005. TIM-1 induces T cell activation and inhibits the development of peripheral tolerance. *Nature Immunol* 6: 447–454
35 Akbari, O., G.J. Freeman, E.H. Meyer, E.A. Greenfield, T.T. Chang, A.H. Sharpe, G. Berry, R.H. DeKruyff, and D.T. Umetsu. 2002. Antigen-specific regulatory T cells develop *via* the ICOS-ICOS-Ligand pathway and inhibit allergen-induced airway hyperreactivity. *Nat Med* 8: 1024–1032

36 Stock, P., O. Akbari, G. Berry, G. Freeman, R. DeKruyff, and D.T. Umetsu. 2004. Induction of TH1-like regulatory cells that express Foxp3 and protect against airway hyperreactivity. *Nature Immunol* 5: 1149–1156
37 Degauque, N., C. Mariat, J. Kenny, D. Zhang, W. Gao, M.D. Vu, S. Alexopoulos, M. Oukka, D.T. Umetsu, R.H. DeKruyff, V. Kuchroo, X.X. Zheng, and T.B. Strom. 2008. Immunostimulatory Tim-1-specific antibody deprograms Tregs and prevents transplant tolerance in mice. *J Clin Invest* 118: 735–741
38 Sizing, I.D., V. Bailly, P. McCoon, W. Chang, S. Rao, L. Pablo, R. Rennard, M. Walsh, Z. Li, M. Zafari, M. Dobles, L. Tarilonte, S. Miklasz, G. Majeau, K. Godbout, M.L. Scott, and P.D. Rennert. 2007. Epitope-dependent effect of anti-murine TIM-1 monoclonal antibodies on T cell activity and lung immune responses. *J Immunol* 178: 2249–2261
39 Xiao, S., N. Najafian, J. Reddy, M. Albin, C. Zhu, E. Jensen, J. Imitola, T. Korn, A.C. Anderson, Z. Zhang et al. 2007. Differential engagement of Tim-1 during activation can positively or negatively costimulate T cell expansion and effector function. *J Exp Med* 204: 1691–1702
40 de Souza, A.J., T.B. Oriss, J. O'Malley K, A. Ray, and L.P. Kane. 2005. T cell Ig and mucin 1 (TIM-1) is expressed on *in vivo*-activated T cells and provides a costimulatory signal for T cell activation. *Proc Natl Acad Sci USA* 102: 17113–17118
41 Binne, L.L., M.L. Scott, and P.D. Rennert. 2007. Human TIM-1 associates with the TCR complex and up-regulates T cell activation signals. *J Immunol* 178: 4342–4350
42 Tami, C., E. Silberstein, M. Manangeeswaran, G.J. Freeman, S.E. Umetsu, R.H. DeKruyff, D.T. Umetsu, and G.G. Kaplan. 2007. Immunoglobulin A (IgA) is a natural ligand of hepatitis A virus cellular receptor 1 (HAVCR1), and the association of IgA with HAVCR1 enhances virus-receptor interactions. *J Virol* 81: 3437–3446
43 Kobayashi, N., P. Karisola, V. Pena-Cruz, D.M. Dorfman, M. Jinushi, S.E. Umetsu, M.J. Butte, H. Nagumo, I. Chernova, B. Zhu et al. 2007. TIM-1 and TIM-4 glycoproteins bind phosphatidylserine and mediate uptake of apoptotic cells. *Immunity* 27: 927–940
44 Meyers, J., S. Chakravarti, D. Schlesinger, D. Illes, H. Waldner, S. Umetsu, J. Kenny, X. Zheng, D. Umetsu, R. DeKruyff, T. Strom, and V. Kuchroo. 2005. Tim-4 is the ligand for Tim-1, and the Tim-1–Tim-4 interaction regulates T cell expansion. *Nature Immunol* 6: 455–464
45 Wilker, P.R., J.R. Sedy, V. Grigura, T.L. Murphy, and K.M. Murphy. 2007. Evidence for carbohydrate recognition and homotypic and heterotypic binding by the TIM family. *Int Immunol* 19: 763–773
46 Shakhov, A.N., S. Rybtsov, A.V. Tumanov, S. Shulenin, M. Dean, D.V. Kuprash, and S.A. Nedospasov. 2004. SMUCKLER/TIM4 is a distinct member of TIM family expressed by stromal cells of secondary lymphoid tissues and associated with lymphotoxin signaling. *Eur J Immunol* 34: 494–503
47 Thery, C., L. Zitvogel, and S. Amigorena. 2002. Exosomes: composition, biogenesis and function. *Nat Rev Immunol* 2: 569–579
48 Santiago, C., A. Ballesteros, L. Martinez-Munoz, M. Mellado, G.G. Kaplan, G.J. Freeman, and J.M. Casasnovas. 2007. Structures of T cell immunoglobulin mucin protein 4

show a metal-ion-dependent ligand binding site where phosphatidylserine binds. *Immunity* 27: 941–951

49 Miyanishi, M., K. Tada, M. Koike, Y. Uchiyama, T. Kitamura, and S. Nagata. 2007. Identification of Tim4 as a phosphatidylserine receptor. *Nature* 450: 435–439

50 Nagata, S. 1997. Apoptosis by death factor. *Cell* 88: 355–365

51 Vaux, D.L., and S.J. Korsmeyer. 1999. Cell death in development. *Cell* 96: 245–254

52 Savill, J., I. Dransfield, C. Gregory, and C. Haslett. 2002. A blast from the past: clearance of apoptotic cells regulates immune responses. *Nat Rev Immunol* 2: 965–975

53 Savill, J., and V. Fadok. 2000. Corpse clearance defines the meaning of cell death. *Nature* 407: 784–788

54 Botto, M., C. Dell'Agnola, A.E. Bygrave, E.M. Thompson, H.T. Cook, F. Petry, M. Loos, P.P. Pandolfi, and M.J. Walport. 1998. Homozygous C1q deficiency causes glomerulonephritis associated with multiple apoptotic bodies. *Nat Genet* 19: 56–59

55 Taylor, P.R., A. Carugati, V.A. Fadok, H.T. Cook, M. Andrews, M.C. Carroll, J.S. Savill, P.M. Henson, M. Botto, and M.J. Walport. 2000. A hierarchical role for classical pathway complement proteins in the clearance of apoptotic cells *in vivo*. *J Exp Med* 192: 359–366

56 Hanayama, R., M. Tanaka, K. Miyasaka, K. Aozasa, M. Koike, Y. Uchiyama, and S. Nagata. 2004. Autoimmune disease and impaired uptake of apoptotic cells in MFG-E8–deficient mice. *Science* 304: 1147–1150

57 Albert, M.L., B. Sauter, and N. Bhardwaj. 1998. Dendritic cells acquire antigen from apoptotic cells and induce class I-restricted CTLs. *Nature* 392: 86–89

58 Scott, R.S., E.J. McMahon, S.M. Pop, E.A. Reap, R. Caricchio, P.L. Cohen, H.S. Earp, and G.K. Matsushima. 2001. Phagocytosis and clearance of apoptotic cells is mediated by MER. *Nature* 411: 207–211

59 Hanayama, R., M. Tanaka, K. Miwa, A. Shinohara, A. Iwamatsu, and S. Nagata. 2002. Identification of a factor that links apoptotic cells to phagocytes. *Nature* 417: 182–187

60 Liu, G., C. Wu, Y. Wu, and Y. Zhao. 2006. Phagocytosis of apoptotic cells and immune regulation. *Scand J Immunol* 64: 1–9

61 Nagata, K., K. Ohashi, T. Nakano, H. Arita, C. Zong, H. Hanafusa, and K. Mizuno. 1996. Identification of the product of growth arrest-specific gene 6 as a common ligand for Axl, Sky, and Mer receptor tyrosine kinases. *J Biol Chem* 271: 30022–30027

Linking lifestyle with microbiota and risk of chronic inflammatory disorders

Fergus Shanahan

Department of Medicine, Alimentary Pharmabiotic Centre, University College Cork, National University of Ireland, Cork, Ireland

Abstract

The inflammatory bowel diseases, Crohn's disease and ulcerative colitis, are among several immune-mediated disorders that consistently increase in incidence and prevalence when a society undergoes transition from 'developing' to 'developed' status. The impact of a changing lifestyle and environment associated with modernisation is greatest during early life. The mechanism may involve an alteration in composition or metabolic activity of the commensal microbiota colonising the host during early life. Since the commensal microbiota influences immunologic maturation and shapes the function of the developing immune system, disturbances in microbial biodiversity may contribute to individual variations in immunologic behaviour during and after childhood. Thus, an environmental influence on the commensal microbiota may underpin much of the changing epidemiology common to several immune-mediated chronic inflammatory disorders.

Conflict of interest statement:
The author is affiliated with a multi-departmental university campus company, Alimentary Health Ltd., which investigates inter alia host-microbe interactions. The content of this paper was neither influenced nor constrained by that fact. The author is supported in part by Science Foundation Ireland.

Introduction

The pivotal contribution of *Helicobacter pylori* to the pathogenesis of chronic peptic ulceration, gastritis and gastric cancer, was rightly acknowledged as a landmark discovery, with a lasting impact on the prevention and treatment of these chronic disorders. However, the bigger lesson of the *Helicobacter pylori* story was that the solution to some complex disorders cannot be found by studying the human host alone, and requires an understanding how the host interacts with its microbial environment. Thus, traditional approaches to human physiology and pathophysiology may be insufficient, too limited to probe the basis of conditions which stem from an abnormal host–microbe interaction. Similarly, traditional approaches to the study of chronic disorders in various populations may be too restrictive and distracted by

The Hygiene Hypothesis and Darwinian Medicine, edited by Graham A.W. Rook

epiphenomena or excessive reliance on 'risk-factor epidemiology' [1]. Epidemiological surveys lack incisiveness if conducted or interpreted without some rapprochement with disease mechanisms.

Here, an attempt is made to reconcile some of more consistent epidemiologic observations regarding inflammatory bowel disease with current concepts of pathogenesis. In particular, the possibility that changes in bowel microbiota might be linked with changing prevalence and disease risk through an influence on immunity will be examined. The inflammatory bowel disorders (IBD), Crohn's disease and ulcerative colitis, will be focussed upon, but the relevance of the gut microbiota to other extra-intestinal inflammatory disorders will also be addressed.

Impact of environmental change and migration on epidemiology of IBD

The most compelling evidence linking the changing epidemiology of IBD with changing lifestyle or environmental influences is the striking increase in frequency of IBD upon transition from 'developing' to developed nation status [2]. As societies develop and IBD emerges, ulcerative colitis appears first, followed after a variable interval by Crohn's disease. This consistent pattern worldwide has also been observed in under-developed sectors of society living within developed countries, such as the original native peoples of North America and New Zealand [2].

Studies of migrants from a low-incidence region to one of high incidence strengthen the evidence for an environmental influence on the risk of developing IBD and suggest that this must occur at an early stage in life. Migrants from India or South Asia to Britain and Canada acquire the same or higher risk of developing IBD as that of natives, but their age at the time of migration is critical. Those with the greatest risk of developing IBD have been the offspring of migrants, born in the new country or have migrated during childhood [3, 4]. What might explain this pivotal role for early life events on the risk of developing IBD? Could it reflect an impact of the environment or lifestyle on the developing immune system? Could the mechanism of such an effect on immunity operate at the level of the microbiota and host–microbe interactions in the gut? It is noteworthy that similar epidemiological trends have been observed with other chronic inflammatory disorders such as asthma, multiple sclerosis and insulin-dependent diabetes, which have an immune-mediated pathogenesis in common [5].

Colliding influences on risk of developing IBD

Three interacting factors contribute to the pathogenesis of most chronic inflammatory disorders, including IBD; these are genetic susceptibility, environmental modifiers, and immune-mediated tissue damage. The molecular nexus of this triad has become much clearer in a subset of patients with the discovery of the first Crohn's disease-related gene (NOD2/CARD15), and this prompted a successful search for other susceptibility genes for both Crohn's and ulcerative colitis [6, 7]. Collectively, this body of work has shown that many of the susceptibility genes code for proteins which are involved at the host–microbe interface, functioning as sensors of the microbial environment or are involved in handling of intracellular bacteria by processes such as autophagy, or participate in the regulation of the host innate and acquired immune response to the microbial environment [6, 7]. Disease heterogeneity arises because of the multiplicity of susceptibility and modifying genes involved and the diversity of lifestyle or environmental exposure, including variable composition of commensal microbiota.

However, while the identification of susceptibility genes has caste light on mechanisms by which gene–environmental interactions might lead to immune-mediated disease, the primacy of environmental factors has been evident from several observations. Firstly, the concordance rate for Crohn's disease in identical twins is less than 50% and that for ulcerative colitis is less than 10% [2]. Secondly, the relative stability of genetic susceptibility genes within the population is at variance with the rapidity of the increase in disease frequency in recent decades. Therefore, to reconcile the changing epidemiology of IBD during periods of socioeconomic transition, the likelihood of environmental-driven disease risk, specifically microbial-driven immunopathology, must be entertained.

The microbiota as an environmental regulator of mucosal and systemic immunity

The immune system is, in essence, the host's sensory mechanism for recognition and disposal of microbial challenge from the environment. In common with other sensory organs, it has receptors for sampling the environment, afferent uptake of environmental information, efferent limbs for dealing with adversity within the environment, rigorous regulation, and the immune system exhibits memory with the capacity to learn. In further similarity to other sensory systems, the immune system is developed at birth but requires education and fine tuning during early childhood. This is achieved by interaction with the local environment. Episodic exposure to childhood infections provides some educational experience, but colonisation with commensals is likely to be far more important because of the huge antigenic diversity of the resident microbiota, which, in the gut, represents the most densely populated ecosystem on the planet [8]. As with all sensory development, deficits in educational development may leave the host vulnerable later to inappropriate interpretation or responses to environmental stimuli. Thus, the commensal microbial biodiversity is central to the robustness of the host immune response to environmental disturbances [9]. Given the impact of the commensal microbiota on mucosal immune development and function [10], it is conceivable that the composition of the colonising microbiota determines individual variations in immunity.

The influence of the commensal microbiota on the development of the immune system is best illustrated by comparative studies of germ-free and conventionally colonised animals. Indeed, it can be deduced that the microbiota must be a source of regulatory signals for mucosal and systemic immune development [11–13]. A molecular basis for this has emerged with the demonstration that a single immunomodulatory polysaccharide derived from *Bacteroides fragilis* can correct maturational defects, not only in the mucosa-associated lymphoid tissue but also in the systemic immune compartment of germ-free mice [14]. The same polysaccharide has also been used to prevent intestinal inflammatory disease in a murine model system [15]. It is likely that other bacterial-derived molecules with regulatory effects on distinct aspects of immune development and function will be found. For example, a peptidoglycan from the microbiota has been shown to be necessary and sufficient for induction of intestinal lymphoid follicles in mice and this involves a NOD1-dependent mechanism within the epithelium [16].

In addition to a regulatory influence on immune maturation, continual signalling by the commensal microbiota with host toll-like receptors (TLRs) under normal steady-state conditions is critical for maintenance of mucosal homeostasis and protection from injury [17, 18]. The adaptation of structure to function within the gut is such that interaction between the contents of the lumen (nutrients and microbes) and the internal milieu of the host is actually favoured over strict maintenance of anatomic barrier function. This is so, because of the large surface area (approximately 200 m^2) that is lined by epithelium of only single cell thickness. Surface enterocytes and mucosal dendritic cells are the host's first line of defence and both are responsible for sampling of the luminal microbiota, with dendritic cells being able to receive material transported across M cells overlying lymphoid follicles or can directly sample the microenvironment of the lumen by extending dendrites directly into the lumen between enterocytes [19]. Dendritic cells ingest and transport live bacteria to the mesenteric lymph node which acts both as a gatekeeper limiting access of commensal microbes to the systemic immune compartment and as a site for induction of mucosal immune responses to the luminal microbiota [20].

The immunosensory role of surface enterocytes and mucosal dendritic cells in distinguishing pathogens from non-pathogens has been reviewed in detail elsewhere [11–13]. The contribution of both cell types to the maintenance of mucosal homeostasis is shaped and modified by the commensal microbiota. For example, the microbiota influences the level of expression and topographic distribution of TLRs on surface enterocytes and may control the expression of regulatory molecules controlling TLR signalling [11]. Furthermore, some components of the commensal microbiota modulate the activation of the transcription factor, NF-κB, within enterocytes by several molecular mechanisms and downregulate inflammatory responses to pathogens [21–23]. Cross-talk among the microbiota, enterocytes, and mucosal dendritic cells helps maintain immune homeostasis. Thus, epithelial cell-intrinsic NF-κB activity is essential for regulation of dendritic cell responses and susceptibility to intestinal inflammation [24–26].

Host–microbe communication is reciprocal

In keeping with other aspects of inter-kingdom signalling, host–microbe interactions in the gut are bi-directional [27]. While there is circumstantial evidence that host genetics may influence the commensal microbiota, it is clear that the immune status of the host influences bacterial composition of the gut. Defects at the effector or regulatory level of mucosal immunity in different species have been associated with aberrant expansion of certain commensal organisms [28, 29]. In addition, the transcription factor T-bet which regulates immune development and function has been shown to have an unexpected influence on commensal populations within the murine intestine. Deletion of T-bet appeared to lead to the emergence of a 'colitogenic' microbiota with the capacity to transfer colitis [30].

The microbiota in inflammatory bowel disease

Several lines of experimental evidence have implicated some but not all components of the microbiota in the pathogenesis of inflammatory bowel disease [31]. The sobering lesson of the discovery of *Helicobacter pylori* as a cause of peptic ulceration prompts the question whether a subset of either Crohn's disease or ulcerative colitis might also be caused by an infectious agent waiting to be discovered. However, there is no conclusive evidence for any single pathogenic cause for either condition and epidemiologic data is at variance with a transmissible agent. Some of the more consistently observed microbial alterations linked with inflammatory bowel disease include a reduction in faecal lactobacilli and bifidobacteria [32], increased detection of adherent-invasive *E. coli* (AIEC) [33], increased detection of *Mycobacterium avium* subsp. *paratuberculosis* (MAP) [34], and reduced bacterial diversity by metagenomic analysis [35], including reductions in the anti-inflammatory commensal, *Faecalibacterium prausnitzii* [36]. Thus, there is increasing evidence for an alteration in the diversity of the microbiota in patients with inflammatory bowel disease; the challenge now is to link specific microbiota with individual variations in the immune response, but progress is occurring here also [37].

The gut microbiota and extra-intestinal inflammatory disorders

While the conditioning influence of the microbiota on intestinal immunity and IBD is well established, increasing evidence suggests that the microbial influence extends to a variety of other disorders beyond the gut. Firstly, the increased incidence and prevalence of IBD in developing countries and societies has been paralleled by similar trends for other chronic disorders, most of which are immunologically-mediated [5]. An environmental influence (including that of the human microbial environment) on the developing immune system may be common to each of these disorders. Secondly, as alluded to earlier, the impact of the microbiota on immune maturation is not limited to gut-associated lymphoid tissue; peripheral lymphoid structures are also influenced [14]. Thirdly, circumstantial support for the role of the microbiota and risk of extra-intestinal inflammatory disease is provided by the apparent efficacy of some probiotics on atopic disorders [32]. Fourthly, the microbiota is an environmental regulator of fat storage and appears to influence the risk of developing obesity and metabolic syndrome, which are chronic disorders associated with a proinflammatory phenotype (reviewed in [38]). Finally, studies of murine Type 1 diabetes have shown that the composition of the gut microbiota may modify the pathogenesis this T cell-mediated destruction of the insulin-producing pancreatic islets. The relationship between the microbiota and risk of developing diabetes is complex, but interaction between the microbiota and the host innate immune system appears to be a critical epigenetic modifying factor [39].

Diet and other lifestyle influences on the microbiota

Many of features of a modern lifestyle may influence the commensal microbiota [2]. These include improved sanitation, urbanisation (reduced exposure to soil microbes), widespread antibiotic usage, refrigeration, reduced family size with delayed exposure to childhood infections, vaccination, decline in endemic parasitism and reduced prevalence of *H. pylori* infection. In addition, a modern sedentary lifestyle and obesity have been linked with alterations in the microbiota [38].

Perhaps the greatest influence on the commensal microbiota is diet and nutrition [2, 40]; reduced consumption of fermented food products and oligosaccharide prebiotics, with increased consumption of refined sugars and saturated fats may have a pivotal influence on the composition or metabolic activity of the microbiota. One commentator has even suggested that the most profound alteration in early immune education during infancy since Neolithic times has been due to changes in the composition of the colonising microbiota over the past century [32]. In this respect, marked reductions in the content of bifidobacteria and lactobacilli within the commensal microbiota of neonates in the modern world, compared with those from developing societies, have been particularly striking.

The possibility that diet could influence the changing epidemiology of inflammatory bowel disease remains speculative and has been addressed in detail elsewhere [2]. However, the increased incidence in both Crohn's disease and ulcerative colitis over recent decades in Japan is especially noteworthy because of remarkably close correlations with changes in dietary fat, particularly animal fat and n-6 polyunsaturated fatty acids [41].

Conclusion

Could a deviation of the gut microbiota from the pattern that probably occurred in hunter-gatherers, contribute to the increases in chronic inflammatory disorders in modern developed countries? There is persuasive circumstantial evidence suggesting that lifestyle and environmental influences account for the changing epidemiology of several chronic inflammatory disorders, in general, and inflammatory bowel disease, in particular. These environmental influences are operative at an early stage in life when initial colonisation with commensal microbiota shapes mucosal and systemic immune maturation. The molecular basis of microbial-induced immune development is beginning to unfold and can be 'mined' for novel therapeutics in the future [42] and also provides a target for therapeutic approaches to inflammatory bowel disease [43, 44].

References

1 Skrabenek P (1999) Risk factor epidemiology: Science or non-science? In: *False Premises False Promises*. Taragon Press, Whithorn, 153–163

2 Bernstein CN, Shanahan F (2008) Disorders of a modern lifestyle: reconciling the epidemiology of inflammatory bowel diseases. *Gut* 57: 1185–91

3 Probert CS, Jayanthi V, Pinder D, Wicks AC, Mayberry JF (1992) Epidemiological study of ulcerative proctocolitis in Indian migrants and the indigenous population of Leicestershire. *Gut* 33: 687–93

4 Pinsk V, Lemberg DA, Grewal K, Barker CC, Schreiber RA, Jacobson K (2007) Inflammatory bowel disease in the South Asian pediatric population of British Columbia. *Am J Gastroenterol* 102: 1077–83

5 Bach JF (2002) The effect of infections on susceptibility to autoimmune and allergic diseases. *N Engl J Med* 347: 911–20

6 Cho JH (2008) The genetics and immunopathogenesis of inflammatory bowel disease. *Nat Rev Immunol* 8: 458–66

7 Mathew CG (2008) New links to the pathogenesis of Crohn disease provided by genome-wide association scans. *Nat Rev Genet* 9: 9–14

8 Marchesi J, Shanahan F (2007) The normal intestinal microbiota. *Curr Opin Infect Dis* 20: 508–13

9 Kitano H, Oda K (2006) Robustness trade-offs and host-microbial symbiosis in the immune system. *Molecular Systems Biology* 2: 2006.0022

10 Cebra JJ (1999) Influences of microbiota on intestinal immune system development. *Am J Clin Nutr* 69: 1046S–1051S

11 O'Hara AM, Shanahan F (2006) The gut flora as a forgotten organ. *EMBO Rep* 7: 688–93

12 Artis D (2008) Epithelial cell recognition of commensal bacteria and maintenance of immune homeostasis in the gut. *Nature Rev Immunol* 8: 411–420

13 Coombes JL, Powrie F (2008) Dendritic cells in intestinal immune regulation. *Nature Rev Immunol* 8: 435–446

14 Mazmanian SK, Liu CH, Tzianabos AO, Kasper DL (2005) An immunomodulatory molecule of symbiotic bacteria directs maturation of the host immune system. *Cell* 122: 107–18

15 Mazmanian SK, Round JL, Kasper DL (2008) A microbial symbiosis factor prevents intestinal inflammatory disease. *Nature* 453: 620–625

16 Bouskra D, Brézillon C, Bérard M, Werts C, Varona R, Boneca IG, Eberl G (2008) Lymphoid tissue genesis induced by commensals through NOD1 regulates intestinal homeostasis. *Nature* 456: 507–510

17 Rakoff-Nahoum S, Paglino J, Eslami-Varzaneh F, Edberg S, Medzhitov R (2004) Recognition of commensal microflora by toll-like receptors is required for intestinal homeostasis. *Cell* 118: 229–41

18 Medzhitov R (2007) Recognition of microorganisms and activation of the immune response. *Nature* 449: 818–826

19 Rescigno M, Urbano M, Valzasina B, Francolini M, Rotta G, Bonasio R, Granucci F, Kraehenbuhl JP, Ricciardi-Castagnoli P (2001) Dendritic cells express tight junction proteins and penetrate gut epithelial monolayers to sample bacteria. *Nat Immunol* 2: 361–367

20 Macpherson AJ, Uhr T (2004) Induction of protective IgA by intestinal dendritic cells carrying commensal bacteria. *Science* 303: 1662–1665

21 Neish AS, Gewirtz AT, Zeng H, Young AN, Hobert ME, Karmali V, Rao AS, Madara JL (2000) Prokaryotic regulation of epithelial responses by inhibition of IkappaB-alpha ubiquitination. *Science* 289: 1560–1563

22 Kelly D, Campbell JI, King TP, Grant G, Jansson EA, Coutts AG, Pettersson S, Conway S (2004) Commensal anaerobic gut bacteria attenuate inflammation by regulating nuclear-cytoplasmic shuttling of PPAR-gamma and RelA. *Nat Immunol* 5: 104–112

23 Petrof EO, Kojima K, Ropeleski MJ, Musch MW, Tao Y, De Simone C, Chang EB (2004) Probiotics inhibit nuclear factor-kappaB and induce heat shock proteins in colonic epithelial cells through proteasome inhibition. *Gastroenterology* 127: 1474–1487

24 Rimoldi M, Chieppa M, Salucci V, Avogadri F, Sonzogni A, Sampietro GM, Nespoli A, Viale G, Allavena P, Rescigno M (2005) Intestinal immune homeostasis is regulated by the crosstalk between epithelial cells and dendritic cells. *Nat Immunol* 6: 507–514

25 Zaph C, Troy AE, Taylor BC, Berman-Booty LD, Guild KJ, Du Y, Yost EA, Gruber AD, May MJ, Greten FR, Eckmann L, Karin M, Artis D (2007) Epithelial-cell-intrinsic IKK-beta expression regulates intestinal immune homeostasis. *Nature* 446: 552–6

26 Nenci A, Becker C, Wullaert A, Gareus R, van Loo G, Danese S, Huth M, Nikolaev A, Neufert C, Madison B, Gumucio D, Neurath MF, Pasparakis M (2007) Epithelial NEMO links innate immunity to chronic intestinal inflammation. *Nature* 446: 557–61

27 Hughes DT, Sperandio V (2008) Inter-kingdom signaling: communication between bacteria and their hosts. *Nat Rev Microbiol* 6: 111–120

28 Ryu JH, Kim SH, Lee HY, Bai JY, Nam YD, Bae JW, Lee DG, Shin SC, Ha EM, Lee WJ (2008) Innate immune homeostasis by the homeobox gene caudal and commensal-gut mutualism in *Drosophila*. *Science* 319: 777–82

29 Suzuki K, Meek B, Doi Y, Muramatsu M, Chiba T, Honjo T, Fagarasan S (2004) Aberrant expansion of segmented filamentous bacteria in IgA-deficient gut. *Proc Natl Acad Sci USA* 101: 1981–6

30 Garrett WS, Lord GM, Punit S, Lugo-Villarino G, Mazmanian SK, Ito S, Glickman JN, Glimcher LH (2007) Communicable ulcerative colitis induced by T-bet deficiency in the innate immune system. *Cell* 131: 33–45

31 Sartor RB (2008) Microbial influences in inflammatory bowel diseases. *Gastroenterology* 134: 577–94

32 Murch SH (2001) Toll of allergy reduced by probiotics. *Lancet* 357: 1057–1059

33 Rhodes JM (2007) The role of *Escherichia coli* in inflammatory bowel disease. *Gut* 56: 610–2

34 Feller M, Huwiler K, Stephan R, Altpeter E, Shang A, Furrer H, Pfyffer GE, Jemmi T, Baumgartner A, Egger M (2007) *Mycobacterium avium* subspecies *paratuberculosis* and Crohn's disease: a systematic review and meta-analysis. *Lancet Infect Dis* 7: 607–13

35 Peterson DA, Frank DN, Pace NR, Gordon JI (2008) Metagenomic approaches for defining the pathogenesis of inflammatory bowel diseases. *Cell Host Microbe* 3: 417–27

36 Sokol H, Pigneur B, Watterlot L, Lakhdari O, Bermúdez-Humarán LG, Gratadoux JJ, Blugeon S, Bridonneau C, Furet JP, Corthier G et al (2008) Faecalibacterium prausnitzii is an anti-inflammatory commensal bacterium identified by gut microbiota analysis of Crohn disease patients. *Proc Natl Acad Sci USA* 105: 16731–6

37 Ivanov II, Frutos Rde L, Manel N, Yoshinaga K, Rifkin DB, Sartor RB, Finlay BB, Littman DR (2008) Specific microbiota direct the differentiation of IL-17-producing T-helper cells in the mucosa of the small intestine. *Cell Host Microbe* 4: 337–49

38 DiBaise JK, Zhang H, Crowell MD, Krajmalnik-Brown R, Decker GA, Rittmann BE (2008) Gut microbiota and its possible relationship with obesity. *Mayo Clin Proc* 83: 460–9

39 Wen L, Ley RE, Volchkov PY, Stranges PB, Avanesyan L, Stonebraker AC, Hu C, Wong FS, Szot GL, Bluestone JA, Gordon JI, Chervonsky AV (2008) Innate immunity and intestinal microbiota in the development of Type 1 diabetes. *Nature* 455: 1109–13

40 Flint HJ, Duncan SH, Scott KP, Louis P (2007) Interactions and competition within the microbial community of the human colon: links between diet and health. *Environ Microbiol* 9: 1101–11

41 Shoda R, Matsueda K, Yamato S, Umeda N (1996) Epidemiologic analysis of Crohn

disease in Japan: increased dietary intake of n-6 polyunsaturated fatty acids and animal protein relates to the increased incidence of Crohn disease in Japan. *Am J Clin Nutr* 63: 741–5

42 Shanahan F, Kiely B (2007) The gut microbiota and disease – an inner repository for drug discovery. *Drug Discov Today: Ther Strategies* 4: 195–200

43 Sartor RB (2008) Therapeutic correction of bacterial dysbiosis discovered by molecular techniques. *Proc Natl Acad Sci USA* 105: 16413–4

44 Mitsuyama K, Sata M (2008) Gut microflora: a new target for therapeutic approaches in inflammatory bowel disease. *Expert Opin Ther Targets* 12: 301–12

Soil bacteria, nitrite and the skin

David R. Whitlock[1] *and Martin Feelisch*[2]

[1]Nitroceutic LLC, Dover, MA 02030, USA
[2]University of Warwick, Warwick Medical School, Coventry, CV4 7AL, UK

Abstract

Little is known about the composition of the skin microbiome and its potential significance for health and disease in the context of the 'hygiene hypothesis'. We here propose that mammals evolved with a dermal microflora that contributed to the regulation of body physiology by providing nitrite from commensal ammonia-oxidising bacteria in response to ammonia released during sweating. We further hypothesise that modern skin hygiene practices have led to a gradual loss of these bacteria from our skin. Together with other lifestyle-related changes associated with an insufficient bodily supply with nitrite and depletion of other nitric oxide(NO)-related species, a condition we here define as 'nitropenia', this has led to a perturbation of cellular redox signalling which manifests as dysregulated immunity and generalised inflammation. If proven correct, this scenario would provide an additional evolutionary rationale and a mechanistic basis for the simultaneous rises in prevalence of a number of seemingly unrelated chronic illnesses over the last 3–4 decades.

Introduction

Birth and a childhood spent in rural areas of the developing world seem to protect against disease acquisition in later life, particularly disorders involving immune system deviation and inflammatory processes such as asthma, allergies, Type 1 diabetes, inflammatory bowel disease, obesity and some degenerative diseases. While this protective effect is most robust for those in poverty [1] in the developing world, poverty in the developed world does not afford protection. In the developed world, particularly in urban areas, asthma and obesity are increasing at an alarming rate, reaching epidemic proportions both in children and adults. The list of disorders exacerbated by obesity is long [2] but the associated mechanisms remain obscure. Regardless of their nature, virtually all of these disorders are also exacerbated by stress. These associations imply involvement of common metabolic pathways and regulatory mechanisms. Considering the ubiquitous importance of nitric oxide (NO) as regulator and effector molecule in a plethora of vastly different cell/organ functions across all levels of organisation, its versatile chemistry and its involvement in the regulation of immune cell function and inflammation we surmise that perturbations of this molecule's signalling pathways provide the missing link.

Increased interest in the human microbiome has revealed that bacterial colonisation of the body plays an important role in the maintenance of health and a robust immune system. The 'hygiene hypothesis' posits that 'hygienic' factor(s) related to changes in lifestyle and environment account for the above health differences, but which factor(s) are important is not well understood. Clearly, the situation is complex and multi-factorial, with involvement of (but not solely due to) bacteria, parasites, and gut bacteria [3]. Reduced exposure to these organisms which accompanied human evolutionary history for millennia ('old friends') are thought to lead to an impaired production of regulatory T cells with inadequate suppression of effector T cells, resulting in exaggerated immune responses [4].

In contrast to the gut, the skin microflora has received relatively little attention, and most studies in the past concentrated on the cultivation of putative pathogens. Only recently have metagenomic studies begun to reveal the diversity of the skin microflora [5, 6], but which organisms are important for human health is unclear. Similarly little is known about how development (in particular urbanisation) changes the interactions of the skin with the environment, especially with regard to its role in modulating human physiology. There are at least two mechanisms by which skin and mucosal flora can affect bodily functions controlled by NO that do not involve immune system activation by direct contact. The first is by reduction of nitrate (NO_3^-) to nitrite (NO_2^-). The second is *via* oxidation of ammonia to nitrite by ammonia oxidising bacteria (AOB).

Regulation of bodily functions by NO – the role of stress and nitrite/nitrate

NO is a pleiotropic signalling and effector molecule that is enzymatically produced from the precursor L-arginine by nitric oxide synthases. The local availability of NO is mainly regulated by the rate of production of reactive oxygen species (ROS). Both, NO and ROS serve a multitude of signalling functions in health and disease, and the balance of NO/ROS formation is a key determinant of cellular redox status and redox signalling [7]. The latter is determined by the extent of nitrosation and oxidation of critical thiol groups of proteins and transcription factors. Perturbations in NO availability and redox signalling are hallmarks of many disease states associated with the 'hygiene hypothesis'. Nitrite and nitrate (NOx), also ubiquitous in dietary sources, are the oxidative decomposition products of NO. These anions were long believed to be devoid of biological activity, but recent research suggests that nitrate is reduced to nitrite not only by bacteria but also by mammalian cells [8] and that nitrite has NO-like signalling properties in its own right [9]. Nitrite has been shown to have effects on platelets, blood pressure [10], protective effects on mucosal function [11], and plays an important role in antimicrobial defense in the stomach [12]. Moreover, nitrite protects a variety of organs against ischemia/reperfusion damage and can be reduced to NO under conditions of low oxygen availability [13].

The majority of nitrate intake in humans is from green leafy vegetables. Many heterotrophic bacteria can use nitrate as a terminal electron acceptor, producing nitrite (this dissimilatory production of nitrite is distinct from assimilatory nitrate reduction to ammonia in which the product is used for the synthesis of amino acids [14]). Such bacteria are commensals on the tongue [15], where they reduce salivary nitrate (concentrated ~10x over plasma) to produce nitrite at levels up to 2 mmol/L. Swallowing lets this nitrite end up in the stomach where part of it is taken up and the remainder is used to generate NO. Thus, endogenous and dietary nitrate represent an important source of nitrite and contribute to the total body pool of NO/NOx (see Fig. 1). The changes in dietary habits over the last couple of decades with increasing use of processed foods, together with markedly reduced levels of NOx in drinking water, has almost certainly led to a reduction in the supply of the body with nitrite.

NO is a rather unusual signalling molecule; it is gaseous and generated on demand at a source, diffuses a distance and, when its local concentration exceeds the activation level, activates an NO sensor. Soluble guanylate cyclase is the best known enzyme stimulated by NO/nitrite, but there are many others. All NO sensors

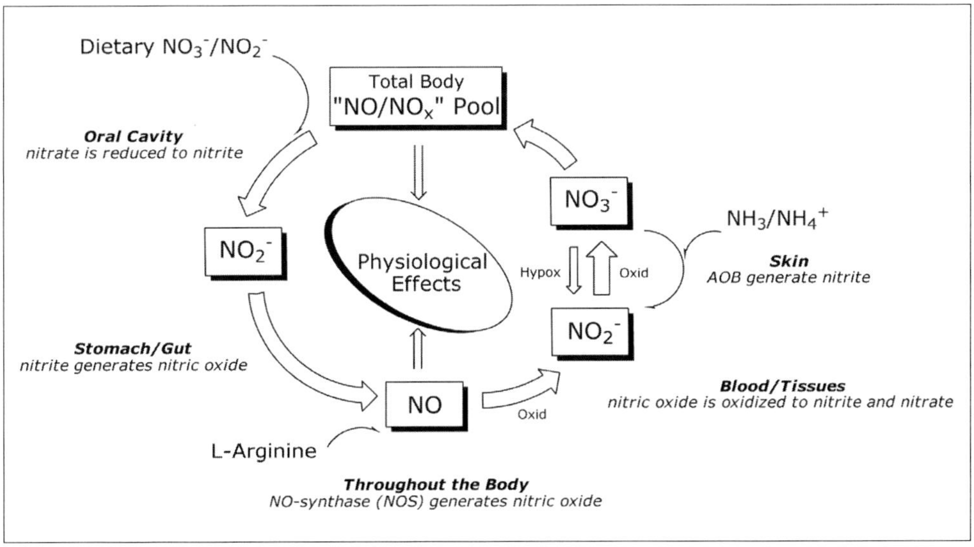

Figure 1
Sources of nitrite and the NO/NOx cycle in the human body

respond to NO from all sources including the basal level. A change in basal NO levels will therefore change the range, onset time and duration of all NO-mediated signalling pathways. When basal NO levels are perturbed by external conditions (as discussed in this chapter) the result can be good regulation (where regulation pathways are working properly) around a bad setpoint (where deviations are skewed in a characteristic direction) rather than distribution around the normal setpoint. This behaviour will occur for all NO-mediated pathways *with no threshold* because these pathways are already being actively regulated by changes in NO (see Fig. 2). This is an important and fundamental, albeit largely neglected concept regarding NO physiology.

NO/NOx physiology is a subject of intense research activity with much cross-talk between signalling pathways. It is possible that the nitrite-based reaction channels of contemporary mammalian cells are a vestige of earlier prokaryotic pathways and

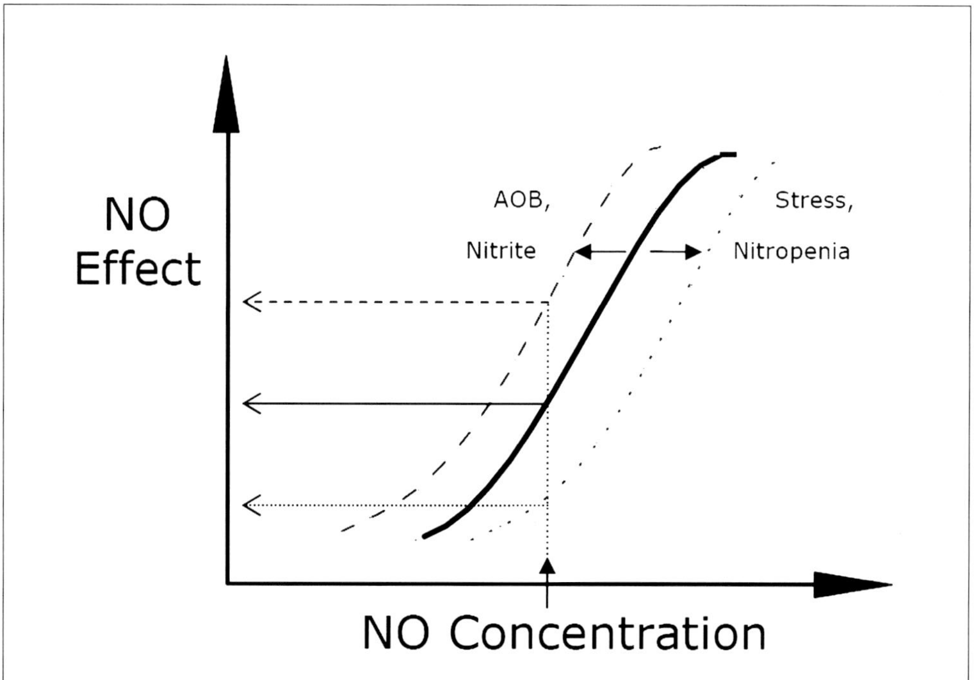

Figure 2
Perturbations in nitric oxide (NO) signalling by stress-induced NO/NOx deprivation ('nitropenia') and reversal by ammonia-oxidising bacteria (AOB) through the delivery of nitrite. When overall NO availability is impaired, the same concentration of NO elicits a lesser effect (see text for details).

that the evolutionarily more recent L-arginine/NO pathway uses signalling cascades originally evolved for nitrite [16]. In any case, the pathways involved are complex, non-linear, highly coupled and operate at multiple time (sub-second to days), length (submicron to centimetre) and concentration (nmol/L to µmol/L) scales in diverse tissue compartments. Nitrite may play an important role in providing a basal NO tone that allows the L-arginine/NO pathway perform more sophisticated functions at the local level. As a corollary, if nitrite supply decreases NO signalling will be affected in ways similar to conditions associated with oxidative stress, requiring higher concentrations of NO to achieve the same effect (essentially resulting in a rightward shift of the concentration-response curve for NO; see Fig. 2). If NO signalling is affected, there is no threshold for dysregulation.

A lowering of NO levels is a near universal stress response [17]. Organisms lower NO levels to disinhibit cytochrome c oxidase and so maximize aerobic ATP production by mitochondria. NO is a generic inhibitor of cytochrome P450 enzymes responsible for xenobiotic substrate metabolism and steroid synthesis [18]. Under most conditions these enzymes are quite uncoupled, generating significant superoxide during cycling [19]. This superoxide lowers NO levels local to the P450 enzyme accelerating its activity. Many compounds are both substrates and products of P450 enzymes, so a change in NO/NOx availability will lead to disruption of P450 physiology, perhaps a mechanism for endocrine disruption independent of xenobiotic chemicals. Low NO regulates the ATP level to a lower setpoint by their combined action on soluble guanylate cyclase. One stress response, ischemic preconditioning, is known to be triggered by low ATP or by oxidative stress, which lowers NO levels *via* consumption by superoxide. The respiratory burst by activated immune cells will lower NO levels by the same mechanism. Thus, as a near universal stress response, deviations in NO/NOx physiology are prime candidate mechanisms for conditions exacerbated by stress. Recent evidence suggests a strong link between oxygen sensing, innate immunity and inflammation *via* NF-κB and other redox- and oxygen-sensitive transcriptional elements [20], which may explain why an impaired NOx availability is often accompanied by a systemic inflammatory component. Both, NO and ROS are involved in the regulation of inflammation at multiple levels. In spite of intense research efforts in this field, however, the conditions which determine whether NO has pro- or anti-inflammatory effects in a particular setting remain poorly defined [21, 22].

Dermal nitrite generation by commensal autotrophic ammonia-oxidising bacteria

An early effect of economic development in modern times is abundant clean water for food preparation and hygiene. The transition in bathing practices due to migration between areas of different economic development times is much more abrupt than was the change during historic development. The 'germ theory' and improved public health education were responsible for the changes in attitudes and hygiene habits in the late 19th and early 20th Centuries; but the timing of the adoption of individual personal hygiene practices was not rapid and is not always well documented. 'The Great Unwashed' were unwashed largely due to lack of opportunity; for example, the number of public bath houses in Britain increased from essentially zero in 1845 to 343 in 1915 [23] and then declined as the poor migrated from fee-based communal bathing facilities to private individual ones [24]. This period also saw the transition of soap from a taxed luxury good to a commodity with annual per capita soap consumption increase from 7 lb in 1851 to 14 pounds in 1881 and 18 pounds in 1912 [25]. Access to clean water occurred slowly. In the Netherlands, the first piped water was for the supply of drinking water on ships leaving port. Individual buckets were sold from taps for additional revenue, and only much later was water directly piped into individual homes. In 1900, about 40% of the Dutch population had access to piped water, increasing to 82% in 1951 [26]. Further dramatic changes in personal hygiene practices have occurred due to the affordability, due to mass production, of sophisticated bathroom equipment and hair/body care products in the last 20–30 years. We hypothesise that modern hygiene practices, in particular the custom of frequent baths or showers with abundant warm water and lavish use of shampoo and liquid soaps, have led to the efficient removal of skin bacteria, including those that were once an important part of the 'normal' commensal microflora of the skin. The very recent practice of using antimicrobial agents in virtually all bathing products can only exacerbate the loss of normal commensals.

An important group of microorganisms, now virtually absent from the skin of most individuals, are AOB, bacteria ubiquitous in soils and natural waters. AOB play an important role in the global nitrogen cycle by catalysing the first step of the nitrification process, the oxidation of ammonia to nitrite [27, 28] which provides their sole source of energy for CO_2 fixation and cell growth. Mammals likely evolved with AOB on their skin, providing their host with nitrite by conversion of ammonia in sweat [29] with scalp, pubic and underarm hair providing a suitable niche due to enhanced sweat production, increased warmth, increased relative humidity and protection from light (the latter is important as ammonia monooxygenase activity is inhibited by light [30]). Low NO increases androgen levels which increases growth of pubic hair, expanding the AOB niche thereby increasing NO/NOx production and absorption in a feedback loop.

The production of a suitable niche for these bacteria provides a rationale for non-thermal sweating (e.g., under stress) (to supply NO/nitrite), the location of body hair (near lymph nodes), why the skin of the scalp is thin and well vascularised (to enhance NO/nitrite absorption), and why sweat glands are most abundant on the feet and palms [31] (for antimicrobial effects of acidified nitrite [32] in surfaces in contact with soil). The pH of normal skin is about 4 [33], and is increased by fungal infection [34], perhaps due to ammonia release by microbial oxidation of amino acids. Ammonia release from amino acid oxidation, or the metabolism of urea would stimulate AOB production of NO/nitrite and limit the growth of other microorganisms, likely through oxidation of quorum sensing compounds [35], a defense strategy utilised by marine algae to keep surfaces free of biofilm. NO and nitrite specifically disrupt biofilms of Staphylococcus [36] and NO is known to cause the transition of Pseudomonas from biofilm to planktonic phenotype [37].

Before adoption of modern hygiene habits and cleanliness standards humans existed without ever bathing other than for spiritual purposes. Bathing in regions without springs, rivers or lakes would have been impossible, bathing in most freshwater sources in tropical regions risky because of the abundance of carnivores and parasites; bathing in cold climates would have been limited to the summer months. Thus, humans must have lived for much of their existence with sweat residues accumulating year round, and inoculation of the skin with AOB would have been impossible to prevent. If true, human physiology would have adapted to utilise this supply of NO/NOx, explaining why a lack of transdermal nitrite delivery due to loss of these commensals would result in a major perturbation of total body nitrite supply and NO/redox tone. The human immune system may also be adapted to the specific lipopolysaccharide produced by these organisms, typically containing only C16 fatty acids [38].

Why have these potentially important commensals been overlooked until now?

AOB are obligate autotrophs [39] incapable of growth on any media used for isolation of pathogens (all known pathogens are heterotrophic [40]); there is not a single reported case of 'infection' by AOB or any other autotrophic bacteria. This is not from lack of exposure because these organisms are universally present in all soils and even drinking water distribution systems [41]. They are incapable of causing infection because they lack the enzyme systems with which to degrade or utilise animal tissues. AOB completely lack virulence factors [42]. AOB are slow growing, with division times ~30 times longer than those of most heterotrophs; alkylbenzene sulfonate detergents are toxic to AOB at ppm levels [43], i.e., >3 orders of magnitude lower than those producing mild skin irritation in humans; modern hygiene practices may have removed these bacteria before we learned to detect them using PCR techniques. They are sufficiently distantly related to heterotrophic bacteria, that some are not detected even by 'universal' 16S primers. Even more distantly related Archaea ammonia oxidisers were first discovered in 2005 [44].

The major nitrogen species taken up by plants is nitrate. When heterotrophic organisms oxidise amino acids, they release ammonia. Essentially all ammonia released into the environment is rapidly oxidised by these bacteria before the ammonia can be taken up by heterotrophic organisms. It would not be surprising if the total biomass in the environment of AOB was comparable to that of heterotrophic organisms.

Could nitrite not be delivered by dermal nitrate-reducing bacteria instead?

Dissimilatory nitrate reduction requires hypoxic conditions. These anoxic conditions can be met in wet biofilms on mucous membranes (as in the oral cavity) where the aqueous layer affects the diffusion of oxygen. However, those conditions are not prevailing on dry skin. In fact, the supply of the external skin with oxygen is almost exclusively *via* the external air [45] (it would become hyperemic if a supply of oxygen from blood were necessary). Mucous membranes are already essentially hyperemic, so biofilms on mucus membranes can be hypoxic without metabolic compromise of the underlying tissues.

A second argument against nitrate reduction is based on the differences in substrate concentrations prevailing. Nitrate is a ubiquitous component of all cells and bodily fluids with typical concentrations of a few tens of µmol/L. Ammonia (or its metabolic equivalent urea) is also present in all bodily fluids, but at ~3 orders of magnitude higher levels, at tens of mmol/L. Thus, assuming comparable absorption and metabolic efficiencies, ammonia oxidation would seem to be considerably more effective in generating physiologically relevant nitrite than nitrate reduction.

What makes the skin attractive as modulator of NO/NOx physiology?

Nitrifiers are found in thermal springs [46], some of which are used as source waters for spas [47]. Part of the therapeutic effects of sauna as originally practiced millennia ago, may have been to release sweat and so nourish a natural biofilm of AOB. While microbial interactions with skin secretions are known to account for the formation of volatile fatty acids responsible for (sometimes unpleasant) body odour, we are not aware of any interaction producing a metabolite that turned out to be beneficial for human health. Nitrite may be the first example of such an entity.

Much of the skin gets oxygen from the external air, and so can be free of blood, and thus free of haemoglobin. All other cells in the body are necessarily diffusively close to oxygenated haemoglobin to ensure an efficient supply of their mitochondria with oxygen. Haemoglobin is a major sink of NO in the body, and destroys it at near diffusion limited rates [48]. Haemoglobin is also an efficient scavenger of nitrite and oxidises it to nitrate, albeit a much lower rates. The outer skin layer would seem to be the only tissue compartment where NO/nitrite reactions can proceed in the absence of haemoglobin. Thus, NO-mediated processes such as, e.g., the priming and regulation of immune cells may proceed with higher efficiency in skin compared to other tissues.

Foreign antigens first come into contact with the immune system in the extravascular space, not the blood. For example, vaccines are typically inoculated subcutaneously, into the extravascular space where dendritic cells engulf foreign particles and process them with major histocompatibility complex molecules. When lymph leaves the skin, it carries with it these and other immune cells, and as they pass through the lymph nodes the immune cells mature and begin presenting antigens on their surfaces. Considering the multitude of effects NO has on T cell function [49], including the production of regulatory T cells [50], it is conceivable that the skin represents a major site of NO-mediated modulation of T cell function/ development.

It is clear that there are very complex interactions between the normal flora of the mucosa and skin involving NO/NOx physiology, including various components of the immune system, and these interactions depend in detail on the conditions occurring in the extravascular space where immune cells first encounter antigen, in the lymph as the cells are transported to the lymph nodes, and in the lymph nodes themselves [51] where T cells encounter antigen presenting cells, differentiate, proliferate, and then migrate into the blood. The level of NO and nitrite, both in the extravascular space where cells first encounter antigens, in the lymph nodes, and in the peripheral tissues, must all be important in T cell physiology.

How can this hypothesis be tested?

Our hypothesis is based on numerous assumptions the validity of which need to be experimentally tested. Many of the regulatory pathways underlying NO/NOx signalling in physiology appear to be far too complex to be appreciated in their entire beauty any time soon. However, other key elements of this hypothesis should not be difficult to verify or reject.

AOB have been shown to live on human skin long-term (years), and produce NO in response to ammonia provided by exercise and non-exercise induced sweating [52]. This demonstrates that sweat provides all nutrients necessary, including ammonia and trace metals, and also shows that sweat is a sufficient substrate for their long-term survival and growth. It further demonstrates that a human colonised with these bacteria is resistant to their loss. However, information about the importance of these bacteria for human health and evidence for their presence as commensals on human skin requires population-based studies including a broad spectrum of individuals from urban and rural areas. Theoretically, the abundance of AOB on their skin should correlate directly with nitrite levels on the skin and inversely with the prevalence for disorders associated with the ‘hygiene hypothesis’. Of similar importance to the production of nitrite is the confirmation that it is actually taken up by the skin to reach compartments important for the regulation of immune cell function.

Conclusions

Considerable evidence points to low NO/NOx in the extravascular space as a contributing factor in immune system disorders and other conditions associated with development. We here coin the term 'nitropenia' to describe adverse health conditions arising from an inadequate availability of NO/NOx and related species. While other factors such as dietary habits also contribute to low NOx levels in the extracellular compartment the skin is likely to represent a major source of these NO species for the following reasons: it is the site for supply of nitrite by commensal AOB, the site for photochemical reactions involving NO/nitrite, and a haemoglobin-free zone where NO reactions can occur more efficiently. Modern living conditions have greatly reduced NO/nitrite supply to the body through the skin. Nitropenia provides an explanation for many observations associated with the 'hygiene/old friends hypothesis', and the greater impact on people with highly pigmented skins, and provides a rational for why the incidence of several seemingly unrelated life-style-related diseases are increasing at this time. Nitropenia also suggests a number of non-invasive therapeutic modalities, including the supplementation of the skin with nitrifying bacteria or application of topical nitrite. Since these are all modest, low cost, low risk treatment modalities, even modest therapeutic effects would seem to be worth investigating.

References

1 Hjern A, Rasmussen F, Johansson M, Aberg N (1999) Migration and atopic disorder in Swedish conscripts. *Pediatr Allergy Immunol* 10: 209–215

2 Daniels SR (2006) The consequences of childhood overweight and obesity. *Future Child* 16: 47–67

3 Kemp A, Björkstén B (2003) Immune deviation and the hygiene hypothesis: A review of the epidemiological evidence. *Pediatr Allergy Immunol* 14: 74–80

4 Rook GA (2009) Review series on helminths, immune modulation and the hygiene hypothesis: the broader implications of the hygiene hypothesis. *Immunology* 126: 3–11

5 Gao Z, Tseng CH, Pei Z, Blaser MJ (2007) Molecular analysis of human forearm superficial skin bacterial biota. *Proc Natl Acad Sci USA* 104: 2927–2932

6 Grice EA, Kong HH, Renaud G, Young AC; NISC Comparative Sequencing Program, Bouffard GG, Blakesley RW, Wolfsberg TG, Turner ML, Segre JA (2008) A diversity profile of the human skin microbiota. *Genome Res* 18: 1043–1050

7 Janssen-Heininger YM, Mossman BT, Heintz NH, Forman HJ, Kalyanaraman B, Finkel T, Stamler JS, Rhee SG, van der Vliet A (2008) Redox-based regulation of signal transduction: principles, pitfalls, and promises. *Free Radic Biol Med* 45(1): 1–17

8 Jansson EA, Huang L, Malkey R, Govoni M, Nihlén C, Olsson A, Stensdotter M,

Petersson J, Holm L, Weitzberg E, Lundberg JO (2008) A mammalian functional nitrate reductase that regulates nitrite and nitric oxide homeostasis. *Nat Chem Biol* 4: 411–417
9 Bryan NS, Fernandez BO, Bauer SM, Garcia-Saura MF, Milsom AB, Rassaf T, Maloney RE, Bharti A, Rodriguez J, Feelisch M (2005) Nitrite is a signaling molecule and regulator of gene expression in mammalian tissues. *Nat Chem Biol* 1: 290–297
10 Webb AJ, Patel N, Loukogeorgakis S, Okorie M, Aboud Z, Misra S, Rashid R, Miall P, Deanfield J, Benjamin N, MacAllister R, Hobbs AJ, Ahluwalia A (2008) Acute blood pressure lowering, vasoprotective, and antiplatelet properties of dietary nitrate *via* bioconversion to nitrite. *Hypertension* 51: 784–790
11 Petersson J, Phillipson M, Jansson EA, Patzak A, Lundberg JO, Holm L (2007) Dietary nitrate increases gastric mucosal blood flow and mucosal defense. *Am J Physiol Gastrointest Liver Physiol* 292: G718–724
12 Dykhuizen RS, Frazer R, Duncan C, Smith CC, Golden M, Benjamin N, Leifert C (1996) Antimicrobial effect of acidified nitrite on gut pathogens: importance of dietary nitrate in host defense. *Antimicrob Agents Chemother* 40: 1422–1425
13 Lundberg JO, Weitzberg E, Gladwin MT (2008) The nitrate-nitrite-nitric oxide pathway in physiology and therapeutics. *Nat Rev Drug Discov* 7: 156–167
14 Stewart V (1998) Nitrate respiration in relation to facultative metabolism in enterobacteria. *Microbiol Rev* 52: 190–232
15 Lundberg JO, Weitzberg E, Cole JA, Benjamin N (2004) Nitrate, bacteria and human health. *Nat Rev Microbiol* 2: 593–602
16 Butler AR, Feelisch M (2008) Therapeutic uses of inorganic nitrite and nitrate: from the past to the future. *Circulation* 117: 2151–2159
17 Esch T, Stefano GB, Fricchione GL, Benson H (2002) Stress-related diseases – a potential role for nitric oxide. *Med Sci Monit* 8: RA103–118
18 Vuppugalla R, Mehvar R (2004) Short-term inhibitory effects of nitric oxide on cytochrome P450-mediated drug metabolism: time dependency and reversibility profiles in isolated perfused rat livers. *Drug Metab Dispos* 32: 1446–1454
19 Zangar RC, Davydov DR, Verma S (2004) Mechanisms that regulate production of reactive oxygen species by cytochrome P450. *Toxicol Appl Pharmacol* 199: 316–331
20 Rius J, Guma M, Schachtrup C, Akassoglou K, Zinkernagel AS, Nizet V, Johnson RS, Haddad GG, Karin M (2008) NF-kappaB links innate immunity to the hypoxic response through transcriptional regulation of HIF-1alpha. *Nature* 453: 807–811
21 Guzik TJ, Korbut R, Adamek-Guzik T (2003) Nitric oxide and superoxide in inflammation and immune regulation. *J Physiol Pharmacol* 54: 469–487
22 Feelisch M (2008) The chemical biology of nitric oxide – an outsider's reflections about its role in osteoarthritis. *Osteoarthritis Cartilage* 16 (Suppl 2): S3–S13
23 A Ramsay. 'The Great Unwashed' – The Working Classes and Personal Hygiene, circa 1840–1914. Leeds History First, Volume 5 2007–08 (Undergraduate Essay) http: // www.leeds.ac.uk/history/studentlife/e-journal/Anna_Ramsay.pdf (accessed 02/23/09)

24 An PG (2004) Helping the poor emerge from "Urban barbarism to civic civilization": Public bathhouses in America, 1890–1915. *Yale J Biol Med* 77: 133–141

25 Edwards HR (1962) Competition and monopoly in the British soap industry, Oxford: Clarendon Press. cited in: Mokyr J, Stein R (1997) *Science, health, and household technology: The effect of the Pasteur Revolution in consumer demand* (Bresnahan TF and Gordon RJ, eds) (1997) The Economics of New Goods, Studies in Income and Wealth, Vol. 58, Chicago and London: The University of Chicago Press

26 Geels F (2005) Co-evolution of technology and society: The transition in water supply and personal hygiene in the Netherlands (1850–1930) – a case study in multi-level perspective. *Technology in Society* 27: 363–397

27 Smith Z, McCaig AE, Stephen JR, Embley TM, Prosser JI (2001) Species diversity of uncultured and cultured populations of soil and marine ammonia oxidizing bacteria. *Microb Ecol* 42: 228–237

28 Francis CA, Roberts KJ, Beman JM, Santoro AE, Oakley BB (2005) Ubiquity and diversity of ammonia-oxidizing archaea in water columns and sediments of the ocean. *Proc Natl Acad Sci USA* 102: 14683–14688

29 Brusilow SW, Gordes EH (1968) Ammonia secretion in sweat. *Am J Physiol* 214: 513–517

30 Hooper AB, Terry KR (1074) Photoinactivation of ammonia oxidation in Nitrosomonas. *J Bacteriol* 119: 899–906

31 Adelman S, Taylor CR, Heglund NC (1975) Sweating on paws and palms: what is its function? *Am J Physiol* 229: 1400–1402

32 Weller R, Ormerod AD, Hobson RP, Benjamin NJ (1998) A randomized trial of acidified nitrite cream in the treatment of tinea pedis. *J Am Acad Dermatol* 38: 559–563

33 Öhman H, Vahlquist A (1998) The pH gradient over the stratum corneum differs in X-linked recessive and autosomal dominant ichthyosis: A clue to the molecular origin of the "Acid Skin Mantle"? *J Invest Dermatol* 111: 674–677

34 Chikakane K, Takahashi H (1995) Measurement of skin pH and its significance in cutaneous diseases. *Clin Dermatol* 13: 299–306

35 Rothfork JM, Timmins GS, Harris MN, Chen X, Lusis AJ, Otto M, Cheung AL, Gresham HD (2004) Inactivation of a bacterial virulence pheromone by phagocyte-derived oxidants: New role for the NADPH oxidase in host defense. *Proc Natl Acad Sci USA* 101: 13867–13872

36 Schlag S, Nerz C, Birkenstock TA, Altenberend F, Götz F (2007) Inhibition of staphylococcal biofilm formation by nitrite. *J Bacteriol* 189: 7911–7919

37 Barraud N, Hassett DJ, Hwang SH, Rice SA, Kjelleberg S, Webb JS (2006) Involvement of nitric oxide in biofilm dispersal of Pseudomonas aeruginosa. *J Bacteriol* 188: 7344–7353

38 Blumer M, Chase T, Watson SW (1969) Fatty acids in the lipids of marine and terrestrial nitrifying bacteria. *J Bacteriol* 99: 366–370

39 Hooper AB (1969) Biochemical basis of obligate autotrophy in Nitrosomonas europaea. *J Bacteriol* 97: 776–779

40 M Schaechter, G Mendoff, D Schlessinger, eds (1989) *Mechanisms of Microbial Disease.* Williams & Wilkins, Baltimore, MD, USA
41 Lipponen MT, Suutari MH, Martikainen PJ (2002) Occurrence of nitrifying bacteria and nitrification in Finnish drinking water distribution systems. *Water Res* 36: 4319–4329
42 Chain P, Lamerdin J, Larimer F, Regala W, Lao V, Land M, Hauser L, Hooper A, Klotz M, Norton J, Sayavedra-Soto L, Arciero D, Hommes N, Whittaker M, Arp D (2003) Complete genome sequence of the ammonia-oxidizing bacterium and obligate chemolithoautotroph *Nitrosomonas europaea. J Bacteriol* 185: 2759–2773
43 Brandt KK, Hesselsoe M, Roslev P, Henriksen K, Sorensen J (2001) Toxic effects of linear alkylbenzene sulfonate on metabolic activity, growth rate, and microcolony formation of Nitrosomonas and Nitrosospira strains. *Appl Environ Microbiol* 67: 2489–2498
44 Könneke M, Bernhard AE, de la Torre JR, Walker CB, Waterbury JB, Stahl DA (2005) Isolation of an autotrophic ammonia-oxidizing marine archaeon. *Nature* 437: 543–546
45 Stucker M, Struk A, Altmeyer P, Herde M, Baumgart H, Lubbers DW (2002) The cutaneous uptake of atmospheric oxygen contributes significantly to the oxygen supply of human dermis and epidermis. *J Physiol* 538: 985–994
46 Zhang CL, Ye Q, Huang Z, Li W, Chen J, Song Z, Zhao W, Bagwell C, Inskeep WP, Ross C, Gao L, Wiegel J, Romanek CS, Shock EL, Hedlund BP (2008) Global occurrence of archaeal amoA genes in terrestrial hot springs. *Appl Environ Microbiol* 74: 6417–6426
47 Weidler GW, Dornmayr-Pfaffenhuemer M, Gerbl FW, Heinen W, Stan-Lotter H (2007) Communities of archaea and bacteria in a subsurface radioactive thermal spring in the Austrian Central Alps, and evidence of ammonia-oxidizing Crenarchaeota. *Appl Environ Microbiol* 73: 259–270
48 Liu X, Miller MJS, Joshi MS, Sadowska-Krowicka H, Clark DA, Lancaster Jr JR (1998) Diffusion-limited reaction of free nitric oxide with erythrocytes. *J Biol Chem* 273: 18709–18713
49 Niedbala W, Cai B, Liew FY (2006) Role of nitric oxide in the regulation of T cell functions. *Ann Rheum Dis* 65 (Suppl 3): iii37–40
50 Niedbala W, Cai B, Liu H, Pitman N, Chang L, Liew FY (2007) Nitric oxide induces $CD4^+CD25^+$ Foxp3 regulatory T cells from $CD4^+CD25$ T cells *via* p53, IL-2, and OX40. *Proc Natl Acad Sci USA* 104: 15478–15483
51 Harris NL, Watt V, Ronchese F, Le Gros G (2002) Differential T cell function and fate in lymph node and nonlymphoid tissues. *J Exp Med* 195: 317–326
52 Whitlock D (2004) NO production on human skin from sweat-derived urea by commensal autotrophic ammonia oxidizing bacteria. *Nitric Oxide* 11: 130 (abstract)

The hygiene hypothesis and allergic disorders

Paolo M. Matricardi[1] *and Eckard Hamelmann*[1,2]

[1]Pediatric Pneumology and Immunology, Charité University Medicine, Berlin, Germany
[2]University Children's Hospital, Ruhr-University Bochum, Germany

Abstract

Allergic diseases are more frequent in the general population than other immune-mediated disorders, such as autoimmune diseases or immunodeficiencies. Therefore, epidemiological studies investigating the hygiene hypothesis in relation to allergic diseases have been performed much more frequently than those investigating the hygiene hypothesis in relation to Crohn's disease, multiple sclerosis, rheumatoid arthritis or other diseases related to dysregulation of the immune system. More than one thousand papers have been written on this subject and it is not possible to condense them in a few pages. In this chapter, we summarise the most important pathways followed by the research on the hygiene hypothesis applied to allergic disorders, i.e., the allergy protective role of foodborne and orofaecal infections, starting from Strachan's initial observations and concluding with the most recent intervention studies.

Introduction

Allergic diseases are more frequent in the general population than other immune-mediated disorders, such as autoimmune diseases or immunodeficiencies. Therefore, epidemiological studies investigating the hygiene hypothesis in relation to allergic diseases are easier to complete and have been performed much more frequently than those investigating the hygiene hypothesis in relation to Crohn's disease, multiple sclerosis, rheumatoid arthritis or other diseases related to dysregulation of the immune system. More than one thousand papers have been written on this subject and it is not possible to condense them in a few pages. In this chapter, we shall therefore try to summarise the most important pathways followed by the research on the hygiene hypothesis applied to allergic disorders, i.e., the allergy protective role of foodborne and orofaecal infections, starting from Strachan's initial observations and concluding with the most recent intervention studies.

The Hygiene Hypothesis and Darwinian Medicine, edited by Graham A. W. Rook

The birth order effect, i.e., the hygiene hypothesis for allergic diseases

In his seminal paper, David Strachan reported that in a British population the risk of having hay fever was directly related to the socioeconomic status (defined by the father's occupation) and inversely related to the overall number of siblings ('sibship size effect'). He also noted that hay fever was less frequent in the presence of older rather than younger siblings ('birth order effect') [1]. Assuming that infections were acquired more frequently in large, less affluent families and earlier in the presence of many older siblings, Strachan proposed that exposure to common infections, especially very early in infancy or even through the mother *in utero*, may 'protect' from hay fever [1]. As a corollary, he hypothesised that the decline in cross-infections within young families due to decreasing family size and to improvement in hygienic standards is, among the set of characteristics of the wealthy lifestyle responsible for the increase of atopy prevalence ('hygiene hypothesis') [1].

The direct association between socioeconomic status (SES) and atopy has been repeatedly observed in developed countries, including USA [2], Italy [3] and German adults [4]. Similarly, the 'sibship-size effect' originally reported by Strachan has been reproduced in more than 20 cross-sectional studies (reviewed in [5]). Finally, this phenomenon was confirmed also by a birth-cohort study involving 1,035 US children. In this study, the risk of wheezing among subjects exposed early in life to other children at home or at day care, steadily declined with age, being high (OR 1.4) at the age of 2, low (OR 0.8) at 6 yrs and very low (OR 0.3) at 13 yrs. These results suggested that bacterial or viral infections (facilitated by early contacts with other children) caused wheezing of infectious aetiology early in life, but in contrast inhibited atopic sensitisation and the subsequent development of allergic wheezing of atopic aetiology (asthma) later in life [6].

HAV serology, birth order and atopy: an early demonstration of the hygiene hypothesis

This hypothesis on the link between atopic sensitisation and birth order was strongly supported by the results of a retrospective study among Italian young military recruits [7]. In this study, young men with antibodies to hepatitis A virus showed a lower prevalence of atopy and atopic respiratory diseases. Interestingly, this was independent of the number of older siblings and other relevant risk factors. Accordingly, the prevalence of atopy among seropositive subjects was always low, whatever the number of older siblings, but among seronegative subjects was only low when they had three or more older siblings. This suggested that common infections acquired either early in life due to the presence of many older siblings (among seronegative subjects) or due to unhygienic living conditions (among seropositive subjects) may have reduced the risk to develop atopy. This study also suggested for the first time that faecal contamination of the environment – a condition which facilitates transmission of the hepatitis A virus – may protect from atopy [7]. Following these observations, it was proposed that foodborne and orofaecal infections, through stimulation of gut-associated lymphoid tissues (GALT), Peyer's patches, and mesenteric lymph nodes, were protective against allergic diseases, autoimmune diseases and other immune-mediated disorders that are on the increase in the developed countries [8]. This hypothesis did not completely exclude that airborne viruses may play a major role in regulating the development of atopy. Nevertheless, it suggested that the gut mucosa is a critical site where microbes may contribute to inhibition of Th2-skewed immune responses against allergens that otherwise show deleterious effects in other mucosal areas (bronchial, nasal, conjunctiva). This hypothesis was in line with the concept that a proportion of the lymphocytes homing to the nasal, bronchial and enteric mucosa belongs to the same recirculating pool [8].

Evidence for an allergy protective role of foodborne and orofaecal infections

The hypothesis that foodborne and orofaecal microbes may protect from atopic sensitisation was further tested by extending the observation on the same Italian military recruits [9]. In a second study, it was found that atopy was inversely related not only to positive HAV serology but also to other orofaecal/foodborne (OF/FB) infections (*Toxoplasma gondii*, TG, and *Helicobacter pylori*, HP), but not to infections mainly acquired through other routes. It was suggested that food hygiene and declining exposure to orofaecal microbes may underlie the allergy and asthma epidemic. Moreover, attention was focused on the gut mucosa and GALT as the sites where microbes may inhibit the development of the atopic phenotype [9]. An inverse association of HAV, TG and HP with allergic diseases was replicated in a study conducted in a representative sample of the Danish general population [10].

Further support for this concept came from an analysis of the public data contained in a large cross-sectional survey of a general population sample in the United States (Third National Health and Nutrition Examination Survey, 1988–1994) [11]. In this study, hay fever and asthma were less frequent in subjects seropositive for HAV, TG, and herpes simplex virus 1 *versus* seronegative subjects, after adjusting for age, sex, race, urban residence, census region, family size, income, and education. Skin sensitisation to peanut and to many airborne allergens was less frequent among HAV-seropositive *versus* -seronegative subjects younger than 40 years of age. This study suggested that, in the United States, serologic evidence of acquisition of certain mainly food-borne and orofaecal infections is associated with a lower probability of having hay fever and asthma. An analysis of the data obtained after stratification of the subjects by year of birth produced a very interesting graph, showing that the epidemic trend of hay fever and asthma was true only in the subset of the

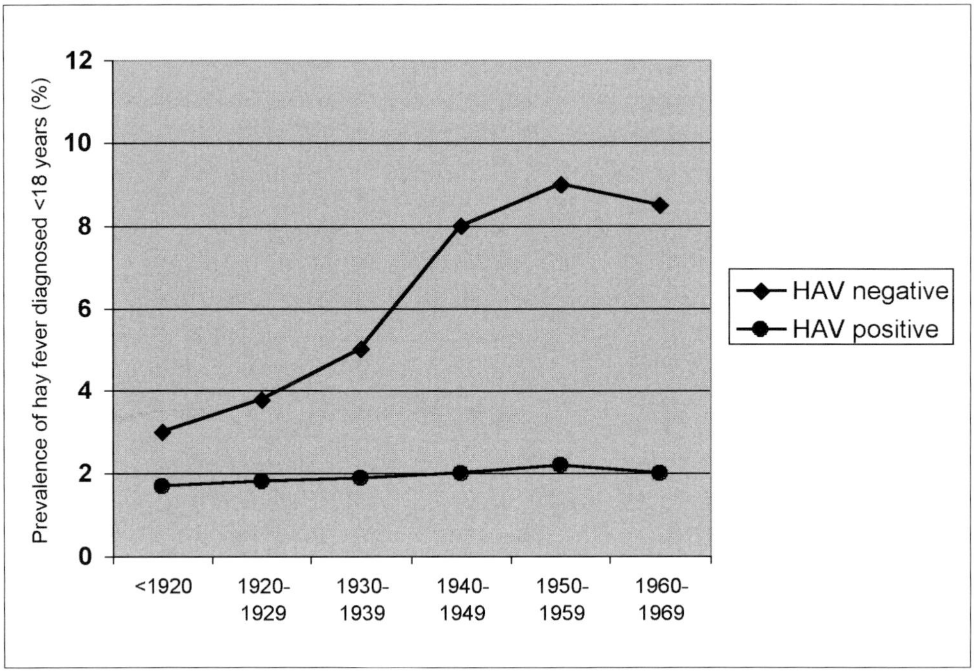

Figure 1
Prevalence of participants with a history of hay fever diagnosed before 18 years of age in relation to serology for HAV by decade of birth (NHANES III, 1988–1994). The prevalence of hay fever diagnosed at or before 18 years of age in HAV-seronegative subjects increased progressively from 2.7%, in cohorts born before 1920 to 8.5%, in cohorts born in the 1960s, whereas they remained constant at around 2% in all cohorts of HAV-seropositive subjects (adapted from [11]).

US population who remained free of HAV infection up to the time of examination (Fig. 1) [11]. Further information on possible mechanisms of action of HAV are discussed in detail in Chapter 4 in this volume.

Studies in other settings supported the concept of an allergy-protective role of foodborne and orofaecal infections. In Norwegian recruits, the serological response to TG, but not to the respiratory encapsulated bacteria, was found to be associated with a lower risk for sensitisation [12]. Similarly, in Germany a positive HP infection status was inversely associated with physician-diagnosed allergy and prescription of anti-allergic medication after adjustment for potential confounders (age, sex, nationality, smoking status and education of the participant) [13]. Among participants in the NHANES 1999–2000 aged 3–19 years, the presence of HP was inversely related to ever having had asthma [14].

Studies in traditional farms confirm an allergy-protective role of microbes

The hypothesis that foodborne and orofaecal microbes may protect from atopic sensitisation was further supported by a long list of studies performed in farming environments. In these studies, direct exposure to stables where livestock is kept showed a strong protective effect against the development of atopy and atopic diseases [15]. Evidence for a lower prevalence of atopic sensitisation among children raised on farms came from studies in Finland [16], Canada [17], Austria [18] and in many other countries (reviewed in [19]).

Studies on the allergy protective role of the farming environment focused then on the components that mediate this phenomenon. The level of environmental endotoxin exposure was measured in homes of farmers' children, children with regular contact to livestock and control children with no contact to farm animals [20]. Endotoxin concentrations were the highest in stables of farming families, but were also significantly higher indoors in dust from kitchen floors and children's mattresses as compared to control children from non-farming families. These data suggested that environmental exposure to endotoxin and/or other bacterial wall components was an important protective factor for the development of atopic diseases in childhood [21]. Similarly, an association of low levels of endotoxin in the house dust with the proportion of peripheral blood CD4 T lymphocytes producing interferon gamma and with skin sensitisation to common allergens had been found in a cross sectional study performed in the United States [22]. The authors had concluded that indoor endotoxin exposure early in life may protect against allergen sensitisation by enhancing type 1 immunity [21]. Direct evidence for a protective, anti-allergic effect of LPS on allergen-induced sensitisation and airway disease was derived from work in animal models of experimental asthma showing that exposure to LPS suppressed IgE production and airway inflammation [23], inhibited the development of allergic airway disease in newborn mice exposed to aerosolised LPS [24], and was even effective in suppressing allergic sensitisation in offspring when LPS exposure took place during pregnancy [25]. As a mechanism for this protective effect it was shown that LPS suppressed allergen-induced Th2 immune responses by induction of Th1 differentiating factors and modulated mucosal tolerance by inducing allergen-specific IgG1 production and effector $CD4^+$ T cells with a mixed regulatory/Th1 phenotype [26].

Further studies tried to disclose what (other) specific environmental factors mediate this protective effect of a farming environment. In a cross-sectional study in Shropshire (UK), farmers' children had significantly less current asthma symptoms and current seasonal allergic rhinitis but not less current eczema symptoms or atopy [22]. However, current unpasteurised milk consumption was associated with significantly less current eczema symptoms and a greater reduction in atopy. Moreover, consumption of unpasteurised milk was associated with a consistent reduction in total IgE levels and higher production of IFN-γ from stimulated whole blood.

The authors concluded that unpasteurised milk was the exposure mediating the protective effect on skin prick test positivity. The effect was independent of farming status and still present with infrequent consumption of unpasteurised milk [21]. Other epidemiological studies confirmed that consumption of unpasteurised milk is a protective factor against asthma, current wheeze, hay fever, current rhinitis, and atopic sensitisation [27] atopy [28] eczema and allergic rhinitis [29], and asthma, rhinoconjunctivitis and sensitisation to pollen, food and horse dander [30].

Which foodborne and orofaecal microbes do the (anti-allergic) job?

The hypotheses that HAV, TG, HP or other OF/FB infections play a causative role in the prevention of allergic sensitisation and diseases is biologically plausible. However, the epidemiological studies generating this hypothesis did not offer any formal demonstration of such a role. Positive serology for these infections could, therefore, be interpreted just as a marker of exposure and acquisition of different categories of infections transmitted through contaminated food or the faecal-oral route. The next question in this component of the hygiene hypothesis was therefore, which kinds of foodborne and faecal-oral infections protect from allergic sensitisation and allergic diseases? In principle, three categories of infections could be taken into consideration: (i) severe pathogens, (ii) commensals, and (iii) mild pathogens. The first category includes microorganisms that can cause fatal diseases such as, for example, the *Vibrio cholerae*. From an evolutionary point-of-view, however, it is very unlikely that fatal infections contribute to the shaping of the immune system to provide a protective influence against immune disorders. Although this hypothesis cannot be totally discarded, it is much more likely that either commensals or mild pathogens, or both play, an important educating role on the immune system. This hypothesis will be discussed in the next two sections.

Do *commensal* bacteria of the GI tract protect from allergic diseases?

It has been known since the seventies that mice reared under sterile conditions do not develop a fully functional immune system and that they are not susceptible to the induction of oral tolerance [31]. The relevance of these observations for the understanding of atopy in humans was for a long time overlooked. A 'commensal' hypothesis emerged in the late nineties when *Lactobacilli* and Eubacteria were observed more frequently in the intestinal microflora of 1-year-old infants living in a country with a low prevalence of atopy (Estonia), while *Clostridia* were more frequent in age-matched infants living in a nearby country with a high prevalence of atopy (Sweden) [32]. It was therefore proposed that the gastrointestinal microflora of westernised children might predispose to atopy because of the stable predominance of bacteria stimulating TH2 activation (e.g., clostridia) or because of the absence of bacteria that stimulate TH1 activation (e.g., Lactobacilli) [33, 34]. Accordingly, we might speculate that atopy may be prevented simply by substituting one colonising species with another. However, from an evolutionary point of view it is unlikely that one single bacterial species has the important task of protecting mammals from atopy. Moreover, translocation through the epithelial barrier and subsequent potent stimulation of the local immune system by a bacterium are only transient and limited to the initial colonisation phases, and they are soon prevented by an IgA response [35]. Otherwise, we probably could not tolerate our own microflora. A stronger hypothesis stated that a high turnover of appropriate bacteria at mucosal level (NALT, BALT, GALT), rather than specific, stable colonisation by certain species, provides the potent, continuous immune stimulation necessary to prevent atopy and atopic diseases [36, 37]. This hypothesis not only has greater evolutionary plausibility (an atopy-preventing effect being attributed to many different bacterial species), but may also explain the 'effects' of sibship size and birth order on atopy. Indeed, cross-infections with new bacterial strains, rather than a specific and stable colonisation, are facilitated in large families living a traditional lifestyle.

A few observational studies investigated the 'gut commensals' hypothesis. In a Swedish cross-sectional study, it was found in stools of allergic participants increased levels of i-caproic acid, a marker of colonisation by *Clostridium difficile*, and lower levels of propionic, i-butyric, butyric, i-valeric and valeric acid. The authors concluded that the composition of gut microflora in allergic subjects may be causally related to their disease [38]. In a small longitudinal study, a reduced ratio of bifidobacteria to clostridia was found in the stools of infants later developing atopy, compared to those remaining non-atopic [39]. The authors concluded that differences in the neonatal gut microflora precede the development of atopy, suggesting a crucial role of the balance of indigenous intestinal bacteria for the maturation of human immunity to a non atopic mode [39].

Data against the 'gut commensal' hypothesis came from a quite large birth cohort study performed in three European cities. The qualitative and quantitative composition of the faecal microflora of over 300 infants born in London, Rome and Göteborg was monitored seven times throughout the first year of age and related to the appearance of sensitisation against food allergens and atopic eczema at 18 months of age [40]. Neither atopic eczema nor food-specific IgE by 18 months of age were associated with time of acquisition of any particular bacterial group. The authors concluded that their study did not support the hypothesis that sensitisation to foods or atopic eczema in European infants in early life is associated with lack or presence of any particular cultivable intestinal commensal bacteria [40]. However, a rather small nested case-control study of the same population, suggested that a reduced diversity in the early faecal microbiota of infants in the first week of life may be causally linked with atopic eczema appearing during the first 18 months of life [41].

Do mild pathogens of the GI tract protect from allergic diseases?

An interesting alternative hypothesis proposed that GI microorganisms, in order to protect from allergy, must stimulate the immune system with a pathogenic effect without causing a fatal disease. This category of microbes would include, by definition, mild pathogens. The analysis of the factors so far reported in the literature and summarised in the previous paragraphs may be of some help in identifying a good candidate to exert an allergy-protective effect within such category. Actually, atopy has been found to be less frequent among subjects who acquire faecal-oral [7] and foodborne [9] infections, in those exposed to stables [15] and to high endotoxin concentrations [19], and consuming unpasteurised milk [21, 22]. Therefore, it was reasoned that an example of atopy-preventing infection may be found among Gram-negative bacteria transmitted by contaminated food and the faecal-oral route, and by animals typical of a farming environment. A number of infectious agents share these properties. Among them, non-typhoid Salmonellae were ideal candidates for an observational study since they cause diseases that are easily diagnosed even in early childhood. The hypothesis that acquisition of infection with Salmonella in the first 4 years of life may counteract the development of respiratory allergic diseases later in childhood was therefore tested with a longitudinal study design [42]. In this study, the incidence of hay fever and asthma was compared in a group of Sardinian children (age 6–18 years) who had been hospitalised at preschool-age (at age <4 years) with salmonellosis (n=148) to that of age-matched children who had been hospitalised (at age <4 years) with acute enteritis of non-bacterial aetiology (n=168) [42]. This study showed that children who had been hospitalised with salmonellosis had a lower prevalence of allergic rhinoconjunctivitis or asthma than controls (Fig. 2). The proportional hazard of salmonellosis for asthma was as low as 0.23 (95% CI: 0.08–0.67; $P<0.01$) and for allergic rhinoconjunctivitis was 0.40 (95% CI: 0.17–0.95; $P=0.04$), after adjusting for a list of relevant confounders. The authors speculated that acquisition of infection early in life by Salmonella may inhibit the development of atopic diseases and in particular of allergic asthma. They concluded that Salmonella may contribute to the prevention of the development of respiratory allergies through a range of mechanisms acting on the innate immune system during a critical period for the maturation of immune response against ubiquitous allergens. Salmonellae, similar to mycobacteria, grow in the endosomes of macrophages and induce a strong activation of $CD4^+$ T helper-1 lymphocytes, which are required to enhance intracellular killing and clearance [43, 44]. This process is highly dependent on production of γ-interferon and interleukin-12, so that patients with selective deficiencies of these cytokines or their receptors develop fatal infections by Salmonella as well as mycobacteria [45]. In addition, Salmonella plays a transient regulatory role on adaptive immunity [46]. Finally, the immune response to Salmonella is controlled by the natural resistance-associated macrophage protein 1 (Nramp-1) gene [47], which plays also a major role in regulation of atopic

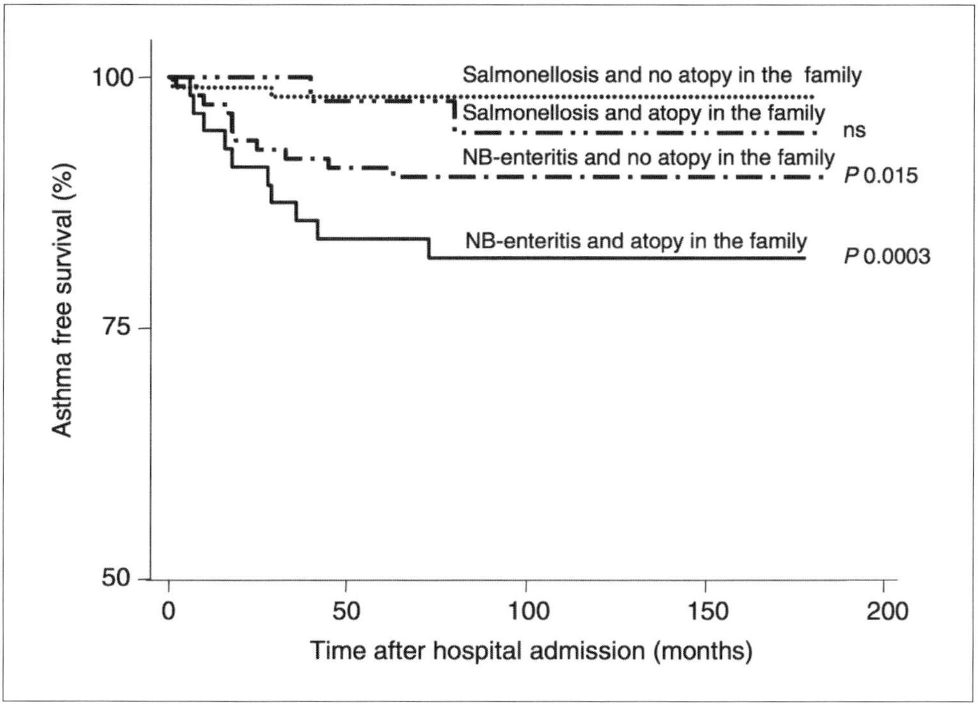

Figure 2
Asthma-free survival in children after hospitalisation for enteritis, by aetiology of enteritis and family history of atopy (adapted from [42]).

responses and airways allergic inflammation in rodents [48]. These properties or other unknown mechanisms provide biologic plausibility to the hypothesis that salmonellosis contributed to protect from allergic asthma and rhinoconjunctivitis the children we examined.

Other mild pathogens protective against allergies: mycobacteria

In a trailblazing paper, Shirakawa et al. [49] showed, in Japanese children immunised with BCG at 3 months of age, an inverse relation between the magnitude of skin responses to intradermal tuberculin injection at 12 years of age and serum levels of total IgE, TH2 cytokines (IL-4, IL-10 and IL-13), and the prevalence of atopy and atopic diseases [49]. These associations suggested that natural exposure to mycobacteria [49], or immunisation with BCG [50] may be protective against atopy. A third explanation proposed that a high skin reactivity to tuberculin and strong immediate reactions to allergens may be mutually excluding phenotypes of the same set of genes regulating predisposition to a TH1 or TH2 predominance in the immune responses [8].

An impressive number of arguments support the concept that daily acquisition of mycobacteria through different routes may be essential to provide protection against allergic diseases and immune disorders [51]. This aspect is in line with the hypothesis that the stimulation of the GI tract by mild pathogens is essential in protecting from allergies [8]. Actually, an important route for the acquisition of environmental mycobacteria is their ingestion through contaminated food, contaminated water, and particles from contaminated soil. There is no doubt that in animal models mycobacteria can both prevent and treat allergic responses either by boosting TH1 or by driving allergen-specific regulatory T cells. However, clinical trials in man using BCG for active treatment or prevention of allergy remain inconclusive (reviewed in [52]).

Helminths and the hypothesis of immunoregulation

Since the seventies it has been proposed that helminths protect from allergy and asthma either through the saturation of high affinity IgE-receptors on mast cells and basophils (by polyclonal IgE) or by induction of blocking IgG antibodies (reviewed in [53]). Helminthic infections were associated with asthma in a few studies, but not in others. This area re-emerged when the immunosuppressive properties of helminths were re-evaluated by combining two new concepts: the 'anti-inflammatory network' (cytokines produced by regulatory T cells: interleukin-10, transforming growth factor-β) and the 'hygiene hypothesis' [54]. Now helminths were considered to prevent allergy and asthma by stimulating the anti-inflammatory network, and the allergy epidemic was attributed to the sharp decline of helminthic infections [54]. Indeed, chronic intestinal helminth infections have been shown to protect children from atopic reactivity in a variety of settings in developing countries [55–59]. Similarly, worm infestation early in life (mainly *Ascaris*, *Oxyuris*) was negatively associated with subsequent eczema in a large population of East German children [60]. In contrast, transient, delayed or milder forms of helminthic infections in children were positively associated with atopic disorders [55, 61]. To explain these new inconsistencies, it was speculated that early, heavy, and chronic helminthic infections protect children in endemic countries against allergies, perhaps by stimulating regulatory T cells and cytokines. This observation was supported by studies in experimental models showing strong induction of regulatory activities subsequently inhibiting allergic immune responses and development of airway inflammation in mice treated with helminths [62] or helminth products [63]. In contrast, sporadic, delayed, or transient infection may potentiate allergy in sensitised children through bystander TH2 stimulation [53, 64].

Cross-sectional studies are clearly insufficient to answer whether helminths promote or suppress allergy and asthma. Theoretically, such a link could be experimentally investigated by monitoring either allergic children during a new helminthic infection or children chronically infected by helminths during de-worming treatment. Indeed, the latter approach showed that Venezuelan children chronically infected mostly with *A. lumbricoides* and *Trichuris trichuria* developed or increased their atopic reactivity to mites after successful treatment with anthelmintic medication [65]. Similarly, Gabonese children chronically infected by *A. lumbricoides* and *T. trichuria* transiently increased their reactivity to locally relevant airborne allergens during a 3-year follow-up after anthelmintic treatment [66]. In contrast, an anthelmintic programme in Ecuadorian schoolchildren reduced helminthiases without promoting atopic sensitisation, allergic symptoms or exercise-induced bronchospasm during a 1-year follow-up [67].

Intervention studies related to the hygiene hypothesis

The hygiene hypothesis has inspired or indirectly supported a new category of intervention strategies based on the use of microbial products for allergy prevention and therapy [68]. They include the use of probiotics, oral bacterial extracts, mycobacteria, LPS derivatives, immunostimulatory sequencies of oligodeoxynucleotides (ISS-ODN), and products derived from helminths. Some of these approaches (oral bacterial extracts, probiotics) have so far given negative or inconsistent results, others (ISS-ODN) seem more promising, but have not reached a level of evidence for efficacy high enough to be recommended by international guidelines.

The one approach that has been most frequently investigated and promoted by the Industry is the attempt to prevent or treat allergic diseases with probiotic bacteria, such as lactobacilli or bifidobacteria [69, 70]. Notwithstanding the lack of a strong rationale [36], the hygiene hypothesis has also been abused as a rationale to 'invent' probiotic functional food for patients with allergic disease [71]. Trials have been performed to evaluate the putative preventive or therapeutic effects of probiotics in allergies. Initial studies, although promising, have been highly criticised for their inadequate design or other weaknesses in data handling or data interpretation [72]. The evidence provided so far is still insufficient to generally advise the use of probiotics for primary prevention or therapy of allergies, and consequently this approach is considered an experimental one by a Task Force of EAACI [68], the GINA group [73], and individual opinion leaders [74].

Since then, further trials using probiotics in the prevention or therapy of allergic diseases have been published. A Finnish intervention trial in over 900 infants demonstrated that probiotic treatment containing Lactobacillus rhamnosus GG (LGG) plus three other probiotic strains had no effect on the incidence of all allergic diseases by age 2 but significantly prevented eczema and especially atopic eczema [75]. By contrast, an Australian intervention trial in over 170 Australian infants demonstrated that early probiotic supplementation with *L. acidophilus* did not reduce the risk of AD in high risk infants, but was associated with increased allergen sensitisation in infants receiving probiotics [76]. A combination of *L. rhamnosus* and *L. reuteri* did not improve significantly SCORAD index of AD patients compared to placebo; nevertheless, the treatment was considered 'beneficial' on the basis of the patients' subjective evaluation during intervention [77]. No differences were observed in infants with atopic eczema and suspected cow milk allergy which were treated with LGG, compared to placebo [78]. No therapeutic effect was observed in a DBPC trial of LGG in German infants with moderate atopic dermatitis [79]. There was no significant difference in the improvement of SCORAD values between the placebo and treatment group in a trial studying the effect of *L. fermentum* in Australian infants with atopic dermatitis [80]. Finally, no clinical or immunological effect of LGG was observed in Dutch infants with atopic dermatitis, compared to placebo [81]. A case report showed that contamination of probiotic preparations with milk allergens can cause anaphylaxis in children with cow's milk allergy [82]. The same report revealed that two out of three probiotics widely used in France contain significant amounts of beta-lactoglobulin. The more recent literature therefore reinforced the statement that probiotic at this stage is not considered a therapeutical option for treatment or prevention of allergy or atopy according to guidelines.

Conclusions

Research lines investigating the hygiene hypothesis by using different models (farming environment, military recruits, general population samples, animal models, etc.) converged to identify the gut-associated immune system as a potential target of an allergy preventive effect of foodborne and orofaecal infections. The infectious agents that may induce such a protective effect through this route are mostly in the category of the mild intracellular pathogens, such as non-typhoid *Salmonellae*, *Toxoplasma gondii*, and Mycobacteria. These infectious agents are included in the broader category of the so called 'old friends' microbes, and they share many characteristics: they are mild intracellular pathogens, stimulate TH1 and T regulatory immunity, are widely spread in the environment and in food/water, and suppress allergy in experimental models. It is to be hoped that from this quite solid basis, new observational and experimental studies will be conceived and designed in order to better define the mechanisms of the allergy-protective role of these infections. Only afterwards can strategies for the safe use of these microbes for the prevention and therapy of allergic diseases be designed [83]. Until then, any further attempts to prevent allergies with products lacking a solid rationale have to be regarded as doubtful, empiric and perhaps non-ethical enterprises.

References

1 Strachan DP. Hay fever, hygiene and household size. *BMJ* 1989; 299,1259–60

2 Gergen PJ, Turkeltaub PC, and Kovar MG. The prevalence of allergic skin test reactivity to eight common aeroallergens in the U.S. population: results from the second National Health and Nutrition Examination Survey. *J Allergy Clin Immunol* 1987; 80: 669–79

3 Matricardi PM, Franzinelli F, Franco A, Caprio G, Murru F et al. Sibship size, birth order, and atopy in 11,371 Italian young men. *J Allergy Clin Immunol* 1998; 101: 439–44

4 Bergmann RL, Edenharter G, Bergmann KE, Lau S, Wahn U et al. Socioeconomic status is a risk factor for allergy in parents but not in their children. *Clin Exp Allergy* 2000; 30: 1740–5

5 Karmaus W, Botezan C. Does a higher number of siblings protect against the development of allergy and asthma? A review. *J Epidemiol Community Health* 2002; 56: 209–217

6 Ball TN, Castro-Rodriguez JA, Griffith KA, Holberg CJ, Martinez FD et al. Siblings, day-care attendance, and the risk of asthma and wheezing during childhood. *N Engl J Med* 2000; 343: 538–43

7 Matricardi PM, Rosmini F, Ferrigno L, Nisini R, Rapicetta M et al. Cross-sectional retrospective study of prevalence of atopy among Italian military students with antibodies against hepatitis A virus. *BMJ* 1997; 314: 999–1003

8 Matricardi PM. Infections preventing atopy: facts and new questions. *Allergy* 1997; 52: 879–82

9 Matricardi PM, Rosmini F, Riondino S, Fortini M, Ferrigno L et al. Exposure to foodborne and orofaecal microbes *versus* airborne viruses in relation to atopy and allergic asthma: epidemiological study. *BMJ* 2000; 320: 412–7

10 Linneberg A, Ostergaard C, Tvede M, Andersen LP, Nielsen NH et al. IgG antibodies against microorganisms and atopic disease in Danish adults: the Copenhagen Allergy Study. *J Allergy Clin Immunol* 2003; 111: 847–53

11 Matricardi PM, Rosmini F, Panetta V, Ferrigno L, Bonini S. Hay fever and asthma in relation to markers of infection in the United States. *J Allergy Clin Immunol* 2002; 110: 381–7

12 Selnes A, Nystad W, Bolle R, Lund E. Diverging prevalence trends of atopic disorders in Norwegian children. Results from three cross-sectional studies. *Allergy* 2005; 60: 894–9

13 Pfefferle PI, Krämer A. *Helicobacter pylori*-infection status and childhood living conditions are associated with signs of allergic diseases in an occupational population. *Eur J Epidemiol* 2008; 23: 635–40

14 Chen Y, Blaser MJ. *Helicobacter pylori* colonization is inversely associated with childhood asthma. *J Infect Dis* 2008; 198: 553–60

15 Kilpelainen M, Terho EO, Helenius H, Koskenvuo M. Farm environment in childhood prevents the development of allergies. *Clin Exp Allergies* 2000; 30: 201–8

16 Ernst P, Cormier Y. Relative scarcity of asthma and atopy among rural adolescents raised on a farm. *Am J Resp Crit Care Med* 2000; 161: 1563–6
17 Riedler J, Eder W, Oberfeld G, Schreuer M. Austrian children living on a farm have less hay fever, asthma and allergic sensitization. *Clin Exp Allergy* 2000; 30: 194–200
18 von Mutius E, Radon K. Living on a farm: impact on asthma induction and clinical course. *Immunol Allergy Clin North Am* 2008 Aug; 28(3): 631–47
19 von Mutius E, Braun-Fahrlander C, Schierl R, Riedler J, Ehlermann S et al. Exposure to endotoxin or other bacterial components might protect against the development of atopy. *Clin Exp Allergy* 2000; 30: 1230–4
20 Gereda JE, Leung DYM, Thatayatikom A, Streib JE, Price MR et al. Relation between house-dust endotoxin exposure, type 1 T-cell development, and allergen sensitisation in infants at high risk of asthma. *Lancet* 2000; 355: 1680–3
21 Perkin MR, Strachan DP. Which aspects of the farming lifestyle explain the inverse association with childhood allergy? *J Allergy Clin Immunol* 2006; 117: 1374–81
22 Riedler J, Braun-Fahrländer C, Eder W, Schreuer M, Waser M et al. Exposure to farming in early life and development of asthma and allergy: a crosssectional survey. *Lancet* 2001; 358: 1129–33
23 Gerhold K, Blümchen K, Bock A, Seib C, Stock P et al. Endotoxins prevent murine IgE production, T(H)2 immune responses, and development of airway eosinophilia but not airway hyperreactivity. *J Allergy Clin Immunol* 2002 Jul; 110(1): 110–6
24 Gerhold K, Bluemchen K, Franke A, Stock P, Hamelmann E. Exposure to endotoxin and allergen in early life and its effect on allergen sensitization in mice. *J Allergy Clin Immunol* 2003 Aug; 112(2): 389–96
25 Gerhold K, Avagyan A, Seib C, Frei R, Steinle J et al. Prenatal initiation of endotoxin airway exposure prevents subsequent allergen-induced sensitization and airway inflammation in mice. *J Allergy Clin Immunol*. 2006 Sep; 118(3): 666–73
26 Gerhold K, Avagyan A, Reichert E, Blumchen K, Wahn U et al. Lipopolysaccharides modulate allergen-specific immune regulation in a murine model of mucosal tolerance induction. *Int Arch Allergy Immunol* 2008; 147(1): 25–34
27 Barnes M, Cullinan P, Athanasaki P, MacNeill S, Hole AM et al. Crete: does farming explain urban and rural differences in atopy? *Clin Exp Allergy* 2001; 31: 1822–8
28 Wickens K, Lane JM, Fitzharris P, Siebers R, Riley G et al. Farm residence and exposures and the risk of allergic diseases in New Zealand children. *Allergy* 2002; 57: 1171–9
29 Waser M, Michels KB, Bieli C, Flöistrup H, Pershagen G et al. Inverse association of farm milk consumption with asthma and allergy in rural and suburban populations across Europe. *Clinical Exp Allergy* 2007; 37: 661–70
30 Von Ehrenstein OS, von Mutius E, Illi S, Baumann L, Bohm O et al. Reduced risk of hay fever and asthma among children of farmers. *Clin Exp Allergy* 2000; 30: 187–93
31 Sudo N, Sawamura S, Tanaka K, Kubo C, Koga Y. The requirement of intestinal bacterial flora for the development of an IgE production system fully susceptible to oral tolerance induction. *J Immunol* 1997; 159: 1739–45

32 Sepp E, Julge K, Vasar M, Naaber P, Bjorksten B et al. Intestinal microflora of Estonian and Swedish infants. *Acta Paediatr Scand* 1997; 86: 956–61
33 Holt PG, Sly PD, Bjorksten B. Atopic *versus* infectious diseases in childhood: a question of balance? *Pediatr Allergy Immunol* 1997; 8: 53–8
34 Bjorksten B, Naaber P, Sepp E, Mikelsaar M. The intestinal microflora in allergic Estonian and Swedish 2-year-old children. *Clin Exp Allergy* 1999; 29: 342–6
35 Shroff KE, Meslin K, Cebra JJ. Commensal enteric bacteria engender a self-limiting humoral mucosal immune response while permanently colonizing the gut. *Infect Immun* 1995; 63: 3904–13
36 Wold AE. The hygiene hypothesis revised: is the rising frequency of allergy due to changes in the intestinal flora? *Allergy* 1998; 53(S46): 20–25
37 Matricardi PM, Bonini S. High microbial turnover rate preventing atopy: a solution to inconsistencies impinging on the Hygiene hypothesis? *Clin Exp Allergy* 2000; 30: 1506–10
38 Bottcher MF, Nordin EK, Sandin A, Midtvedt T, Bjorksten B. Microflora-associated characteristics in faeces from allergic and nonallergic infants. *Clin Exp Allergy* 2000; 30: 1590–6
39 Kalliomaki M, Kirjavainen P, Eerola E, Pentti K, Salminen S et al. Distinct patterns of neonatal gut microflora in infants in whom atopy was and was not developing. *J Allergy Clin Immunol* 2001; 107: 129–34
40 Adlerberth I, Strachan DP, Matricardi PM, Ahrné S, Orfei L et al. Gut microbiota and development of atopic eczema in 3 European birth cohorts. *J Allergy Clin Immunol* 2007; 120: 343–50
41 Wang M, Karlsson C, Olsson C, Adlerberth I, Wold AE et al. Reduced diversity in the early faecal microbiota of infants with atopic eczema. *J Allergy Clin Immunol* 2008; 121: 129–34
42 Pelosi U, Porcedda G, Tiddia F, Tripodi S, Tozzi AE et al. The inverse association of salmonellosis in infancy with allergic rhinoconjunctivitis and asthma at school-age: a longitudinal study. *Allergy* 2005; 60: 626–30
43 Mizuno Y, Takada H, Nomura A, Jin CH, Hattori H et al. Th1 and Th1-inducing cytokines in *Salmonella* infection. *Clin Exp Immunol* 2003; 131: 111–17
44 Mastroeni P, Menager N. Development of acquired immunity to *Salmonella*. *J Med Microbiol* 2003; 52: 453–9
45 de Jong R, Altare F, Haagen IA, Elferink DG, Boer T et al. Severe mycobacterial and *Salmonella* infections in interleukin-12 receptor-deficient patients. *Science* 1998; 280: 1435–8
46 Rotta G, Edwards EW, Sangaletti S, Bennett C, Ronzoni S et al. Lipopolysaccharide or whole bacteria block the conversion of inflammatory monocytes into dendritic cells *in vivo*. *J Exp Med* 2003; 198: 1253–63
47 Soo SS, Villareal-Ramos B, Anjam Khan CM, Hormaeche CE, Blackwell JM. Genetic control of immune response to recombinant antigens carried by attenuated *Salmonella*

typhimurium vaccine strain: Nramp1 influences T-helper subset responses and protection against leishmanial challenge. *Infect Immun* 1998; 66: 1910–17

48 Smit JJ, van Loveren H, Hoekstra MO, Nijkamp FP, Bloksma N. Influence of the macrophage bacterial resistance gene, Nramp1 (Slc11a1), on the induction of allergic asthma in the mouse. *FASEB J* 2003; 17: 958–60

49 Shirakawa T, Enomoto T, Shimazu S, Hopkin JM. The inverse association between tuberculin responses and atopic disorders. *Science* 1997; 275: 77–9

50 Silverman M. BCG vaccination and atopy – unfinished business? *Lancet* 1997; 350: 380–1

51 Rook GA, Stanford JL. Give us this day our daily germs. *Immunol Today* 1998; 19: 113–6

52 Rook GA, Hamelmann E, Brunet LR. Mycobacteria and allergies. *Immunobiology* 2007; 212(6): 461–73

53 Grove DI. What is the relationship between asthma and worms? *Allergy* 1982; 37: 139–48

54 Yazdanbakhsh M, Kremsner PG, Van Ree R. Allergy, parasites and the hygiene hypothesis. *Science* 2002; 296: 490–544

55 Lynch NR, Lopez RI, Di Prisco-Fuenmayor MC, Hagel I, Medouze L et al. Allergic reactivity and socio-economic level in a tropical environment. *Clin Allergy* 1987; 17: 199–207

56 Scrivener S, Yemaneberhan H, Zebenigus M, Tilahan D, Girma S et al. Independent effects of intestinal parasite infection and domestic allergen exposure on risk of wheeze in Ethiopia: a nested case-control study. *Lancet* 2001; 358: 1493–9

57 Nyan OA, Walraven GE, Banya WA, Milligan P, Van Der Sande M et al. Atopy, intestinal helminth infection and total serum IgE in rural and urban adult Gambian communities. *Clin Exp Allergy* 2001; 31: 1672–8

58 Cooper PJ, Chico ME, Bland M, Griffin GE, Nutman TB. Allergic symptoms, atopy, and geohelminth infections in a rural area of Ecuador. *Am J Respir Crit Care Med* 2003; 168: 313

59 van den Biggelaar AH, van Ree R, Rodrigues LC et al . Decreased atopy in children infected with *Schistosoma haematobium*: a role for parasite-induced interleukin-10. Lancet 2000; 356: 1723–7

60 Schafer T, Meyer T, Ring J, Wichmann HE, Heinrich J. Worm infestation and the negative association with eczema (atopic/nonatopic) and allergic sensitization. *Allergy* 2005; 60: 1014–20

61 Buijs J, Borsboom G, van Gemund JJ, Hazeboek A, van Dongen PA et al. Toxocara seroprevalence in 5-year-old elementary schoolchildren: relation with allergic asthma. *Am J Epidemiol* 1994; 140: 839–47

62 Dittrich AM, Erbacher A, Specht S, Diesner F, Krokowski M et al. Helminth infection with *Litomosoides sigmodontis* induces regulatory T cells and inhibits allergic sensitization, airway inflammation, and hyperreactivity in a murine asthma model. *J Immunol* 2008; 180(3): 1792–9

63 Schnoeller C, Rausch S, Pillai S, Avagyan A, Wittig BM et al. A helminth immunomodulator reduces allergic and inflammatory responses by induction of IL-10-producing macrophages. *J Immunol* 2008; 180(6): 4265–72
64 Cooper PJ. Can intestinal helminth infections (geohelminths) affect the development and expression of asthma and allergic disease? *Clin Exp Immunol* 2002; 128: 398–404
65 Lynch NR, Hagel I, Perez M, Di Prisco MC, Lopez R et al. Effect of anthelmintic treatment on the allergy reactivity of children in a tropical slum. *J Allergy Clin Immunol* 1993; 92: 404–11
66 Van den Biggelaar AH, Rodrigues LC, van Ree R, van der Zee JS, Hoeksma-Kruize YC et al. Long-term treatment of intestinal helminths increases mite skin-test reactivity in Gabonese schoolchildren. *J Infect Dis* 2004; 189: 892–900
67 Cooper PJ, Chico ME, Vaca MG, Moncayo AL, Bland JM et al. Effect of albendazole treatments on the prevalence of atopy in children living in communities endemic for geohelminth parasites: a cluster-randomised trial. *Lancet* 2006; 367(9522): 1598–603
68 Matricardi PM, Bjorksten B, Bonini S, Bousquet J, Djukanovic R et al. EAACI Task Force 7. Microbial products in allergy prevention and therapy. *Allergy* 2003; 58: 461–71
69 Salminen S, Isolauri E. Opportunities for improving the health and nutrition of the human infant by probiotics. *Nestle Nutr Workshop Ser Pediatr Program* 2008; 62: 223–33
70 Isolauri E, Salminen S; Nutrition, Allergy, Mucosal Immunology, and Intestinal Microbiota (NAMI) Research Group Report. Probiotics: use in allergic disorders: a Nutrition, Allergy, Mucosal Immunology, and Intestinal Microbiota (NAMI) Research Group Report. *J Clin Gastroenterol* 2008; 42 Suppl 2: S91–6
71 Laiho K, Ouwehand A, Salminen S, Isolauri E. Inventing probiotic functional foods for patients with allergic disease. *Ann Allergy Asthma Immunol* 2002; 89(6 Suppl 1): 75–82
72 Matricardi PM. Probiotics against allergy: data, doubts, and perspectives. Allergy. 2002; 57: 185–7
73 Global Strategy for Asthma Management and Prevention. Global Initiative for Asthma, 2004
74 Flohr C, Pascoe D, Williams HC. Atopic dermatitis and the ‘hygiene hypothesis’: too clean to be true? Br J Dermatol 2005; 152: 202–16
75 Kukkonen K, Savilahti E, Haahtela T, Juntunen-Backman K, Korpela R et al. Probiotics and prebiotic galacto-oligosaccharides in the prevention of allergic diseases: a randomized, double-blind, placebo-controlled trial. *J Allergy Clin Immunol* 2007; 119: 192–8
76 Taylor AL, Dunstan JA, Prescott SL. Probiotic supplementation for the first 6 months of life fails to reduce the risk of atopic dermatitis and increases the risk of allergen sensitization in high-risk children: a randomized controlled trial. *J Allergy Clin Immunol* 2007; 119: 184–91
77 Rosenfeldt V, Benfeldt E, Nielsen SD, Michaelsen KF, Jeppesen DL et al. Effect of pro-

biotic *Lactobacillus* strains in children with atopic dermatitis. *J Allergy Clin Immunol* 2003; 111: 389–95

78 Viljanen M, Savilahti E, Haahtela T, Juntunen-Backman K, Korpela R et al. Probiotics in the treatment of atopic eczema/dermatitis syndrome in infants: a double-blind placebo-controlled trial. *Allergy* 2005; 60: 494–500

79 Grüber C, Wendt M, Lau S, Kulig M, Wahn U et al. Randomized placebo-controlled trial of *Lactobacillus rhamnosus* GG as treatment of mild to moderate atopic dermatitis in infancy. *J Allergy Clin Immunol* 2005; 117: S239

80 Weston S, Halbert A, Richmond P, Prescott SL. Effects of probiotics on atopic dermatitis: a randomised controlled trial. *Arch Dis Child* 2005; 90: 892–7

81 Brouwer ML, Wolt-Plompen SA, Dubois AE, van der Heide S, Jansen DF et al. No effects of probiotics on atopic dermatitis in infancy: a randomized placebo-controlled trial. *Clin Exp Allergy* 2006; 36(7): 899–906

82 Lee TT, Morisset M, Astier C, Moneret-Vautrin DA, Cordebar V et al. Contamination of probiotic preparations with milk allergens can cause anaphylaxis in children with cow's milk allergy. *J Allergy Clin Immunol* 2007; 119: 746–7

83 Hamelmann E, Herz U, Holt P, Host A, Lauener RP et al. New visions for basic research and primary prevention of pediatric allergy: an iPAC summary and future trends. *Pediatr Allergy Immunol* 2008; Suppl 19: 4–16

Multiple sclerosis

Jorge Correale

Institute for Neurological Research Dr. Raúl Carrea, FLENI, and School of Biomedical Sciences, Austral University, Buenos Aires, Argentina

Abstract

Multiple sclerosis (MS) is an inflammatory demyelinating disease of the Central Nervous System (CNS). Although its etiology remains unknown, several lines of evidence support autoimmunity as playing a major role in the development of the disease. MS incidence has significantly increased during the second half of the 20th century. This has been attributed to improved sanitation and reduced exposure to infection. The hygiene hypothesis is not new and is currently used to explain the increasing incidence of allergies and other autoimmune diseases. Because helminths are powerful modulators of the host immune system, it has also been suggested that reduced exposure to helminths due to improved hygiene conditions may favor MS development. In this chapter epidemiological, experimental and clinical data supporting the protective role of helminths in MS are reviewed. Better understanding of host–parasite interactions, as well as identification of specific parasite molecules causing immunomodulatory modulation will help combat allergies and autoimmune diseases without having to pay the price of undesired infectious side-effects.

Introduction

Multiple sclerosis (MS) is an inflammatory demyelinating disease of the central nervous system (CNS), affecting an estimated 2 million people worldwide and representing the second cause of nervous system disability in young adults after traumatic brain injuries. The disease affects mainly young adults between 20–40 years of age, and is approximately twice as frequent in females as in males. The course of MS is highly variable, but most classically characterized by a relapsing-remitting (RR) pattern in which acute exacerbations are followed by periods of stability (remissions). However, in up to 50% of patients this pattern evolves to a secondary progressive course characterized by relentless neurological function deterioration over a period of years, or it can also, in a minority of patients (~15%), progress from onset (primary progressive course) [1].

Although the etiology of MS remains unknown, several lines of evidence support the notion that autoimmunity plays a major role in disease susceptibility and development [2]. It is generally accepted that both MS and experimental allergic encephalomyelitis (EAE), an animal model resembling MS, involve T helper type-

The Hygiene Hypothesis and Darwinian Medicine, edited by Graham A. W. Rook

1 (Th1) and T helper type-17 (Th17) cells which recognize several components of the myelin sheath, including myelin basic protein, myelin oligodendrocyte glycoprotein, or proteolipic protein, as immunologically relevant antigens (Ag). Such autoreactive T cells are thought to be crucial for the initiation and maintenance of the CNS inflammatory response leading to myelin destruction and axonal loss [3]. However, although inflammation is generally believed to be a primary feature of demyelination in MS, myelin destruction has recently been reported to occur prior to inflammation [4]. Thus, endogenous glia, such as microglia or astrocytes, might be a source of injury mediators.

Autoimmune diseases are currently considered to arise from a combination of genetic susceptibility and environmental factors [5–7]. The genetic component of MS is believed to result from the action of common allelic variants in several genes. Some loci may be involved in early pathogenic events, while others could influence disease development and progression. Whole-genome screens for linkage and/or associations have now been completed in over 30 datasets, supporting the long-held view that MS is a polygenic disorder. Comparative analysis reveals only partial replication of results across the various screens, with the exception of the MHC region mapping to the short arm of chromosome 6, where strong linkage and association signals consistently indicate the presence of a major susceptibility gene or genes. This signal segregates primarily with the HLA-DR15 haplotype DQB1*0602,DQA1*0102*, DRB1*1501, DRB5*0101. The association of MS with the HLA-Class II locus has been a reproducible finding across nearly all populations studied. In addition, a recent combined analysis from more than 12,000 samples identified a number of genes involved in disease susceptibility, including the IL-2 receptor on chromosome 10p15 and the IL-7 receptor on chromosome 5p13 [8]. However, studies of identical twins in which one has MS have shown that only 30% of the second twins develop disease [9]. The discordance of MS among monozygotic twins suggests that additional factors, such as environmental modulators, could be involved.

MS epidemiology

There are clear differences in the geographic distribution of MS. Although the disease occurs worldwide, it is most common in Caucasian individuals from Northern Europe, and rare among Asians and Africans [10]. The disease has low prevalence in tropical regions, increasing in those north of the equator. In the United States, prevalence below parallel 37 is reported to be 35.5/100,000, while above parallel 37 it is 68.8/100,000. A similar north-south distribution gradient has been observed in west European countries. Several studies have demonstrated that African-American patients in United States develop MS less frequently than Caucasians. However, both races show similar north-south incidence decline, indicating the importance of environmental factors [11].

Migration studies may help determine the role of environmental and genetic factors. Adult immigrants leaving Europe for South Africa have been shown to present three-fold higher risk of developing MS than those migrating at age 15 or younger. Risk of MS is higher among children of immigrants to the UK from India, Africa and the Caribbean (all areas of very low prevalence) than among their parents, and similar to that of children born in the UK [12]. The relevant role of age at migration was demonstrated in a study conducted on Ashkenazi (from northern Europe) and Sephardic Jews (from Asia and Africa) living in Israel, which resulted in increased risk in Ashkenazi who were older at time of migration (after adolescence), suggesting age is perhaps related but probably only during the first two decades of life [13]. Earlier migration studies suggest risk of acquiring MS is largely established before the age of 15, however more recent studies show risk may develop over several years and not be restricted to childhood or early adult life.

Infectious agents and MS

Several studies implicate environmental factors in childhood and probably young adulthood as strong determinants of MS risk, but do not identify the nature of these factors. Both infectious and noninfectious factors deserve consideration [5, 14–16]. Microbial infections can act as environmental triggers inducing or promoting autoimmunity, resulting in clinical manifestations of autoimmune disease in genetically predisposed individuals [17]. Alternatively, infectious diseases might accelerate established but subclinical autoimmune processes. To examine to what extent an infectious agent could provide an explanation for MS epidemiology, two hypotheses have been proposed. The first one has been called the 'polio' hypothesis, postulates that a viral infection acquired in late childhood or adulthood increases the risk for MS, but is less harmful and confers protective immunity if acquired in infancy [18]. The second hypothesis supported by Kurtzke is the prevalence hypothesis, based on investigations of the Faroe Islands epidemic, which postulates that MS is caused by a pathogen that is more common in regions of high MS prevalence. This agent is widespread and in most individuals causes an asymptomatic and persistent infection, which years later might cause neurological symptoms [19].

A recent meta-analysis of EBV infections supports the role of EBV in the genesis of MS. Like most herpes viruses, EBV infections determine continuous stimulation of the immune system. Using individuals in early childhood as reference, the risk is about 10-fold less among EBV-negative individuals, and about two- to three-fold more among those infected with EBV later in life. Thus, there is at least a 20-fold increase in risk among individuals with a history of mononucleosis compared with those who are EBV negative [20]. Moreover, an important contribution to understanding the relationship between EBV and MS has come from longitudinal studies of antibodies in serum of healthy individuals who later developed MS. Increasing anti-EBV antibodies become significant five or more years before MS symptom onset. This increase is characterized by significant anti-EBNA 1 antibody elevation accompanied by less prominent increase in anti-EBNA 2 expressed in latently infected cells. These findings suggest a more severe primary infection or reactivation of infection accompanied by a vigorous cellular immune response [21–22]. Likewise, T cells specific for EBNA 1 are present significantly more often in MS patients compared to healthy subjects, indicating a correlation between cellular immune response against EBV and MS incidence [23].

Other infectious agents have been proposed as putative candidates to increase MS risk. One of these organisms is *Chlamydia pneumoniae*, a Gram-negative bacterium. Earlier enthusiasm over *C. pneumoniae* however now appears to be fading after original findings of *C. pneumoniae* DNA and antibodies in CSF of patients with MS could not be confirmed [16]. Likewise, during the last decade a role for herpes virus 6 (HHV-6) in the genesis of MS has been proposed. Evidence of HHV-6 involvement in MS includes presence of viral DNA detected in post-mortem MS tissue, increased viral DNA in blood cells during disease exacerbations, as well as increased IgG and IgM levels against HHV-6 in serum and CSF of MS patients compared to healthy controls. However, these findings do not explain migration data observed for MS patients. Furthermore, viral DNA and antibodies against HHV-6 are also observed in other neurological diseases, suggesting DNA or viral antigen detection in brain tissue might reflect HHV-6 reactivation from latency in peripheral blood T cells trafficking through the brain of patients with inflammatory CNS diseases [14, 16].

Despite decades of research, it has not been possible to establish a clear causal relationship between an infectious agent and MS. Different factors may explain this situation. First, acute infections may induce autoreactive cells, but most infections are cleared within weeks, and by the time MS is diagnosed (months or years later) the infectious particles have been eliminated. Second, antibodies and autoreactive T cells against infectious agents can be found in almost every individual, most of whom will not develop MS. Third, under certain circumstances molecular mimicry phenomena may induce cells that protect against subsequent autoimmune disease development. Likewise, duration of exposure to the infectious agent may cause disease acceleration when occurring early in the disease process, but may dampen autoimmunity when expressed relatively late [24].

Infectious agents and the hygiene hypothesis in MS

Several epidemiological and experimental studies support the hygiene hypothesis, which postulates that infections will protect rather than induce/accelerate autoimmune diseases, such as MS [25–26]. Different factors have been proposed to explain this scenario [17, 25]. First, infectious agents can shift the immunological balance toward a more immunosuppressive state. Second, inflammation can cause massive autoreactive cell hyperactivation which may lead to activation-induced cell death, and dimish the load of aggressive cells. Third, infection at another location might keep autoreactive cells from reaching the site of autoimmune destruction. Fourth, competitive mechanisms may exist between anti-infectious responses and other immune responses. This putative mechanism may be considered at different levels: a) competition for Ag processing by phagocytic cells; b) competition over Ag binding to MHC molecules; c) competition for essential cytokines necessary for lymphocyte differentiation and homeostasis. Finally, the presence of superantigens within the structure of different infectious agents may induce deletion or activation of different T cell populations expressing a particular TCR Vβ region. Evidence of this particular mechanism is demonstrated through EAE inhibition by enterotoxin B.

Improvement in living conditions and reduced exposure to childhood infections in particular, have been suggested as contributing to increases observed in atopy and autoimmunity. In 1966 Leibowitz and co-workers suggested prevalence of MS correlated with childhood environments characterized by high levels of sanitation [27]. This was the first time the hygiene hypothesis was linked to MS. Recent findings consistent with this hypothesis have come from studies in epidemiology, immunology, and animal models. Epidemiology data provide strong evidence of a steady rise in MS incidence in developed countries during recent decades. Concomitantly, there has been a decrease in the incidence of many infectious diseases in these countries as a result of the use of antibiotics, vaccination or improved hygiene and better socioeconomic conditions. For example, in Lower Saxony, Germany, the incidence of MS doubled from 1969–1986 [28], and Mexican studies have shown a 29-fold increase over 20 years, since the original reports in 1970 [29]. Likewise, in the region of Sassari, Sardinia, MS incidence increased from 1.1/100,000 inhabitants between 1965–1969 to 5.8/100,000 between 1995–1999 [30]. Although this increase could be due to improvement in diagnosis, recent studies have demonstrated that using similar study methods during comparable time periods, MS incidence did not change in other regions such as Ferrara (northern Italy) or Iceland, ruling out diagnostic accuracy improvement as responsible for the greater number of MS cases in Sardinia [31]. In developed countries, industrialization has strongly contributed to human migration from rural areas to the cities. One of the consequences of resettlement has been changing from ecosystem to which dweller immune systems had adapted since prehistoric times to areas with different pathogens. The fact that infections were no longer present has led to emergence of autoimmune diseases [25]. Supporting these findings recent investigations demonstrate a dichotomous relationship between global distribution of MS and the parasite *Trichuris trichiura*, a common human helminth present worldwide. MS prevalence appears to fall steeply once a critical threshold of *T. trichiura* (about 10%) is exceeded [32]. Longitudinal and migratory studies evaluating the prevalence of MS, such as those conducted in the French west Indies between 1878 and 1994, have been also consistent with the hygiene hypothesis, showing MS incidence has increased in the region in association with significant reduction of parasite infections during the same period of time [33].

Investigations in animal models also support the hygiene hypothesis in MS. A pre-established infection with the parasitic helminth *Schistosoma mansoni* or pre-treatment of mice with *S. mansoni* ova significantly reduced incidence and delayed onset of EAE in mice [34–35]. This altered disease progression was associated with decreased IFN-γ, TNF-α, and IL-12 production, as well as with increased production of IL-10 and TGF-β. Moreover, in parasite-infected animals, infiltrating macrophages were absent from inflammatory lesions, and although $CD4^+$ T cell levels were similar during disease peak, they were significantly reduced 6 weeks post-stimulation, suggesting that schistosomiasis may affect the course of EAE by targeting the macrophage compartment during induction and by downregulating effector functions mediated by T cells.

Recently we identified a group of 12 patients in our MS clinic, presenting eosinophilia subsequently shown to be caused by mild, asymptomatic intestinal parasitosis [36]. Infected patients were matched with 12 uninfected MS patients, and no significant differences in clinical characteristics were found between either group in the 24 months preceding the study or at baseline, defined as eosinophilia onset. Following standard medical practice, as well as tropical medicine expert recommendations indicating that asymptomatic adults patients usually do not require treatment, infected MS patients were not given anthelmintic medication. Both groups were followed for approximately 4.5 years and compared with respect to clinical, magnetic resonance imaging (MRI) and immunological parameters. Parasite-infected MS patients showed a significantly lower number of relapses, minimal changes in disability scores and significantly lower MRI activity compared to uninfected MS individuals. Parasite-driven protection was associated with induction of regulatory T cells secreting suppressive cytokines IL-10 and TGF-β, as well as $CD4^+CD25^+FoxP3^+$T cells displaying significant suppressive function. These findings provide evidence to support autoimmune downregulation secondary to parasite infections in MS patients through regulatory T cell action, with effects extending beyond response to an invading agent. Evidence of regulatory T cells present during parasite infections is now emerging, offering a potential explanation on the mechanism through which infected hosts exhibit altered immune responses to bystander Ag [37]. Thus, parasites may lead to increased regulatory T cell numbers or activity, either by generating new cells or by activating or expanding existing cells.

In addition to the development of regulatory T cells, helminth infections in MS patients also induce regulatory B cells capable of dampening the immune response through production of IL-10 [38]. Existence of immunoregulatory B cells exerting a key role in immune regulation *via* the production of IL-10 has already been demonstrated in animal models in which IL-10 produced by B cells were able to protect mice against the development of collagen-induced arthritis, promoting disease resolution of experimental colitis and EAE [39–41]. IL-10 is essential for regulatory function development in this subset of B cells, and B cells isolated from IL-10 knock-out mice fail to show this protective function [41]. In agreement with our findings, recent observations indicate that B cells from MS patients exhibit relative deficiency in their capacity to produce IL-10 [42]. Our results demonstrate that this defect is entirely restored after intestinal-helminth infections [38]; also worthy of note was the fact that production of IL-10 by B cells was restricted to helminth-infected individuals exclusively. B cells from patients infected by other parasites such as *Trypanosoma cruzi* exhibited IL-10 production levels similar to those observed in uninfected MS patients, indicating that intracellular parasites are not able to downmodulate harmful autoimmune responses in the same way that helminths do [38]. Indeed, previous investigations have demonstrated that intracellular parasites such as *T. cruzi* generally elicit a Th1 type of response [43]. Likewise, B cells from *Paracoccidioides brasiliensis*-infected patients, although expressing a Th2 immune response after Ag-stimulation of peripheral blood mononuclear cells, exhibit B-cell IL-10 production levels similar to those observed in uninfected MS patients, suggesting that increased production of IL-10 by B cells found in helminth-infected MS patients is not determined by the Th2 profile observed in these individuals [38]. In addition, IL-10 producing B cells isolated from helminth-infected MS patients expressed the MHC Class Ib molecule CD1d, which aside from Ag presentation, is also involved in immunoregulation. Thus, glycolipids presented by CD1d cells expressing the invariant Vα14 (in mice), or Vα24 (in humans) T cell receptor, activate natural killer T cells, a mechanism that has been found to prevent autoimmune responses in different animal models [44]. Furthermore, the cytoplasmic tail of CD1d is linked to signaling cascades associated to IL-10 transcription, suggesting another mechanism of immunoregulation [45]. Supporting this view, studies in a murine model of intestinal inflammation have demonstrated that upregulation of CD1d expression confers ability to suppress inflammation through IL-10 production by B cells [40].

Infectious agents may also downmodulate the immune response through mechanisms that do not involve specific immune response induction against their constituent Ags. The first mechanism to be considered in this context was that of Toll-like receptor (TLR) stimulation [46]. One might assume that TLR stimulation leading to proinflammatory cytokines production would accelerate MS progression. However, glycosylated molecules expressed and secreted by parasites such as *S. mansoni*, bind to TLR and C-type lectin receptors (CLRs) expressed on dendritic cells and B cells (particularly TLR2), antagonizing proinflammatory pathways [46–47]. Indeed, *S. mansoni* products as well as TLR2 agonists induced IL-10 production by B cells and dendritic cells leading to a direct anti-inflammatory effect [48]. The figure accompanying this text summarizes the most important mechanisms involved in helminth modulation of the immune response during the course of MS (Fig. 1).

The growing body of epidemiological and experimental evidence detailed above suggests that helminths applied in a controllable clinical setting could relieve or attenuate immune responses, therefore providing a starting point for possible

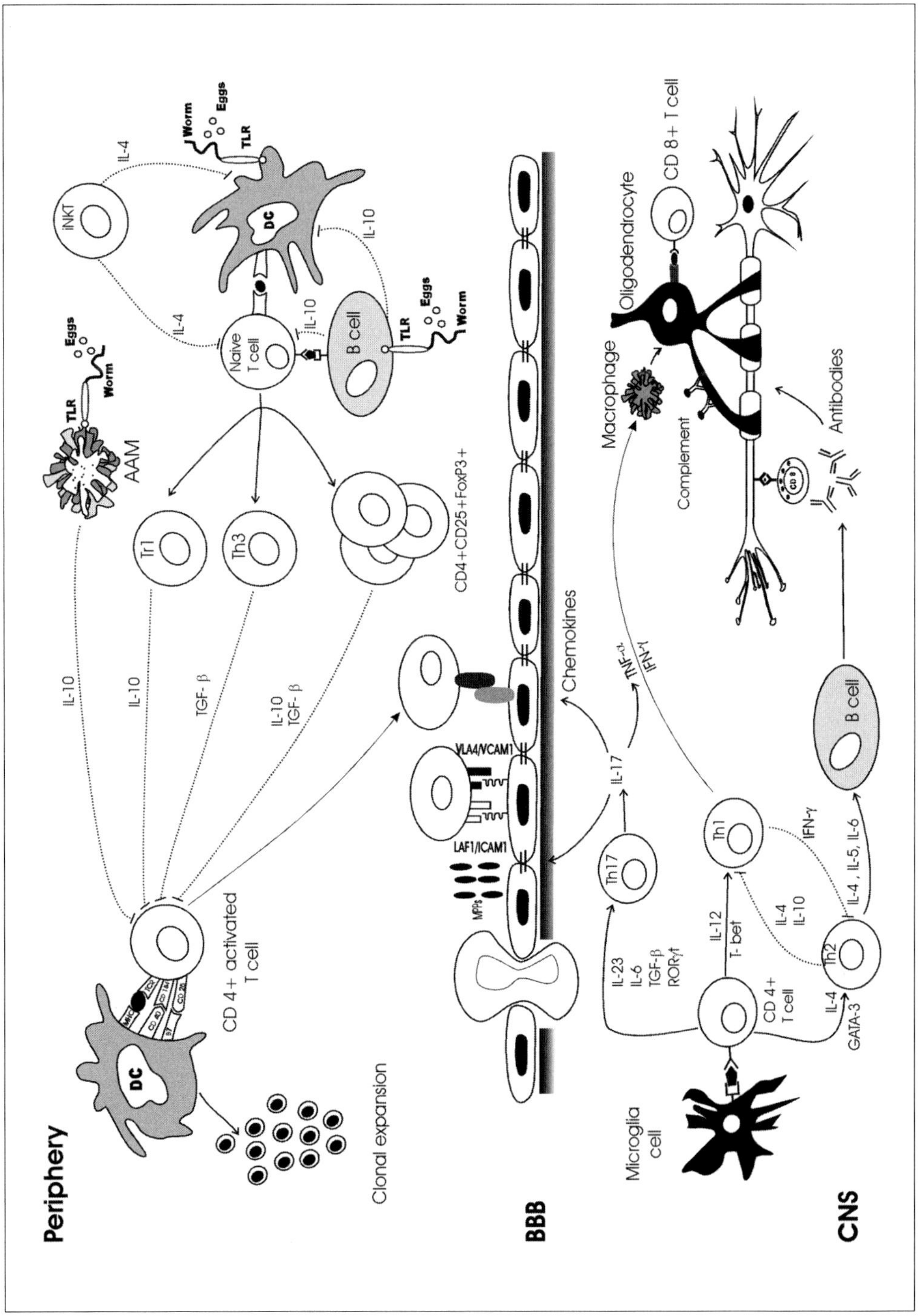
Periphery
BBB
CNS
DC
CD 4+ activated
T cell
Clonal expansion
IL-10
IL-10
TGF- β
IL-10
TGF- β
Tr1
Th3
CD4+CD25+FoxP3+
AAM
TLR
Worm
Eggs
Naive
T cell
iNKT
IL-4
IL-4
IL-10
IL-10
B cell
MPPs
LAF1/ICAM1
VLA4/VCAM1
Chemokines
IL-17
TNF-α
IFN-γ
Th17
Th1
Th2
IL-23
IL-6
TGF-β
RORγt
IL-12
T- bet
IL-4
IL-10
IFN-γ
IL-4 , IL-5, IL-6
CD 4+
T cell
IL-4
GATA-3
Microglia
cell
Macrophage
Oligodendrocyte
CD 8+ T cell
Complement
Antibodies

Figure 1
Induction of autoimmunity. Helminths modulate the immune response in MS. Genetic and environmental factors induce peripheral activation of myelin-reactive T cells after recognition of specific antigens. Following these activating signals, T cells interact with ligands expressed by endothelial cells along the blood-brain-barrier, including selectin, vascular adhesion molecule-1 (VCAM-1), and intracellular adhesion molecule (ICAM-1), ultimately crossing the endothelium into the CNS through diapedesis. Matrix metalloproteinases (MMP) may further enhance this process. Binding of putative MS antigens triggers and enhances the immune response, resulting in proinflammatory cytokine secretion. Multiple mechanisms of myelin and oligodendrocyte destruction have been postulated including direct injury by $CD4^+$ *and* $CD8^+$ *T cells, complement-mediated injury, antibody-mediated injury and myelin digestion by activated macrophages. Injury of the myelin membrane results in denuded axons susceptible to further damage.*
Induction of Immunosuppression. Helminth infections modify innate immune system cells through interactions with TLRs and CRLs, arresting inflammatory mediator production and eliciting release of immunosuppressive cytokines such as IL-10 and TGF-β, which in turn inhibit autoimmune response development. The number of $CD4^+CD25^+FoxP3^+$ *regulatory T cells is also increased by helminth infections. Naïve B cells can also produce IL-10 in response TLR binding. The ultimate effect of IL-10 production by B cells is to constrain pathology driven by Th1 and Th17 cells. This may be mediated by a direct effect on the* $CD4^+$ *T cells themselves, or through reduction in immune priming by dendritic cells. As helminths induce invariant natural killer T cells (iNKT), these cells might also play a role in modulating autoimmune responses. Helminth infections have also been shown to induce activated macrophages which alternatively secrete small amounts of inflammatory mediators and inhibit T cell proliferation. Dotted lines indicate inhibitory pathways. AAM: alternately activated macrophage; BBB: blood-brain-barrier; CNS: central nervous system; DC: dendritic cell; iNKT: invariant NK T cell; MMP: matrix metalloproteinases; TLR: toll-like receptor.*

immunomodulatory therapy development. Indeed, several models of autoimmunity have validated such an approach. Limited clinical trials have already been carried out assessing the effects of oral administration of the porcine whipworm *Trichuris suis* and *Necator americanus* on patients with active inflammatory bowel disease refractory to conventional therapies [49–50]. These initial studies revealed effective relief in Crohn's disease patients, and modest effects in ulcerative colitis. Also, an exploratory trial of helminth treatment for asthma has begun at the University of Nottingham, and a MRI-controlled Phase II study of oral *T. suis* administration for MS has been recently approved by the FDA [51]. A recent study proposed at the University of Nottinghan will explore whether controlled infection with a clinically safe number of larvae of hookworm results in a protective immune response in relapsing MS [52]. The caveat of a live parasite approach is evident, even if the chosen parasite is unable to productively infect the host patient, as is the case of *T. suis*, there may be still some side-effects, particularly in immunosuppressed patients. More desirable would be the administration of purified, well-characterized immunomodulatory helminth-derived products to avoid active infection associated risks. Nevertheless, the parasitic product armamentarium is complex and involves mixtures of immunoactive molecules, each with varying optimal concentration and duration of effect.

It is important to note that although investigations previously described have supported the hygiene hypothesis concept, evidence from other studies has been equivocal or contradictory. Most the indications in favor of the hygiene hypothesis are based on observational, cross-sectional studies, and not randomized controlled interventional ones. The problem with cross-sectional studies is that causal relationship cannot be assessed with sufficient validity. Thus, the negative relationship observed between the decline in incidence of major infectious diseases and increase in MS is suggestive, but not by any means demonstrative of a causal relationship [53]. Another limitation of the hygiene hypothesis in MS arises from the EBV paradox, namely the extremely low risk for MS among individuals who are EBV-seronegative. According to the hygiene hypothesis these individuals should have high MS risk; paradoxically their MS risk is much lower than that of their EBV-positive peers [14] (but see discussion of the issue of virus infections in Chapters 1 and 2 of this volume). Other epidemiological studies in certain populations also contradict the hygiene hypothesis. For example, contrary to the expected, some studies show MS prevalence in Sassari, Sardinia, is significantly higher in genetically archaic rural areas where less pronounced western influence exists, suggesting aside from hygiene-related factors that genetics might be a determinant of high susceptibility to MS observed in Sardinia [31]. Moreover, family size has been used to indicate overcrowded living conditions, also linked to poor hygiene and greater exposure to infections. Smaller family size and lower exposure to infections have been associated with decreased risk of autoimmune disease. Likewise, inverse correlation between number of older siblings (presumed a surrogate for degree of early exposure to childhood disease) and prevalence of autoimmune and allergic diseases has been described in different studies across the developed world. However, these observations were not confirmed by The Canadian Collaborative study on MS which investigated more than 10,000 individuals. It was found that birth order had no effect on MS risk in most families, and no support for the hypothesis that having older siblings protected against MS was found [54]. Finally, observations in some animal models also may render the hygiene hypothesis in MS vulnerable. Spontaneous EAE may develop in transgenic mice housed in a non-sterile facility, but not in those maintained in sterile pathogen-free conditions [55]. Likewise, administration of a mixture of non-absorbed antibiotics ameliorated EAE development in mice. This observation was associated with a reduction in proinflammatory cytokine production from regional lymph nodes, and reduction of mesenteric Th17 cells, a process dependent on invariant NK T cells [56].

Overall, many unanswered questions regarding the hygiene hypothesis in MS remain. It is evident that future studies in this area will be required to establish whether certain infections, particularly those produced by helminths during critical periods of infancy exert a protective effect against MS. Parasite infections are often long-lived and inhabit immunocompetent hosts; consequently it is not surprising that they would have acquired modulatory molecules that ameliorate host responses thus enhancing their survival. Understanding host–parasite interactions and identifying different parasite molecules possessing immunomodulatory effects will help combat allergies and autoimmune diseases, without paying the price of infectious side-effects. Future research should also be directed towards discovering and validating better markers for hygiene related factors, in order to rule out the possibility that the link between MS and a hygiene-related factor is an indirect and non-causal effect.

References

1 Lublin FD, Reingold SC (1996) Defining the clinical course of multiple sclerosis: results of an international survey. *Neurology* 46: 907–911

2 Sospedra M, Martin R (2005) Immunology of multiple sclerosis. *Annu Rev Immunol* 23: 683–747

3 Trapp BD, Peterson J, Ransohoff RM, Rudick R, Mörk S, Bö L (1998) Axonal transection in the lesions of multiple sclerosis. *N Engl J Med* 338: 278–285

4 Barnett MH, Prineas JW (2004) Relapsing and remitting multiple sclerosis: pathology of newly forming lesions. *Ann Neurol* 55: 458–468

5 Marrie RA (2004) Environmental risk factors in multiple sclerosis aetiology. *Lancet Neurol* 3: 709–718

6 Oksenberg JR, Baranzini SE, Sawcer S, Hauser SL (2008) The genetics of multiple sclerosis: SNP to pathways to pathogenesis. *Nat Rev Genet* 9: 516–526

7 Dyment DA, Ebers GA, Sadovnick AD (2004) Genetics of multiple sclerosis. *Lancet Neurol* 3: 104–110

8 The International Multiple Sclerosis Genetics Consortium, Hafler DA, Compstom A, Sawcer S, Lander ES, Daly MJ, De Jager PL, de Bakker PI, Gabriel SB, Mirel DB, Ivinson AJ et al (2007) Risk alleles for multiple sclerosis identified by a genome wide study. *N Engl J Med* 357: 851–862

9 Sadovnick AD, Armstrong H, Rice GPA, Bulman D, Hashimoto L, Paty DW, Hashimoto SA, Warren S, Hader W, Murray TJ et al (1993) A population-based study of multiple sclerosis in twins: update. *Ann Neurol* 33: 281–285

10 Rosati G (2001) The prevalence of multiple sclerosis in the world: an update. *Neurol Sci* 22: 117–139

11 Kurtzke JF (1997) The epidemiology of multiple sclerosis. In: Raine CS, McFarland HF, Tourtellotte WW (eds): *Multiple Sclerosis Clinical and Pathogenic Basis*. Chapman & Hall Medical, London, 91–140

12 Elian M, Nigthingale S, Dean G (1990) Multiple Sclerosis among United Kingdom-born children of immigrants from the Indian subcontinent, Africa and the West Indies. *J Neurol Neurosurg Psychiatry* 53: 906–911

13 Kahana E, Zilber N, Abramson JH, Biton V, Libowitz Y, Abramsky O (1994) Multiple Sclerosis: genetic versus environmental aetiology: epidemiology in Israel updated. *J Neurol* 241: 341–346

14 Ascherio A, Munger KL (2007) Environmental risk factors for multiple sclerosis. Part I: The role of infection. *Ann Neurol* 61: 288–299

15 Ascherio A, Munger KL (2007) Environmental risk factors for multiple sclerosis. Part II: Noninfectious factors. *Ann Neurol* 61: 504–513

16 Gilden DH (2005) Infectious causes of multiple sclerosis. *Lancet Neurol* 4: 195–202

17 Christen U, von Herrath MG (2005) Infections and autoimmunity – Good or bad? *J Immunol* 174: 7481–7486

18 Poskanzer DC, Walker AM, Yonkondy J, Sheridan JL (1976) Studies in the epidemiology of multiple sclerosis in the Okney and Shetland Islands. *Neurology* 26: 14–17

19 Kurtzke JF (1993) Epidemiological evidence for multiple sclerosis as an infection. *Clin Microbiol Rev* 6: 382–427

20 Rickinson AB, Kieff E (1996) Epstein-Barr virus. In. Fields BN, Knipe DM, Howley PM (eds) *Fields virology*, 3rd edition. Lippincott-Raven Publishers, Philadelphia, 2397–2446

21 Kasunoki Y, Huang H, Fukuda Y, Ozaki K, Saito M, Hirai Y, Akiyama M (1993) A positive correlation between the precursors frequency of cytotoxic lymphocytes to autologous Epstein-Barr virus-transformed B cells and antibody titer level against Epstein-Barr virus-associated nuclear antigen in healthy seropositive individuals. *Microbiol Immunol* 37: 461–469

22 Henle W, Henle G, Niederman JC, Klemola E, Haltia K (1971) Antibodies to early antigens induced by Epstein-Barr virus in infectious mononucleosis. *J Infect Dis* 124: 58–67

23 Lunemann JD, Edwards N, Muraro PA, Hayashi S, Cohen JI, Münz C, Martin R (2006) Increased frequency and broadened specificity of latent EBV nuclear antigen-1 specific T cells in multiple sclerosis. *Brain* 129: 1493–1506

24 von Herrath MG, Fujinami RS, Whitton JL (2003) Microorganisms and autoimmunity: making the barren field fertile? *Nature Rev Microbiol* 1: 151–157

25 Bach JF (2002) The effect of infections on susceptibility to autoimmune and allergic diseases. *N Engl J Med* 347: 911–920

26 Starchan DP (1989) Hay fever, hygiene and household size. *Br Med J* 299: 1259–1260

27 Leibowitz U, Atanovsky A, Medalie JM, Smith HA, Halpern L, Alter M (1966) Epidemiological study of multiple sclerosis in Israel. II. Multiple Sclerosis and level of sanitation. *J Neurol Neurosurg Psychiatry* 29: 60–68

28 Poser S, Stickel B, Krsch U, Burckhardt D, Nordman B (1989) Increasing incidence of multiple sclerosis in South Lower Saxony, Germany. *Neuroepidemiology* 8: 207–213

29 Gonzalez O, Sotelo J (1995) Is the frequency of multiple sclerosis increasing in México? *J Neurol Neurosurg Psychiatry* 59: 528–530

30 Pugliatti M, Riise T, Sotgiu MA, Sotgiu S, Satta WM, Mannu L, Sanna G, Rosati G (2005) Increasing incidence of multiple sclerosis in the province of Sassari, northern Sardinia. *Neuroepidemiology* 25: 129–134

31 Sotgiu S, Pugliatti M, Sotgiu A, Sanna A, Rosati G (2003) Does the "hygiene hypothesis" provide an explanation for the high prevalence of multiple sclerosis in Sardinia? *Autoimmunity* 36: 257–260

32 Fleming JO, Cook TD (2006) Multiple sclerosis and the hygiene hypothesis. *Neurology* 67: 2085–2086

33 Cabre P, Signate A, Olindo S, Merle H, Caparros-Lefebvre D, Bera O, Smadja D (2005) Role of return migration in the emergence of multiple sclerosis in the French West Indies. *Brain* 128: 2899–2910

34 Sewell D, Qing Z, Reinke E, Elliot D, Weinstock J, Sandor M, Fabry Z (2003) Immunomodulation of experimental autoimmune encephalomyelitis by helminth ova immunization. *Int Immunol* 15: 59–69

35 La Flamme AC, Ruddenklau K, Bäckstrom BT (2003) Schistosomiasis decreases central nervous system inflammation and alters the progression of experimental autoimmune encephalomyelitis. *Infect Immun* 71: 4996–5004

36 Correale J, Farez M (2007) Association between parasite infection and immune responses in Multiple Sclerosis. *Ann Neurol* 61: 97–108

37 Maizels RM, Balic A, Gomez-Escobar N, Nair M, Taylor MD, Allen JE (2004) Helminth parasites-masters of regulation. *Immunol Rev* 201: 89–116

38 Correale J, Farez M, Razzitte G (2008) Helminth infections associated with multiple sclerosis induce regulatory B cells. *Ann Neurol* 64: 187–199

39 Mauri C, Gray D, Mushtaq N, Londei M (2003) Prevention of arthritis by interleukin 10-producing B cells. *J Exp Med* 197: 489–501

40 Mizoguchi A, Mizoguchi E, Takedatsu H, Blumberg RS, Bhank AK (2002) Chronic intestinal inflammatory conditions generate IL-10 producing regulatory B cells characterized by CD1d upregulation. *Immunity* 16: 219–230

41 Fillatreau S, Sweenie CH, McGeachy MJ, Gray D, Anderton SM (2002) B cells regulate autoimmunity by provision of IL-10. *Nat Immunol* 3: 944–950
42 Duddy M, Niino M, Adatia F, Hebert S, Freedman M, Artkins H, Kim HJ, Bar-Or A (2007) Distinct effector cytokine profiles of memory and naïve human B cell subsets and implication in multiple sclerosis. *J Immunol* 178: 6092–6099
43 Jankovic D, Steinfleder S, Kullberg MC, Sher A (2006) Mechanisms underlying helminth-induced Th2 polarization: default, negative or positive pathways. *Chem Immunol Allergy* 90: 65–81
44 Godfrey DI, Kronenberg M (2004) Going both ways: immune regulation *via* CD1d-dependent NKT cells. *J Clin Invest* 114: 1379–1388
45 Colgan SP, Hersberg RM, Furuta GT, Blumberg RS (1999) Ligation of intestinal epithelial CD1d induces bioactive IL-10: critical role of cytoplasmic tail in autocrine signaling. *Proc Natl Acad Sci USA* 96: 13938–13943
46 Kane CM, Cervi L, Sun J, McKee AS, Masek KS, Shapira S, Hunter CA, Pearce EJ. Helminth antigens modulate TLR-initiated dendritic cell activation (2004) *J Immunol* 173: 7454–7461
47 Fillatreau S, Gray D, Anderton SM (2008) Not always the bad guys: B cells as regulators of autoimmune pathology. *Nat Rev Immunol* 8: 391–397
48 Correale J, Farez M (2008) Toll-like receptors induce regulatory B cells during helminth infections associated with multiple sclerosis. *J Neuroimmunol* 203: 132–133
49 Summers RW, Elliot DE, Urban JF Jr, Thompson R, Weinstock JV (2005) *Trichuris suis* therapy for active ulcerative colitis: a randomized trial. *Gastroenterology* 128: 825–832
50 Croese J, O'Neil J, Masson J, Cooke S, Melrose W, Pritchard D, Spare R (2006) A proof of concept study establishing *Necator americanus* in Crohn's patients and reservoir donors. *Gut* 55: 136–137
51 Fleming JO (2007) The hygiene hypothesis and multiple sclerosis. *Ann Neurol* 61: 85–89
52 Constantinescu C. Immunoregulation by controlled parasite exposure in Multiple Sclerosis. Available at: http: //www.clinicaltrials.gov/ct2/show/NCT00630383 (accessed 10 November 2008)
53 van Shayck CP, Knottnerus JA (2004) No clinical evidence base to support the hygiene hypothesis. *Prim Care Respir J* 13: 76–79
54 Sadovnick AD, Yee IM, Ebers GC, Canadian Collaborative Study Group (2005) Multiple sclerosis and birth order: a longitudinal cohort study. *Lancet Neurol* 4: 611–617
55 Goverman J, Woods A, Larson L, Weiner L, Hood L, Zaller DM (1993) Transgenic mice that express a myelin basic protein-specific T cell receptor develop spontaneous autoimmunity. *Cell* 72: 551–560
56 Yokote H, Miyake S, Croxford JL, Oki S, Mizusawa H, Yamamura T (2008) NKT cell-dependent amelioration of a mouse model of multiple sclerosis by altering gut flora. *Am J Pathol* 173: 1–10

Inflammatory bowel disease and the hygiene hypothesis: an argument for the role of helminths

David E. Elliott[1] *and Joel V. Weinstock*[2]

[1]Division of Gastroenterology and Hepatology, Department of Internal Medicine, University of Iowa, Roy J. and Lucille A.Carver College of Medicine, Iowa City, IA and VAMC, Iowa City, IA, USA
[2]Division of Gastroenterology and Hepatology, Department of Internal Medicine, Tufts Medical Center, Boston, MA, USA

Abstract

Variations in more than 40 genetic loci can alter the risk for developing inflammatory bowel disease (IBD). However, the epidemiology of ulcerative colitis and Crohn's disease suggest that a recent environmental change accounts for most of the disease risk. In this chapter we will introduce IBD and outline its dramatic rise in prevalence over the last 70 years. We will consider the effective eradication of helminths during this time period and the effects of helminths on immunity. We will review the current evidence that helminths induce regulatory immune circuits that suppress aberrant inflammation and may be useful clinically to treat immune-mediated disease.

Inflammatory bowel disease

Inflammatory bowel disease (IBD) is a set of chronic remitting/relapsing idiopathic inflammatory disorders of the gastrointestinal tract that segregate into two major disease phenotypes named ulcerative colitis (UC) and Crohn's disease (CD). Either can be debilitating with persistent cramping abdominal pain, intractable diarrhea, urgent (run don't walk) need to defecate, night sweats, fever, anemia due to gastrointestinal blood loss, and severe fatigue. However, each has distinguishing clinical, pathological, and endoscopic features. Inflammation in UC always involves the rectum (proctitis), and depending on the person, contiguously extends proximal until involving the entire colon (pancolitis). The inflammation is mainly limited to the mucosa (the superficial lining) of the colon but this can include extensive ulceration and granular erythema with friability (easy bleeding). Inflammation in CD can involve any part of the GI tract ('gum to bum') but usually involves the portion of the small bowel that attaches to the colon (terminal ileum, ileitis). CD causes deep sharply demarcated non-contiguous ulcerations that can become transmural involving all layers of the GI tract. If the ulcers penetrate the wall they can cause

The Hygiene Hypothesis and Darwinian Medicine, edited by Graham A. W. Rook

abdominal perforation, abscess, and/or tunnels (fistula) to another organ or skin. The chronic inflammation in UC eventually results in a high risk for colon cancer. The chronic inflammation in CD can cause colon cancer but also causes fibrotic narrowing (stricture) of the intestinal lumen with obstruction and severe malnutrition.

Treatment of IBD consists of suppressing the immune and/or inflammatory response (Tab. 1). Although these drugs can cause significant toxicity, they allow IBD patients to achieve remission and pursue normal lifestyles. The requirement for treatments based on immune suppression places IBD squarely alongside multiple sclerosis and rheumatoid arthritis in the group of immune-mediated inflammatory diseases.

The current belief is that IBD results from dysregulated immune responses driven in part by bacterial products. Immune responses result from the coordinated activity of multiple cell types which comprise the innate and adaptive immune systems. Coordination occurs through signals mediated through cell contact and cytokine

Table 1 - Medications frequently used to treat IBD

Agent	Use	Examples	Mechanism of action
mesalmine (5-ASA)	UC > CD	Sulfasalazine	inhibits cyclooxygenase, lipoxygenase, and acts as a PPAR-γ agonist inhibiting NF-κB pathways [142]
corticosteroid	UC = CD	hydrocortisone, prednisolone, budesonide	inhibits multiple pathways [143]
azathioprine, 6-mercaptopurine	UC = CD	NA	interferes with proliferation and promotes apoptosis of activated lymphocytes [144]
methotrexate	CD	NA	inhibits dihydrofolate reductase and thymidylate synthase to suppress lymphocyte proliferation
cyclosporine	UC	NA	blocks cytokine production by inhibiting calcineurin.
anti-TNF-α monoclonal antibody	CD > UC	infliximab, adalimumab, certolizumab	blocks proinflammatory TNF-α signaling and may cause apoptosis of activated T cells
anti-β4 integrin monoclonal antibody	CD	natalizumab	blocks β4-based adhesion and cell trafficking by inflammatory cells

release. The profiles of surface marker and cytokine expression displayed by cells involved in an immune response usually can be grouped into one of four major categories. Type 1 responses (Th1) are characterized by IFN-γ release and are targeted toward control of intracellular pathogens. Type 2 responses (Th2) are characterized by IL-4 release and are targeted toward control of toxins and helminths. Type 17 responses (Th17) are characterized by IL-17 release and are targeted toward control of extracellular pathogens. Regulatory responses (Treg) are characterized by IL-10 and TGF-β production and are targeted toward modulation of the other three responses. Although each response type is referred to by a $CD4^+$ T cell subtype, in actuality multiple other cell types can make signature cytokines within a focal response. Mucosal samples taken from patients with active inflammation suggest that CD is associated with a Th1/Th17 cytokine profile while UC is associated with a modified Th2 (IL-5 without IL-4)/Th17 profile [1–5]. The percentage of $CD4^+$ $CD25^+$ regulatory T cells is increased in UC as compared to CD but is below that seen in diverticulitis [6–8]. Regulatory invariant NKT are decreased in CD and less so in UC as compared to healthy controls [9]. Loss of these cell types is seen in other autoimmune/inflammatory diseases. A recent episode of infectious gastroenteritis increases the risk of developing IBD [10]. Although no particular bacteria identified to date appears to be causal for IBD, patents with CD have a more restricted fecal biodiversity and many bacteria cannot be easily identified [11]. Patients with CD have increased antibodies to bacterial flagellin [12, 13] but this may reflect augmented bacterial penetration of the mucosal barrier (translocation) rather than a specific bacterial product that drives CD [14, 15].

There are several animal models of IBD [16, 17]. None are perfect parallels to human disease but they permit study of predisposing conditions, triggers and responses that produce intestinal inflammation. Colitis can be induced in normal mice by exposure to mucosal toxins. The two main models of toxin-induced colitis are: 1) DSS (dextran sulphate sodium) colitis in which the agent induces inflammation from intestinal epithelial cell injury without a pathogenic lymphocyte response; 2) TNBS (trinitrobenzenesulfonic acid) colitis in which the agent induces inflammation from an excessive Th1/Th17 response [18]; and colitis occurs spontaneously in mice with immunologic defects such as lack of IL-10, TGF-β-signaling, or STAT3-signaling. Colitis in these mice is due to dysregulated Th1 or Th17 activity. Colitis can also be caused by transfer of naïve T cells into immune deficient mice in the absence of regulatory T cells. Most of these models require the presence of normal bacterial flora for inflammation to occur. In general, these animal models demonstrate that intestinal inflammation occurs when there is dysregulation of adaptive and/or innate immune responses.

The genetic basis for IBD is being explored by genome-wide association studies. These studies compare genetic variation across the entire human genome to identify associations with the presence or absence of disease. Currently, about 40 discrete loci appear to contribute at least some risk for IBD [19, 20]. Table 2 lists loci that

Table 2 - Genetic loci with identified or candidate genes associated with the risk of developing Crohn's disease

Locus	Gene/Candidate	Function
Loci with confirmed or with highly probable contribution to CD risk		
16q12	NOD2/CARD15	MDP sensor (recognition of bacteria)
2q37	ATG16L1	Autophagy
5q33	IRGM	Initiates autophagy of intracellular bacteria
22q12	XBP1	ER stress response
18p11	PTPN2	Cell signaling (tyrosine kinase), growth factor stimulation
1q44	NLRP3	Cyropyrin, Inflammasome component
12q12	MUC19, LRRK2	Mucus protein, Unknown (autophagy)
5p13	PTGER4 (EP4)	Prostaglandin receptor
9q32	TNFSF15	Induces endothelial cell apoptosis
10q21	ZNF365	Unknown (zinc-finger protein)
1p13	PTPN22	Cell signaling (tyrosine kinase) associates with CBL
1q23	ITLN1	Galactose-binding lectin (recognition of bacteria)
6p22	CDKAL1	Unknown (regulation of a cyclin-dependent kinase)
6q27	CCR6	Chemokine receptor
9p24	JAK2	Cell signaling (tyrosine kinase), cytokine stimulation
11q13	C11orf30	Unknown (oncogene)
17q21	ORMDL3	Unknown (also associated with asthma risk)
17q21	STAT3	Cell signaling, cytokine stimulation
21q22	ICOSLG	ICOS ligand (costimulation)
Loci with confirmed or with highly probable contribution to CD and UC risk		
1p31	IL23R	IL23 receptor
5q33	IL12B (IL12p40)	Component of IL12 and IL23
6p21	MHC genes	Epitope selection
10q24	Nkx2-3	Regulates epithelial cell and lymphocyte development
3p21	MST1	Cytokine, macrophage stimulatory protein
21q22	PSMG1	Proteasome assembly
20q13	TNFRSF6B	Decoy receptor protects against FasL-induced cell death

Table 2 (continued)

Locus	Gene/Candidate	Function
Loci with likely contribution to CD risk		
2p23	GCKR	Cell signaling (e.g., Wnt signaling)
2p16	PUS10	Unknown (tRNA pseudouridine synthesis)
17q12	CCL2, CCL7	C-C chemokines, macrophage recruitment
6p25	LYRM4	Unknown (protein folding)
6p25	SLC22A23	Organic ion transporter
2q11	IL18RAP	IL18 receptor component
10p12	C10ORF67	Unknown, shared with sarcoidosis

have identified or candidate CD genes. Many of these loci provide evidence that altered innate immune responses to bacteria are important in CD.

The first gene variation identified to confer risk for Crohn's disease was in *CARD15/NOD2* [21, 22]. *NOD2* codes for a cytosolic pathogen-associated molecular pattern-recognition (PAMP) protein that identifies bacterial protein muramyl dipeptide (MDP). In Caucasian populations, compound heterozygotes and homozygotes for *NOD2* mutations carry an odds ratio of 17.1 for CD, while simple heterozygosity carries an odds ratio of 2.4 [23]. The exact mechanism by which *NOD2* variants contribute to CD risk is not clear. NOD2 is part of a larger class of innate immune receptors (toll- and nod-like receptors, (TLR, NLR)) that trigger responses to bacteria, virus, and fungi. NOD2 is expressed by monocytes and intestinal Paneth cells. Paneth cells release antibacterial peptides into the intestinal lumen. Paneth cells are numerous in the terminal ileum and NOD2 mutations increase the risk of ileal CD compared to CD involving only non-ileal sites [23]. *NOD2* variants may increase the risk of CD by causing hyporeactivity of certain innate responses, thereby forcing excess responses in other pathways disrupting homeostatic mechanisms. Alternatively, CD-related NOD2 mutations may directly impair negative regulation of TLR-mediated responses to enteric microflora, promoting excessive inflammation [24]. Although NOD2 mutations increase the risk for CD, most patients with CD have normal NOD2 and most people with variant NOD2 do not develop CD.

Further evidence implicating innate immune mechanisms in CD is the identification of autophagy-related gene *ATG16L1* as a susceptibility variant for CD [25]. *ATG16L1* codes for part of a multimeric protein complex that helps degrade senescent cytoplasmic material. In addition to this homeostatic function, autophagy is an evolutionarily conserved innate defense mechanism against invading viruses and intracellular bacteria [26]. The exact role of autophagosomal mutations in CD pathogenesis is not yet clear, but initial evidence from experiments using *S. typhimurium* and intestinal epithelial cells have shown that altered gene products from *ATG16L1* impair pathogen clearance and elimination of intracellular bacteria [27]. IRGM is another autophagy gene product involved in the isolation and degradation of intracellular bacteria [28] that is associated with CD risk.

The NOD2 and autophagy gene findings highlight the role of innate mechanisms in mucosal immune homeostasis. Innate cells affect adaptive (lymphocyte) cell function by display of costimulatory molecules and cytokines. Dendritic cells and macrophages make IL-23 which promotes Th17 function. Genome-wide association searches found IL-23 receptor gene variations that protect from or increase risk for CD [29, 30]. A locus close to the IL-12B, the 'p40' component shared by IL-12 and IL-23, has also been correlated to IBD risk [19, 20]. Loci that contain genes for immune cell signaling cascades and chemokine receptors are also implicated in CD (Tab. 2). For example, the region 6q27 contains CCR6 which is a chemokine receptor expressed on Th17 cells that assists in trafficking to sites of inflammation [31, 32].

Although many loci contribute to the predisposition of developing IBD, neither CD nor UC are genetically determined diseases. The contribution of identified loci explains about 20% of the risk for developing CD [19]. Monozygotic twins begin life with identical genes, but long-term concordance rates (i.e., both twins developing disease) is 50% for CD and 19% for UC [33]. This suggests that much of the risk for IBD is contributed by environmental factors.

The rise of IBD

The epidemiology of IBD demonstrates that environmental factors dramatically affect prevalence. Prior to the twentieth century IBD was largely unknown. In 1909 a special meeting by the Royal Society of Medicine was held to discuss whether ulcerative colitis existed as a non-infectious form of dysentery. No cases had been seen in London prior to the 1880s. Between 1884 and 1909 the hospitals in London were averaging two UC cases a year, mostly in young adults from 'well to do' families [34]. Singular case reports of rare small bowel inflammation with autopsy findings consistent for CD were reported in European literature as early as 1813 [35]. Crohn's disease became recognized as a entity in 1932 [36] separating regional ileitis from tubercular ileitis. By 1952 it became clear that CD could affect the colon and was different from UC [37]. During the first half of the twentieth century, IBD was sporadic, seen in usually in individuals from the upper classes, urban areas and northern latitudes [38]. During the second half of the twentieth century, IBD gained in scope and prevalence [38, 39]. Figure 1 shows the increase in CD incidence that occurred in different locals [40]. Analysis of U.S. military records suggests that being raised in the rural South afforded protection [41] and IBD occurred less

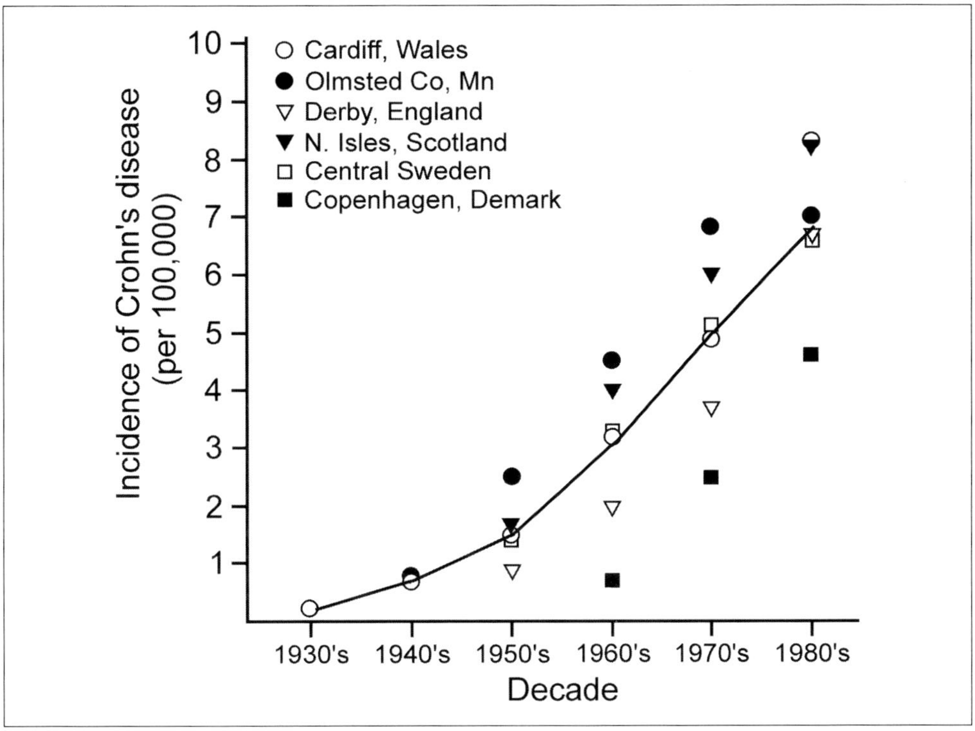

Figure 1
The increased incidence of Crohn's disease in different locales over time

frequently in people with blue collar jobs involving exposure to dirt and physical exercise [42].

Currently, the prevalence of IBD in the United States is about 440/100,000 or about 1 to 1.7 million people [43]. There is still a North/South gradient but it has declined over time. The prevalence of IBD in Manitoba Canada is about 510/100,000 and still occurs most frequently in patients of higher socioeconomic status [44]. The IBD prevalence in northern England, in 1995, was 412/100,000 [45]. The current estimate is that 2.2 million people in Western Europe and the United Kingdom have IBD [39]. Although once thought to have stabilized, the incidence of CD continues to gradually rise in England [46], France [47], and Sweden [48].

While IBD is prevalent in Australia [49], Canada, Western Europe, the United Kingdom, and the United States; it remains less so in Eastern Europe, Asia, Africa, and South America. However, as countries in these regions develop socioeconomically, the incidence of IBD increases [39]. The incidence of UC and CD in Hungary increased from 1.7 and 0.4/100,000 in 1977–1981 to 11.0 and 4.7/100,000 in

1997–2001 [50]. Although low by U.S. and European standards, the prevalence of IBD is increasing dramatically in Japan (Fig. 2 [51]) and South Korea [52]. This suggests that an environmental influence promotes IBD in developed countries or protects form IBD in lesser developed countries. Congruent with this is the observation that when people move from a country with low prevalence to a country with high prevalence of IBD, their children acquire a higher risk of developing IBD [53, 54].

The dramatic rise in IBD that occurred in the last half of the 20th century in highly industrialized countries could not be due to genetic change within the population. Instead it has to result from an environmental change. The increase in IBD that occurs as Eastern European and Asian countries develop reinforces this conclusion. Many lifestyle and environmental changes occur as countries develop socioeconomically. For example, the polyunsaturated/saturated fatty acid ratio in the American diet increased from 0.16 in the 1940s to 0.46 in the 1980s [55]. Smoking prevalence increased dramatically during this time frame and is now increasing in developing countries [56]. However, smoking appears to protect individuals from

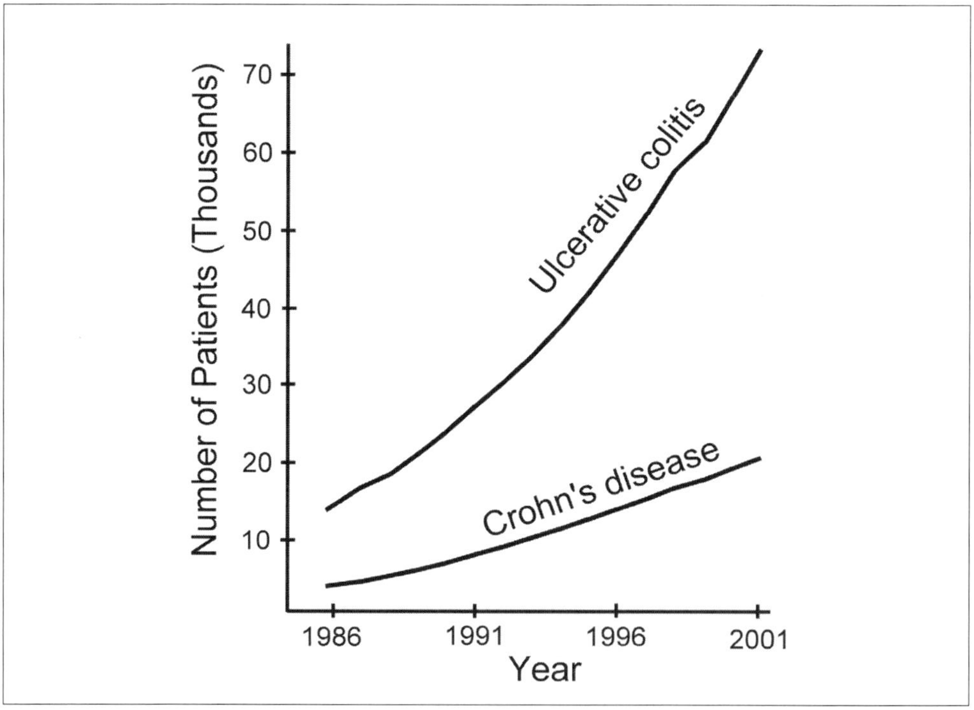

Figure 2
The increasing prevalence of IBD in Japan

developing ulcerative colitis rather than put them at risk [57]. The question then, is what environmental change explains the epidemic rise in IBD? Beginning in the mid 1990s we began working on the hypothesis that loss of exposure to helminthic parasites increases the risk of developing IBD [58].

Helminths are old acquaintances

Helminths are parasitic worms and comprise a successful group of organisms. Every mammal living in a natural environment is colonized with helminths. Indeed, helminths parasitize arthropods, mollusks, fish, amphibians, reptiles and birds. Worms do not fossilize well but amber-entrapped helminths in host insects date back 110 million years [59]. The conviction is that animals diversified with their parasitic worms. This co-evolutionary process explains why helminths usually have restricted host ranges, often host species specific.

While each species of helminth usually infects a narrow range of hosts, there are many species of helminths. Helminths are divided into two phyla; platyhelminths (flat worms) and nematodes (round worms). Platyhelminths are further divided into trematodes (flukes) and cestodes (tapeworms). Although all are called 'worms' there is huge genetic disparity between platyhelminths and nematodes. The phylogenetic distance between platyhelminths and nematodes is like that between mollusks and arthropods [60]. Even within nematodes, the variation is vast with multiple origins for establishment of parasitism [61]. Thus, although they look similar, different helminths have distinct adaptations for parasitic lifestyles. Table 3 lists some of the most common helminths that all together currently infect more than 1 billion people. A complete description of each worm, its unique method for finding a host, and possible pathologic effects of infection is beyond the scope of this chapter. However, certain generalizations are warranted. First, people can be infected with more than one species of helminth that utilize different niches within the host. Second, many helminth infections are long-lived (decades) or rapidly recur so individuals in endemic areas have persistent infections. Third, while many individuals in a locale are infected with a helminth, most of these will have few or no symptoms. Fourth, many helminths are killed by freezing or depend on intermediate hosts that require tropical conditions. Therefore, many helminths are limited to the tropics or have decreased geographic prevalence as climate cools. Fifth, many helminths spread by inundating host excretions with their offspring. New hosts are infected by direct or indirect contact with these excretions.

Evidence for human infections with helminths dates back many thousands of years [62]. For example eggs from *Ascaris lumbricoides*, *Trichuris trichiura*, and *Fasciola hepatica* were found in 30,000 year old fossilized stool specimens (coprolites) in France. Tissue from a 5,000 year old mummified Egyptian contained schistosoma circulating antigen showing active infection at the time of death [63]. Pre-Colombian

Table 3 - A short list of common helminths of people with their 'common names', and (anatomic location)

Platyhelminths 'flatworms'
- Trematodes 'flukes'
  - *Clonorchis sinensis* (biliary epithelium)
  - *Fasciola hepatica* (biliary epithelium)
  - *Fasciola gigantica* (intestinal epithelium)
  - *Fasciolopsis buski* (intestinal epithelium)
  - *Schistosoma* sp. (venous blood vessels)
    - *S. hematobium* (vesicular plexus)
    - *S. mansoni* (mesenteric plexus)
    - *S. japonicum* (mesenteric plexus)
    - *S. mekongi* (mesenteric plexus)
- Cestodes 'tapeworms'
  - *Diphyllobothrium latum* 'fish tapeworm' (intestinal epithelium)
  - *Taenia saginata* 'beef tapeworm' (intestinal epithelium)
  - *Taenia solium* 'pig tapeworm' (intestinal epithelium + systemic cysts)
  - *Hymenolepsis nana* 'dwarf tapeworm' (intestinal epithelium)
  - *Hymenolepsis diminuta* 'rat tapeworm' (intestinal epithelium)
  - *Echinococcus* sp. 'hydatid cysts' (systemic cysts)

Nematodes 'round worms'
- *Ascaris lumbricoides* (intestinal epithelium)
- *Enterobius vermicularis* 'pinworm/threadworm' (intestinal epithelium)
- *Trichinella* sp. (intestinal epithelium and systemic cysts)
- *Trichuris trichiura* 'whipworm' (intestinal epithelium)
- *Necator americanus* 'hookworm' (intestinal epithelium)
- *Ancylostoma duodenale* 'hookworm' (intestinal epithelium)
- *Strongyloides stercoralis* 'threadworm' (intestinal epithelium)
- *Paracapillaria philippinensis* (intestinal epithelium)
- *Brugia malayi* 'filaria' (lymphatic vessels)
- *Wuchereria bancrofti* 'filaria' (lymphatic vessles)
- *Onchocerca volvulus* 'river blindness' (subcutaneous tissues)

coprolites demonstrate that *A. lumbricoides*, *T. trichiura*, and *Enterobius vermicularis* were common guests for indigenous Americans [62]. This long history of ubiquitous poly-helminth exposure likely influenced human genetic variation.

Helminths are limited by hygiene and modern lifestyles

Prior to the 20th century, helminth eradication was relegated to ineffectual patent medications (Fig. 3). However, in 1909 John D. Rockefeller established the 'Rockefeller Sanitary Commission for Eradication of Hookworm Disease' which grew into the Rockefeller Foundation [64, 65]. Although the foundation failed to achieve full control of hookworm, the effort helped establish public health and sanitation

Figure 3
Many patent medicines claimed to help get rid of helminths

Figure 4
The Rockefeller Foundation began as a program to eradicate hookworm. It set the framework for many public health programs.

programs (Fig. 4). After World War II, anthelmintic drug treatment evolved from minimally effective and highly toxic to highly effective and minimally toxic, prompting greater acceptance. In a 1968 lecture Professor Harold W. Brown commented

> *"Well-nourished persons often harbor helminths without apparent damage and one may question the wisdom of treating such infections, especially with chemotherapeutic agents with toxic qualities. On the other hand, if a safe and effective drug is available, treatment should be given until evidence is in hand that these helminth infections do not harm or are beneficial to man. Unfortunately, reinfection is common and prompt unless sanitary measures are instituted" [66].*

A major impetus for developing modern hygienic practices was the eradication of helminths. The conditions under which most people lived remained fairly uniform for millennia. A war here, a disaster there, did not affect the ubiquity of helminth infection. Helminths enjoyed high prevalence until the industrial revolution which brought rapid economic development and concurrent improvement in sanitation and hygienic practice. Modern sanitation has proved disastrous to most helminths. Indoor plumbing and modern sewage treatment spirit away *A. lumbricoides* and *T. trichiura* eggs before they can spread infection. Frequent baths and rapidly laundered clothing disinfect fomites blocking transmission of *E. vermicularis* or *H. nana*. Sidewalks and shoes obstruct *Necator americanus*, *Ancylostoma duodenale* and *Strongyloides stercoralis*. Easily available cooking utensils and modern food processing kill *Diphyllobothrium*, *Taenia*, and *Trichinella* larva. These changes have all but eradicated helminths from industrialized countries.

Initial loss of significant helminth exposure probably began in northern latitude urban centers among the more wealthy members of society [67]. As infrastructure developed in the 20th century, helminth exposure in rural areas declined. The prevalence of hookworm dropped in the Southeastern United States (Fig. 5) [68]. By 1987 these same Southeastern states reported that less than 2% of evaluated stool specimens were positive for hookworm [69] and most of the people harboring hookworm were recent immigrants from endemic countries. Other helminths show the same trend (Tab. 4) [67].

Another good example is trichinosis. *Trichinella* sp. are intestinal helminths that produce larva which migrate to and encyst in host muscles (trichinosis). People acquire the helminth by eating poorly cooked contaminated pork. Until the 1960s trichinosis was endemic in the Northeastern and Western United States. Surveillance autopsy studies examining prevalence of *Trichinella* cysts throughout the USA found a decline in trichinosis from 16.1% (7.2% recently infected) in 1936–1941 to 4.2%

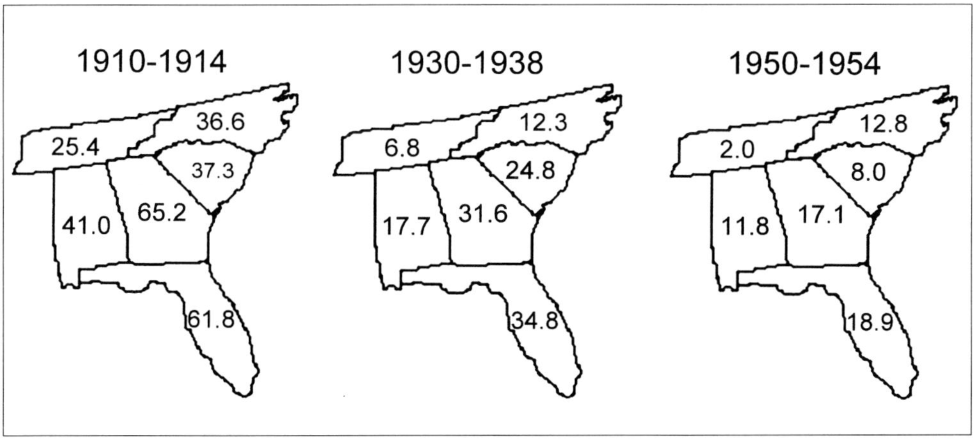

Figure 5
The prevalence (% positive stool tests) of hookworm infection declined in the Southeastern United States.

Table 4 - Percent of stool specimens evaluated at Charity Hospital New Orleans, Louisiana, USA showing helminth eggs. Data from [67]

Helminth	**1942**	**1972**
Ascaris lumbricoides	4.2%	2.2%
Trichuris trichiura	7.5%	4.7%
Enterobius vermicularis[a]	1.4%	0.4%
Taenia sp.	0.2%	Not detected

[a]*Includes cellophane tape tests*

(0.5% recently infected) in samples taken from 1966–1968 [70]. Now trichinosis is exceedingly rare averaging less than 25 cases/year in the United States [71].

A similar eradication of helminths occurred in post-war Western Europe. Estimates in 1947 were that Europe (including the UK) had 62, 34, and 32 million active infections with *E. vermicularis*, *T. trichiura*, and *A. lumbricoides* (respectively) or about 36% of the population harbored helminths [72]. Now even pinworm (*E. vermicularis*) has become a rarity in Europe [73].

As countries socioeconomically develop, helminth carriage declines dramatically. The prevalence of *T. trichiura* infections in South Korean schoolchildren fell from 74.2% in 1969 to 0.02% in 2004 [74]. During this time frame the incidence of UC increased in Seoul (Fig. 6) [52]. Although prevalent exposure remains in tropical areas with poor sanitation and high levels of poverty [75], in the industrialized

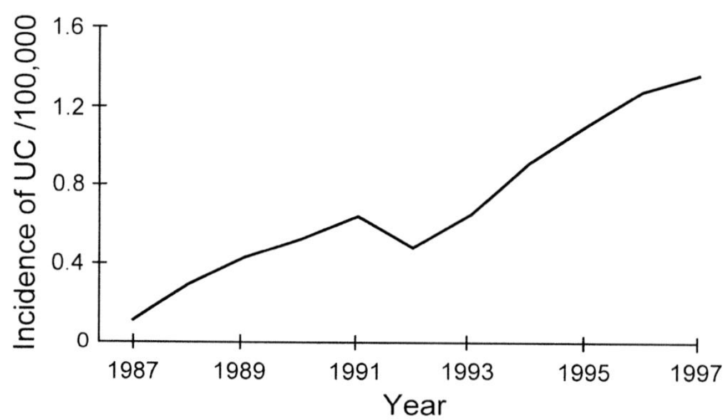

Figure 6
The incidence of ulcerative colitis increased dramatically in Seoul, South Korea during the time that helminths were being eradicated. Modified from [52].

nations over the last half of the 20th century our longstanding acquaintance with helminths has atrophied.

Helminths affect immunity

Helminths are complex organisms with long-life spans that survive within people by evading and/or altering host immune responses. For example, peripheral blood lymphocytes (PBL) from patients with filariasis make immune regulatory IL-10 and TGF-β in response to filarial antigens [76]. Filarial infections also cause PBL to have increased expression of CTLA4 and PD-1 (inhibitory receptors), increased FoxP3 expression (a regulatory T cell transcription factor) and diminished T-bet and GATA3 (Th1 and Th2 transcription factors) expression [77]. Anergy to filaria was associated with induction of inhibitory E3 ubiquitin ligases due to augmented CTLA4 and TGF-β signaling [77].

Helminths also modify immune responses beyond those directed against the worms. People exposed to helminths have Th2 and IL-10-skewed peripheral blood lymphocyte (PBL) cytokine responses to T cell mitogen stimulation compared to groups without exposure [78]. People with helminths respond to tetanus vaccination with an immune bias away from a Th1 cytokine profile [79–81]. Anthelmintic drug treatment to remove worms improves IFN-γ responses to BCG vaccination [82, 83] and antibody responses to low dose cholera vaccine [84]. Ethiopian immigrants to Israel with chronic helminth exposure show increased PBL CTLA4 expression, decreased intracellular phosphorylation of various tyrosine kinases, and impaired degradation of phosphorylated IkB in response to mycobacterial antigens all of which slowly recover after treatment with anthelmintics [85]. A recent study of Cameroon school age children evaluating peripheral blood lymphocyte cytokine production found that levels of regulatory cytokines IL-10 and TGF-β in response to mitogens correlated directly to intensity of infection with *Ascaris lumbricoides* and *Trichuris trichiura* [86].

Studies comparing helminth-colonized to helminth-naïve people groups are complicated by difficult to control variables including socioeconomic, dietary, nutritional, and educational differences between villages or regions. Studies in inbred mice permit control of such variables. Like people, mice harboring helminths have depressed Th1 and augmented Th2 responses to test antigens [87, 88] and mycobacteria [89]. Colonization of mice with *B. malayi* (filaria) recruits alternatively activated macrophages that inhibit lymphocyte proliferation [90]. Mice infected with the nematode *Heligmosomoides polygyrus*, have immune responses with pronounced Th2 and Treg activity that actively suppress Th1 and Th17 function [91–94].

Helminths alter their host's immune system through multiple mechanisms [95] including release of compounds that trigger IL-10 [96] or prevent IL-12 and TNF-α [97] production, release of actual mediators such as prostaglandin E2 [98], and release of cytokine mimics such as TGF-β-like molecules[99]. This makes our recent loss of helminth exposure a strong environmental candidate to explain the increased prevalence of IBD. If this is the case, then colonization with helminths should limit pathogenic inflammation [58, 100, 101].

Helminths suppress aberrant inflammation

The protective effects of helminth colonization have been characterized in mouse models of colitis. Mice colonized with the male *S. mansoni* worms are protected from DSS colitis [102]. Colitis in the DSS model results from barrier dysfunction rather than an immune hyper-response. The protection mediated by *S. mansoni* results from induction of protective macrophages rather than T regulatory cell function [102]. Mice exposed to a range of helminths (e.g., *S. mansoni*, *T. spiralis*, *Hymenolepis diminuta* or *H. polygyrus*) are protected from TNBS-type colitis [103–107]. Helminth exposure inhibits the expression of the proinflammatory cytokines IFN-γ, IL-12p40, and IL-17A, while augmenting expression of regulatory factors such as IL-10, TGF-β, FoxP3$^+$ T cells, and regulatory CD8$^+$ T cells [91, 93, 103, 105, 106]. Colonization with *Trichuris muris* or *H. polygyrus* inhibits development of colitis in IL-10-deficient mice [58]. Furthermore, colonization of IL-10-deficient mice with *H. polygyrus* [108] or exposure to *S. mansoni* antigens (Elliott, unpublished results) after colitis is fully established results in resolution of the otherwise chronic inflammation. Improvement in IL-10$^{-/-}$ mice is associated with inhibition of gut-associated immune cell production of IFN-γ, IL-12p40, and IL-17 [91, 108]. Helminth exposure augments T cell FoxP3 mRNA expression and CD8$^+$ T cell regulatory function even in the absence of T cell IL-10 production [93, 108].

Specific critical regulatory pathways induced by helminth exposure may differ by helminth species, mouse strain, and inflammatory model. Although the induction of multiple regulatory circuits permits suppression of IL-10-deficient colitis, congenital loss of STAT6 [103] or T cell TGF-β signaling (Ince, unpublished observation) or acute loss of IL-10 signaling [104] circumvents helminth-associated protection from other types of colitis. Thus, at least some helminth-augmented regulatory circuits are non-redundant. Helminths also protect from disease in murine models of Type 1 diabetes, reactive airway disease, Grave's disease, anaphylaxis, and multiple sclerosis [109–117]. Protection in these disease models is associated with augmentation of IL-10, TGF-β, and Treg circuits.

Natural helminth exposure may afford protection from inflammatory disease

Immunomodulatory effects similar to those found in colonized mice have been observed in people with helminth infections [78–86]. It is not surprising that infection with helminths is associated with decreased inflammatory disease expression or activity in people. Examples include case reports and studies involving ulcerative colitis, multiple sclerosis, allergen reactivity, and infantile eczema.

Büning et al. reported on a case where a young girl with ulcerative colitis developed more intense disease after eradication of *E. vermicularis* [118]. Importantly, they had performed colonic biopsies before and after treatment with pyrantel to remove the pinworms. Mucosal IL-10 and TGF-β mRNA expression and the percentage of $FoxP3^+$ lamina propria cells all decreased dramatically after loss of the helminths. In a small case-control observational study, Correale and Farez reported on the course of remitting-relapsing MS in 12 patients with intestinal helminths compared to 12 matched MS patients without helminths. Over the 4.6 years of observation, three relapse episodes occurred in the patients with helminths while 56 episodes occurred in the patients without helminths ($p < 0.0001$). Peripheral blood lymphocytes (PBL) from infected patients made significantly more IL-10 and TGF-β and less IFN-γ and IL-12 in response to myelin basic peptide than did PBL from uninfected patients [119]. A study of 1,055 children from northern Brazil found that those with *T. trichiura* (OR = 0.58, $p = 0.02$) or *A. lumbricoides* (OR = 0.63, $p = 0.01$) infections had reduced skin test reactivity against a panel of common allergens [120]. A placebo-controlled study, involving 341 Gabonese schoolchildren treated to eradicate *A. lumbricoides* and *T. trichiura* infections, showed that children who received anthelmintic drugs were at higher risk for developing skin test reactivity to dust mite allergens (OR = 2.51, $p < 0.001$) than children who received placebo [121]. A study of 103 women to evaluate the benefit of helminth eradication in pregnancy found that children of worm-free women were far more likely to develop infantile eczema (39% *versus* 9%, $p = 0.002$) than those born to women with active helminth infections [122]. These observations provide strong evidence that colonization with helminths hinders development or activity of disease due to dysregulated immunity.

Clinical use of helminths

Because natural helminth infection appears to protect from or modify the course of inflammatory disease, clinically controlled helminth exposure could have therapeutic application. Some helminths cause significant disease that negates their therapeutic application. Other helminths, with minimal or no known pathogenicity are being studied clinically.

Hookworm (*N. americanus*) is being investigated for therapeutic use. Hookworms can cause disease [123] but very light infections with *N. americanus* are probably safe [124]. People acquire hookworm infections by applying infective larvae to their skin. Currently, these larvae are cultured from the stool of human volunteer donors that are actively colonized with *N. americanus*. These donors are extensively screened to reduce the risk of co-transmitting other infections. A feature of *N. americanus* that makes it an attractive candidate is that infection from a single application can last over 6 years. Therefore, repeat dosing may not be needed.

A small open-label trial tested *N. americanus* in nine patients with Crohn's disease [125]. Two of these patients had moderately active disease when they received 50 larvae. Both showed improvement in their symptom scores. The other seven patients in the initial trial had inactive or very mild disease which did not significantly change with helminth exposure. A small dose-ranging study of *N. americanus* [126] evaluated 12 volunteers who were blinded to the dose of helminths they received. Two persons withdrew (one given 100 and one given 50 larvae) due to symptoms of diarrhea and vomiting or abdominal pain. However, the lowest (10 larvae) dose was well tolerated and resulted in patent colonization. Larger studies using patients with Crohn's disease and asthma are underway.

Most of the therapeutic helminth trials to date used the porcine whipworm *Trichuris suis*. *T. suis* is closely related to *T. trichiura* (human whipworm) and can briefly colonize people [127]. However, *T. suis* has never been documented to cause human disease. *Trichuris* are good candidates for clinical use. People acquire whipworm infections by ingesting microscopic embryonated parasite eggs. Whipworms do not migrate beyond the intestines, do not multiply within their host, and cannot be directly transmitted from host to host. While the human whipworms (*T. trichiura*) can be obtained only from colonized primates, *T. suis* is obtained from pigs raised in pathogen-free environments. For the published studies, adult *T. suis* worms were collected from porcine colons, washed, and cultured *in vitro* to collect freshly deposited ova.

Initially, the effect of *T. suis* colonization was studied in a small open-label trial of seven patients with IBD (four Crohn's disease, three ulcerative colitis). The patients ingested 2,500 embryonated ova and were observed. All had improvement in their symptoms [128]. A second study tested repeated dosing (2,500 ova every 3 weeks) of *T. suis* in 29 patients with active Crohn's disease [129]. At week 24, 79% had a significant reduction in symptoms. A third study was a double blind placebo-controlled trial of *T. suis* in 54 patients with active ulcerative colitis [130]. The patients received either 2,500 *T. suis* ova or placebo every 2 weeks for 12 weeks. Many of the patients given *T. suis* improved compared to those given placebo (43.3% *versus* 16.7%, $p<0.04$). The study also included a blinded crossover limb where patients originally on placebo where switched to *T. suis* and those on *T. suis* were switched to placebo. In the crossover limb, 56.3% of the patients given *T. suis* improved compared to 13.3% of patients given placebo ($p=0.02$) [131]. Trials are now underway studying the effect of *T. suis* in allergic rhinitis and multiple sclerosis. Additional studies in Crohn's disease, ulcerative colitis and other immunological diseases are planned for the near future.

Arguments against an immunoregulatory or therapeutic role for helminths

Although there is epidemiologic, experimental and clinical evidence that helminth exposure suppresses aberrant inflammation like IBD, there are arguments that helminth eradication is not the etiologic change responsible for the rise in immune-mediated disease. For example, many other changes have occurred with development and the prevalence of IBD continues to increase in countries long after loss of helminth colonization [132]. However, the most dramatic increase in prevalence of immune-mediated disease occurs soon after eradication of helminths as is now happening in Korea, South America and India. After this rapid rise, the increase slows and may even plateau in highly industrialized countries [39]. Any continued slow increase could reflect more accurate diagnosis of minimally active disease due to improvements in imaging and laboratory tests. Furthermore, if the continued increase in prevalence is real, it may reflect loss of generational epigenetic patterning. Maternal helminth colonization can influence infant immunity. The PPD-stimulated cytokine response profile of PBL from children born to *Wuchereria bancrofti*-colonized women showed 26 fold less IFN-γ production ($p = 0.008$) than PBL from children of worm-free mothers [133]. Children of worm-free mothers are more likely to develop infantile eczema than children of colonized mothers [122]. Helminth-modified maternal gene expression may be epigenetically conferred to offspring [134] and explain increasing disease prevalence in the second worm-free generation.

The more common concern is whether it is safe to use helminths clinically to treat immune-mediated diseases. In mice, colonization with *H. polygyrus* increases susceptibility to *Citrobacter rodentium*, a murine bacterial pathogen [135]. This increased susceptibility is due to induction of IL-10 producing dendritic cells that suppress an exuberant immune response [136] and permits expansion of the bacteria. Although helminth exposure may augment Citrobacter infections, clearance of mycobacterial infections does seem impaired [137]. Currently, we use immunosuppressive medications to treat IBD and most other immune-mediated inflammatory diseases (Tab. 1). Many of these medications have a high potential for dangerous adverse effects. Unlike immunosuppressive drugs, helminths do not seem to increase susceptibility to opportunistic infections. To date, no significant adverse effects attributable to the helminths have been noted in patients treated with *T. suis*. It is likely that physician-supervised helminth exposure will be very safe.

Another concern is that helminths may become pathogenic in patients treated with immune suppressive medications. Most helminths are biologically unable to multiply in their host. There is no alteration in host immunity that would permit an increase in helminth (e.g., *T. suis*) number without re-exposure to the infective stage. Only two helminths (*Strongyloides stercoralis* and *Paracapillaria philippinensis*) are known to circumvent this rule. Neither *S. stercoralis* nor *P. philippinensis* are proposed for therapeutic use. Other helminths neither multiply nor expand their niche in immunosuppressed patients such as those with HIV-AIDS [138]. If a helminth does seem to cause pathology, it can be killed with available anthelmintic medications. Therefore, use of helminths in patients treated with immune suppressive medications is not necessarily problematic.

Identification of diseases due to loss of helminth exposure

Helminths may not inhibit or modulate all types of inflammation. Mice with oxazalone colitis, another model of intestinal inflammation, develop worse disease if colonized with *H. diminuta* [139]. Yet, exposure to this helminth protects from TNBS-like colitis [104]. There is a spectrum of over 40 different immune-mediated diseases which together affect greater than 10% of the population in highly developed countries. Each immune-mediated disease is unique. However, Th1/Th17-type diseases like multiple sclerosis or Crohn's disease, and Th2/Th17-type diseases like asthma appear to be suppressed by helminths.

Many of the immune-mediated diseases share predisposing genes (Tab. 5). If helminth exposure offsets a genetic predisposition toward aberrant inflammation, then eradication of helminths would result increased prevalence for that disease. Genomic surveys suggest that variation in IL-12p40, IL-23R, and CCR6 (expressed on Th17 cells [31]) contribute to the risk of IBD. In mice, helminth colonization suppresses mucosal IL-12p40, IL-23 and IL-17 expression [91]. If this effect occurs in people, then helminth exposure could obscure any genetic tendency toward IBD conferred by variation in these genes. Helminths induce alternatively activated macrophages [140] as does MST1 [141]. Helminthic induction of alternatively-activated macrophages could offset effects due to variation in MST1. Variation in IL-2R5α (CD25) contribute to multiple sclerosis and type-1 diabetes risk. Helminth colonization increases expression of $CD25^+$ regulatory T cells [94, 119] which could negate a proinflammatory variation in the IL-2Rα gene. As we learn more about the specific immune dysregulation that occurs in each disease and the different immune regulatory pathways helminths exploit, we may be able to predict which diseases are attributable to helminth eradication and which patients may benefit from helminth exposure.

Acknowledgements
Supported by grants from the Department of Veterans Affairs Office of Research and Development, the National Institutes of Health, the Crohn's and Colitis Foundation of America, and the Broad Medical Research Program of the Eli and Edythe L. Broad Foundation.

Table 5 - Gene variants suspected of conferring risk for more than one autoimmune disease

Gene	Crohn's	UC	MS	T1D	RA	Psoriasis	SLE	Celiac	Asthma	AITD	Sarcoid
PTPN22	X			X	X		X				
IRF5	X	X			X		X				
IL23R	X	X				X					
PTPN2	X			X							
IL18RAP	X							X			
TRAF1/C5					X				X		
ORMDL3	X								X		
TNFAIP3					X		X	X			
STAT4					X		X				
IL12B	X	X				X					
IL2RA			X	X							
IL7RA			X	X							
MST1	X	X									
Nkx2-3	X	X									
PSMG1	X	X									
CLEC16A			X	X							
IL2/IL21				X				X			
SH2B3				X				X			
TNFRSF6B	X	X									
CTLA4				X						X	
C10ORF67	X										X

UC, ulcerative colitis; MS, multiple sclerosis; T1D, autoimmune (Type 1) diabetes; RA, rheumatoid arthritis, SLE, systemic lupus erythematosus; AITD, autoimmune thyroid disease (Grave's disease and Hashimoto's thyroiditis)

References

1 Fuss IJ, Neurath M, Boirivant M et al. Disparate CD4+ lamina propria (LP) lymphokine secretion profiles in inflammatory bowel disease. Crohn's disease LP cells manifest increased secretion of IFN-gamma, whereas ulcerative colitis LP cells manifest increased secretion of IL-5. *J Immunol* 1996;157: 1261–70

2 West GA, Matsuura T, Levine AD et al. Interleukin 4 in inflammatory bowel disease and mucosal immune reactivity. *Gastroenterology* 1996; 110: 1683–95

3 Raddatz D, Bockemuhl M, Ramadori G. Quantitative measurement of cytokine mRNA in inflammatory bowel disease: relation to clinical and endoscopic activity and outcome. *Eur J Gastroenterol Hepatol* 2005; 17: 547–57
4 Fujino S, Andoh A, Bamba S et al. Increased expression of interleukin 17 in inflammatory bowel disease. *Gut* 2003; 52: 65–70
5 Nielsen OH, Kirman I, Rudiger N et al. Upregulation of interleukin-12 and -17 in active inflammatory bowel disease. *Scand J Gastroenterol* 2003; 38: 180–5
6 Berndt U, Bartsch S, Philipsen L et al. Proteomic analysis of the inflamed intestinal mucosa reveals distinctive immune response profiles in Crohn's disease and ulcerative colitis. *J Immunol* 2007; 179: 295–304
7 Holmen N, Lundgren A, Lundin S et al. Functional $CD4^{+}CD25$high regulatory T cells are enriched in the colonic mucosa of patients with active ulcerative colitis and increase with disease activity. *Inflamm Bowel Dis* 2006; 12: 447–56
8 Maul J, Loddenkemper C, Mundt P et al. Peripheral and intestinal regulatory $CD4^{+}$ CD25(high) T cells in inflammatory bowel disease. *Gastroenterology* 2005; 128: 1868–78
9 Grose RH, Thompson FM, Baxter AG et al. Deficiency of invariant NK T cells in Crohn's disease and ulcerative colitis. *Dig Dis Sci* 2007; 52: 1415–22
10 Porter CK, Tribble DR, Aliaga PA et al. Infectious gastroenteritis and risk of developing inflammatory bowel disease. *Gastroenterology* 2008; 135: 781–6
11 Sokol H, Lay C, Seksik P et al. Analysis of bacterial bowel communities of IBD patients: what has it revealed? *Inflamm Bowel Dis* 2008; 14: 858–67
12 Lodes MJ, Cong Y, Elson CO et al. Bacterial flagellin is a dominant antigen in Crohn disease. *J Clin Invest* 2004; 113: 1296–306
13 Sitaraman SV, Klapproth JM, Moore DA III et al. Elevated flagellin-specific immunoglobulins in Crohn's disease. *Am J Physiol Gastrointest Liver Physiol* 2005; 288: G403–G406
14 Adams RJ, Heazlewood SP, Gilshenan KS et al. IgG antibodies against common gut bacteria are more diagnostic for Crohn's disease than IgG against mannan or flagellin. *Am J Gastroenterol* 2008; 103: 386–96
15 Ziegler TR, Luo M, Estivariz CF et al. Detectable serum flagellin and lipopolysaccharide and upregulated anti-flagellin and lipopolysaccharide immunoglobulins in human short bowel syndrome. *Am J Physiol Regul Integr Comp Physiol* 2008; 294: R402–R410
16 Strober W, Fuss IJ, Blumberg RS. The immunology of mucosal models of inflammation. *Ann Rev Immunol* 2002; 20: 495–549
17 Xavier RJ, Podolsky DK. Unravelling the pathogenesis of inflammatory bowel disease. *Nature* 2007; 448: 427–34
18 Zhang Z, Zheng M, Bindas J et al. Critical role of IL-17 receptor signaling in acute TNBS-induced colitis. *Inflammatory Bowel Diseases* 2006; 12: 382–8
19 Barrett JC, Hansoul S, Nicolae DL et al. Genome-wide association defines more than 30 distinct susceptibility loci for Crohn's disease. *Nat Genet* 2008; 40: 955–62
20 Fisher SA, Tremelling M, Anderson CA et al. Genetic determinants of ulcerative colitis

include the ECM1 locus and five loci implicated in Crohn's disease. *Nat Genet* 2008; 40: 710–12
21 Hugot JP, Chamaillard M, Zouali H et al. Association of NOD2 leucine-rich repeat variants with susceptibility to Crohn's disease. *Nature* 2001; 411: 599–603
22 Ogura Y, Bonen DK, Inohara N et al. A frameshift mutation in NOD2 associated with susceptibility to Crohn's disease. *Nature* 2001; 411: 603–6
23 Economou M, Trikalinos TA, Loizou KT et al. Differential effects of NOD2 variants on Crohn's disease risk and phenotype in diverse populations: a metaanalysis. *Am J Gastroenterol* 2004; 99: 2393–404
24 Cho JH. The genetics and immunopathogenesis of inflammatory bowel disease. *Nat Rev Immunol* 2008; 8: 458–66
25 Hampe J, Franke A, Rosenstiel P et al. A genome-wide association scan of nonsynonymous SNPs identifies a susceptibility variant for Crohn disease in ATG16L1. *Nat Genet* 2007; 39: 207–11
26 Levine B, Deretic V. Unveiling the roles of autophagy in innate and adaptive immunity. *Nat Rev Immunol* 2007; 7: 767–77
27 Rioux JD, Xavier RJ, Taylor KD et al. Genome-wide association study identifies new susceptibility loci for Crohn disease and implicates autophagy in disease pathogenesis. *Nat Genet* 2007; 39: 596–604
28 Singh SB, Davis AS, Taylor GA et al. Human IRGM induces autophagy to eliminate intracellular mycobacteria. *Science* 2006; 313: 1438–41
29 Duerr RH, Taylor KD, Brant SR et al. A genome-wide association study identifies IL23R as an inflammatory bowel disease gene. *Science* 2006; 314: 1461–3
30 Tremelling M, Cummings F, Fisher SA et al. IL23R variation determines susceptibility but not disease phenotype in inflammatory bowel disease. *Gastroenterology* 2007; 132: 1657–64
31 Acosta-Rodriguez EV, Rivino L, Geginat J et al. Surface phenotype and antigenic specificity of human interleukin 17-producing T helper memory cells. *Nat Immunol* 2007; 8: 639–46
32 Yamazaki T, Yang XO, Chung Y et al. CCR6 regulates the migration of inflammatory and regulatory T cells. *J Immunol* 2008; 181: 8391–401
33 Halfvarson J, Bodin L, Tysk C et al. Inflammatory bowel disease in a Swedish twin cohort: a long-term follow-up of concordance and clinical characteristics. *Gastroenterology* 2003; 124: 1767–73
34 Allchin WH. A Discussion on "Ulcerative Colitis.": Introductory Address., 1909 v2 Edn 1909
35 Combe C, Saunders W. *A singular case of stricture and thickening of the ileum*, 4 Edn 1813
36 Crohn BB, Ginzburg L, Oppenheimer GD. *Regional Ileitis: A Pathologic and Clinical Entity*, 99 Edn 1932
37 Wells C. Ulcerative colitis and Crohn's disease. *Ann R Coll Surg Engl* 1952; 11: 105–20

38 Kirsner JB. The historical basis of the idiopathic inflammatory bowel diseases. *Inflamm Bowel Dis* 1995; 1: 2–26
39 Loftus EV, Jr. Clinical epidemiology of inflammatory bowel disease: Incidence, prevalence, and environmental influences. *Gastroenterology* 2004; 126: 1504–17
40 Elliott DE, Summers RW, Weinstock JV. *Helminths and the Modulation of Mucosal Inflammation*, 21 Edn 2005
41 Sonnenberg A, Wasserman IH. Epidemiology of inflammatory bowel disease among U.S. military veterans. *Gastroenterology* 1991; 101: 122–30
42 Sonnenberg A. Occupational distribution of inflammatory bowel disease among German employees. *Gut* 1990; 31: 1037–40
43 Kappelman MD, Rifas-Shiman SL, Kleinman K et al. The prevalence and geographic distribution of Crohn's disease and ulcerative colitis in the United States. *Clin Gastroenterol Hepatol* 2007; 5: 1424–9
44 Green C, Elliott L, Beaudoin C et al. A population-based ecologic study of inflammatory bowel disease: searching for etiologic clues. *Am J Epidemiol* 2006; 164: 615–23
45 Rubin GP, Hungin AP, Kelly PJ et al. Inflammatory bowel disease: epidemiology and management in an English general practice population. *Aliment Pharmacol Ther* 2000; 14: 1553–9
46 Gunesh S, Thomas GA, Williams GT et al. The incidence of Crohn's disease in Cardiff over the last 75 years: an update for 1996–2005. *Aliment Pharmacol Ther* 2008; 27: 211–9
47 Molinie F, Gower-Rousseau C, Yzet T et al. Opposite evolution in incidence of Crohn's disease and ulcerative colitis in Northern France (1988–1999) *Gut* 2004; 53: 843–8
48 Lapidus A. Crohn's disease in Stockholm County during 1990–2001: an epidemiological update. *World J Gastroenterol* 2006; 12: 75–81
49 McDermott FT, Whelan G, St John DJ et al. Relative incidence of Crohn's disease and ulcerative colitis in six Melbourne hospitals. *Med J Aust* 1987; 146: 525, 528–5, 529
50 Lakatos L, Mester G, Erdelyi Z et al. Striking elevation in incidence and prevalence of inflammatory bowel disease in a province of western Hungary between 1977–2001. *World Journal of Gastroenterology* 2004; 10: 404–9
51 Ouyang Q, Tandon R, Goh KL et al. The emergence of inflammatory bowel disease in the Asian Pacific region. *Curr Opin Gastroenterol* 2005; 21: 408–13
52 Yang SK, Hong WS, Min YI et al. Incidence and prevalence of ulcerative colitis in the Songpa-Kangdong District, Seoul, Korea, 1986–1997. *Journal of Gastroenterology & Hepatology* 2000; 15: 1037–42
53 Carr I, Mayberry JF. The effects of migration on ulcerative colitis: a three-year prospective study among Europeans and first- and second- generation South Asians in Leicester (1991–1994.) *American Journal of Gastroenterology* 1999; 94: 2918–22
54 Jayanthi V, Probert CS, Pinder D et al. Epidemiology of Crohn's disease in Indian migrants and the indigenous population in Leicestershire. *Quarterly Journal of Medicine* 1992; 82: 125–38

55 Stephen AM, Wald NJ. Trends in individual consumption of dietary fat in the United States, 1920–1984. *Am J Clin Nutr* 1990; 52: 457–69
56 Yach D, Wipfli H. A century of smoke. *Ann Trop Med Parasitol* 2006; 100: 465–79
57 Birrenbach T, Bocker U. Inflammatory bowel disease and smoking: a review of epidemiology, pathophysiology, and therapeutic implications. *Inflamm Bowel Dis* 2004; 10: 848–59
58 Elliott DE, Urban JFJ, Argo CK et al. Does the failure to acquire helminthic parasites predispose to Crohn's disease? *FASEB Journal* 2000; 14: 1848–55
59 Poinar G Jr, Buckley R. Nematode (Nematoda: Mermithidae) and hairworm (Nematomorpha: Chordodidae) parasites in Early Cretaceous amber. *J Invertebr Pathol* 2006; 93: 36–41
60 Philippe H, Lartillot N, Brinkmann H. Multigene analyses of bilaterian animals corroborate the monophyly of Ecdysozoa, Lophotrochozoa, and Protostomia. *Mol Biol Evol* 2005; 22: 1246–53
61 Mitreva M, Blaxter ML, Bird DM et al. Comparative genomics of nematodes. *Trends Genet* 2005; 21: 573–81
62 Goncalves ML, Araujo A, Ferreira LF. Human intestinal parasites in the past: new findings and a review. *Memorias do Instituto Oswaldo Cruz* 2003; 98 Suppl 1: 103–18
63 Deelder AM, Miller RL, de Jonge N, Krijger FW. Detection of schistosome antigen in mummies. *Lancet* 1990; 335: 724–5
64 Warren KS. The control of helminths: nonreplicating infectious agents of man. *Annu Rev Public Health* 1981; 2: 101–5
65 Anonymous. The Rockefeller Commission for the Eradication of Hookworm Disease. *Science* 1909; 30: 635–6
66 Brown HW. Anthelmintics, new and old. *Clin Pharmacol Ther* 1969; 10: 5–21
67 Hubbard DW, Morgan PM, Yaeger RG et al. Intestinal parasite survey of kindergarten children in New Orleans. *Pediatric Research* 1974; 8: 652–8
68 Wright WH. Current Status of Parasitic Diseases. *Public Health Reports* 1955; 70: 966–75
69 Kappus KD, Lundgren RGJ, Juranek DD et al. Intestinal parasitism in the United States: update on a continuing problem. *Am J Trop Med Hyg* 1994; 50: 705–13
70 Zimmermann WJ, Steele JH, Kagan IG. The changing status of trichiniasis in the U.S. population. *Public Health Reports* 1968; 83: 957–66
71 McNabb SJ, Jajosky RA, Hall-Baker PA et al. Summary of notifiable diseases--United States, 2006. *MMWR Morb Mortal Wkly Rep* 2008; 55: 1–92
72 Stoll NR. This wormy world. *J Parasitol* 1947; 33: 1–18
73 Gale EA. A missing link in the hygiene hypothesis? *Diabetologia* 2002; 45: 588–94
74 Hong ST, Chai JY, Choi MH et al. A successful experience of soil-transmitted helminth control in the Republic of Korea. Korean *J Parasitol* 2006; 44: 177–85
75 Bethony J, Brooker S, Albonico M et al. Soil-transmitted helminth infections: ascariasis, trichuriasis, and hookworm. *Lancet* 2006; 367: 1521–32
76 Doetze A, Satoguina J, Burchard G et al. Antigen-specific cellular hyporesponsiveness

in a chronic human helminth infection is mediated by T(h)3/T(r)1-type cytokines IL-10 and transforming growth factor-beta but not by a T(h)1 to T(h)2 shift. *International Immunology* 2000; 12: 623–30

77 Babu S, Blauvelt CP, Kumaraswami V et al. Regulatory networks induced by live parasites impair both Th1 and Th2 pathways in patent lymphatic filariasis: implications for parasite persistence. *J Immunol* 2006; 176: 3248–56

78 Bentwich Z, Weisman Z, Moroz C et al. Immune dysregulation in Ethiopian immigrants in Israel: relevance to helminth infections? *Clinical & Experimental Immunology* 1996; 103: 239–43

79 Sabin EA, Araujo MI, Carvalho EM et al. Impairment of tetanus toxoid-specific Th1-like immune responses in humans infected with *Schistosoma mansoni*. *Journal of Infectious Diseases* 1996; 173: 269–72

80 Nookala S, Srinivasan S, Kaliraj P et al. Impairment of tetanus-specific cellular and humoral responses following tetanus vaccination in human lymphatic filariasis. *Infect Immun* 2004; 72: 2598–604

81 Cooper PJ, Espinel I, Paredes W et al. Impaired tetanus-specific cellular and humoral responses following tetanus vaccination in human onchocerciasis: a possible role for interleukin-10. *J Infect Dis* 1998; 178: 1133–8

82 Elias D, Wolday D, Akuffo H et al. Effect of deworming on human T cell responses to mycobacterial antigens in helminth-exposed individuals before and after bacille Calmette-Guerin (BCG) vaccination. *Clin Exp Immunol* 2001; 123: 219–25

83 Elias D, Britton S, Aseffa A et al. Poor immunogenicity of BCG in helminth infected population is associated with increased *in vitro* TGF-beta production. *Vaccine* 2008; 26: 3897–902

84 Cooper PJ, Chico ME, Losonsky G et al. Albendazole treatment of children with ascariasis enhances the vibriocidal antibody response to the live attenuated oral cholera vaccine CVD 103-HgR. *J Infect Dis* 2000; 182: 1199–206

85 Borkow G, Leng Q, Weisman Z et al. Chronic immune activation associated with intestinal helminth infections results in impaired signal transduction and anergy. *J Clin Invest* 2000; 106: 1053–60

86 Turner JD, Jackson JA, Faulkner H et al. Intensity of intestinal infection with multiple worm species is related to regulatory cytokine output and immune hyporesponsiveness. *J Infect Dis* 2008; 197: 1204–12

87 Kullberg MC, Pearce EJ, Hieny SE et al. Infection with *Schistosoma mansoni* alters Th1/Th2 cytokine responses to a non-parasite antigen. *J Immunol* 1992; 148: 3264–70

88 Pearlman E, Kazura JW, Hazlett FEJ et al. Modulation of murine cytokine responses to mycobacterial antigens by helminth-induced T helper 2 cell responses. *J Immunol* 1993; 151: 4857–64

89 Sacco R, Hagen M, Sandor M et al. Established T(H1) granulomatous responses induced by active *Mycobacterium avium* infection switch to T(H2) following challenge with *Schistosoma mansoni*. *Clin Immunol* 2002; 104: 274–81

90 Loke P, MacDonald AS, Robb A et al. Alternatively activated macrophages induced by

nematode infection inhibit proliferation *via* cell-to-cell contact. *Eur J Immunol* 2000; 30: 2669–78

91 Elliott DE, Metwali A, Leung J et al. Colonization with *Heligmosomoides polygyrus* suppresses mucosal IL-17 production. *J Immunol* 2008; 181: 2414–9

92 Ince MN, Elliott DE, Setiawan T et al. *Heligmosomoides polygyrus* induces TLR4 on murine mucosal T cells that produce TGFbeta after lipopolysaccharide stimulation. Journal of Immunology 2006; 176(2): 726–9

93 Metwali A, Setiawan T, Blum AM et al. Induction of $CD8^{+}$ regulatory T cells in the intestine by *Heligmosomoides polygyrus* infection. *Am J Physiol Gastrointest Liver Physiol* 2006; 291: G253–G259

94 Finney CA, Taylor MD, Wilson MS et al. Expansion and activation of CD4(+)CD25(+) regulatory T cells in *Heligmosomoides polygyrus* infection. *Eur J Immunol* 2007; 37: 1874–86

95 Maizels RM, Yazdanbakhsh M. Immune regulation by helminth parasites: cellular and molecular mechanisms. *Nature Reviews* 2003; *Immunology* 3: 733–44

96 Thomas PG, Harn DA Jr. Immune biasing by helminth glycans. *Cellular Microbiology* 2004; 6: 13–22

97 Goodridge HS, Wilson EH, Harnett W et al. Modulation of macrophage cytokine production by ES-62, a secreted product of the filarial nematode *Acanthocheilonema viteae*. *J Immunol* 2001; 167: 940–5

98 Liu LX, Buhlmann JE, Weller PF. Release of prostaglandin E2 by microfilariae of *Wuchereria bancrofti* and *Brugia malayi*. *Am J Trop Med Hyg* 1992; 46: 520–3

99 Gomez-Escobar N, Gregory WF, Maizels RM. Identification of tgh-2, a filarial nematode homolog of *Caenorhabditis elegans* daf-7 and human transforming growth factor beta, expressed in microfilarial and adult stages of *Brugia malayi*. *Infection & Immunity* 2000; 68: 6402–10

100 Elliott DE, Summers RW, Weinstock JV. Helminths as governors of immune-mediated inflammation. *International Journal for Parasitology* 2007; 37: 457–64

101 van Riet E, Hartgers FC, Yazdanbakhsh M. Chronic helminth infections induce immunomodulation: consequences and mechanisms. *Immunobiology* 2007; 212: 475–90

102 Smith P, Mangan NE, Walsh CM et al. Infection with a helminth parasite prevents experimental colitis *via* a macrophage-mediated mechanism. *J Immunol* 2007; 178: 4557–66

103 Elliott D, Li J, Blum A et al. Exposure to schistosome eggs protects mice from TNBS-induced colitis. *Am J Physiol* 2003; 284: G385–G391

104 Hunter MM, Wang A, Hirota CL et al. Neutralizing anti-IL-10 antibody blocks the protective effect of tapeworm infection in a murine model of chemically induced colitis. *J Immunol* 2005; 174(11): 7368–75

105 Khan WI, Blennerhasset PA, Varghese AK et al. Intestinal nematode infection ameliorates experimental colitis in mice. *Infection & Immunity* 2002; 70: 5931–7

106 Setiawan T, Metwali A, Blum AM et al. *Heligmosomoides polygyrus* promotes regula-

tory T cell cytokine production in normal distal murine intestine. *Infect Immun* 2007; 75: 4655–63

107 Sutton TL, Zhao A, Madden KB et al. Anti-Inflammatory mechanisms of enteric *Heligmosomoides polygyrus* infection against trinitrobenzene sulfonic acid-induced colitis in a murine model. *Infect Immun* 2008; 76: 4772–82

108 Elliott DE, Setiawan T, Metwali A et al. *Heligmosomoides polygyrus* inhibits established colitis in IL-10-deficient mice. *Eur J Immunol* 2004; 34: 2690–8

109 Zaccone P, Fehervari Z, Jones FM et al. *Schistosoma mansoni* antigens modulate the activity of the innate immune response and prevent onset of type 1 diabetes. *Eur J Immunol* 2003; 33: 1439–49

110 Saunders KA, Raine T, Cooke A et al. Inhibition of autoimmune type 1 diabetes by gastrointestinal helminth infection. *Infect Immun* 2007; 75: 397–407

111 Kitagaki K, Businga TR, Racila D et al. Intestinal helminths protect in a murine model of asthma. *J Immunol* 2006; 177: 1628–35

112 Wilson MS, Taylor MD, Balic A et al. Suppression of allergic airway inflammation by helminth-induced regulatory T cells. *J Exp Med* 2005; 202: 1199–212

113 Mangan NE, van Rooijen N, McKenzie AN et al. Helminth-modified pulmonary immune response protects mice from allergen-induced airway hyperresponsiveness. *J Immunol* 2006; 176: 138–47

114 La Flamme AC, Ruddenklau K, Backstrom BT. Schistosomiasis decreases central nervous system inflammation and alters the progression of experimental autoimmune encephalomyelitis. *Infection & Immunity* 2003; 71: 4996–5004

115 Sewell D, Qing Z, Reinke E et al. Immunomodulation of experimental autoimmune encephalomyelitis by helminth ova immunization. *International Immunology* 2003; 15: 59–69

116 Nagayama Y, Watanabe K, Niwa M et al. *Schistosoma mansoni* and alpha-galactosylceramide: prophylactic effect of Th1 Immune suppression in a mouse model of Graves' hyperthyroidism. *J Immunol* 2004; 173: 2167–73

117 Mangan NE, Fallon RE, Smith P et al. Helminth infection protects mice from anaphylaxis *via* IL-10-producing B cells. *J Immunol* 2004; 173: 6346–56

118 Büning J, Homann N, von Smolinski D et al. Helminths as governors of inflammatory bowel disease. *Gut* 2008; 57: 1182–3

119 Correale J, Farez M. Association between parasite infection and immune responses in multiple sclerosis. *Ann Neurol* 2007; 61: 97–108

120 Rodrigues LC, Newcombe PJ, Cunha SS et al. Early infection with *Trichuris trichiura* and allergen skin test reactivity in later childhood. *Clin Exp Allergy* 2008; 38: 1769–77

121 van den Biggelaar AH, Rodrigues LC, van Ree R et al. Long-term treatment of intestinal helminths increases mite skin-test reactivity in Gabonese schoolchildren. *J Infect Dis* 2004; 189: 892–900

122 Elliott AM, Mpairwe H, Quigley MA et al. Helminth infection during pregnancy and development of infantile eczema. *JAMA* 2005; 294: 2032–4

123 Hotez PJ, Brooker S, Bethony JM et al. Hookworm infection. *N Engl J Med* 2004; 351: 799–807
124 Pritchard DI, Brown A. Is *Necator americanus* approaching a mutualistic symbiotic relationship with humans? *Trends in Parasitology* 2001; 17: 169–72
125 Croese J, O'Neil J, Masson J et al. A proof of concept study establishing *Necator americanus* in Crohn's patients and reservoir donors. *Gut* 2006; 55: 136–7
126 Mortimer K, Brown A, Feary J et al. Dose-ranging study for trials of therapeutic infection with *Necator americanus* in humans. *Am J Trop Med Hyg* 2006; 75: 914–20
127 Beer RJ. The relationship between *Trichuris trichiura* (Linnaeus 1758) of man and *Trichuris suis* (Schrank 1788) of the pig. *Research in Veterinary Science* 1976; 20: 47–54
128 Summers RW, Elliott DE, Qadir K et al. *Trichuris suis* seems to be safe and possibly effective in the treatment of inflammatory bowel disease. *Am J Gastroenterol* 2003; 98: 2034–41
129 Summers RW, Elliott DE, Urban JF Jr et al. *Trichuris suis* therapy in Crohn's disease. *Gut* 2005; 54: 87–90
130 Summers RW, Elliott DE, Urban JF Jr et al. *Trichuris suis* therapy for active ulcerative colitis: a randomized controlled trial. *Gastroenterology* 2005; 128: 825–32
131 Elliott DE, Summers RW, Weinstock JV. Helminths and the modulation of mucosal inflammation. *Current Opinion in Gastroenterology* 2005; 21(1): 51–8
132 de Silva HJ, de Silva NR, de Silva AP et al. Emergence of inflammatory bowel disease 'beyond the West': do prosperity and improved hygiene have a role? *Trans R Soc Trop Med Hyg* 2008; 102: 857–60
133 Malhotra I, Mungai P, Wamachi A et al. Helminth- and Bacillus Calmette-Guerin-induced immunity in children sensitized *in utero* to filariasis and schistosomiasis. *J Immunol* 1999; 162: 6843–8
134 Jirtle RL, Skinner MK. Environmental epigenomics and disease susceptibility. *Nat Rev Genet* 2007; 8: 253–62
135 Chen CC, Louie S, McCormick B et al. Concurrent infection with an intestinal helminth parasite impairs host resistance to enteric *Citrobacter rodentium* and enhances *Citrobacter*-induced colitis in mice. *Infect Immun* 2005; 73: 5468–81
136 Chen CC, Louie S, McCormick BA et al. Helminth-primed dendritic cells alter the host response to enteric bacterial infection. *J Immunol* 2006; 176: 472–83
137 Erb KJ, Trujillo C, Fugate M et al. Infection with the helminth *Nippostrongylus brasiliensis* does not interfere with efficient elimination of *Mycobacterium bovis* BCG from the lungs of mice. *Clin Diagn Lab Immunol* 2002; 9: 727–30
138 Karp CL, Auwaerter PG. Coinfection with HIV and tropical infectious diseases. II. Helminthic, fungal, bacterial, and viral pathogens. *Clin Infect Dis* 2007; 45: 1214–20
139 Hunter MM, Wang A, McKay DM. Helminth infection enhances disease in a murine TH2 model of colitis. *Gastroenterology* 2007; 132: 1320–30
140 Reyes JL, Terrazas LI. The divergent roles of alternatively activated macrophages in helminthic infections. *Parasite Immunol* 2007; 29: 609–19
141 Morrison AC, Correll PH. Activation of the stem cell-derived tyrosine kinase/RON

receptor tyrosine kinase by macrophage-stimulating protein results in the induction of arginase activity in murine peritoneal macrophages. *J Immunol* 2002; 168: 853–60
142 Rousseaux C, Lefebvre B, Dubuquoy L et al. Intestinal antiinflammatory effect of 5–aminosalicylic acid is dependent on peroxisome proliferator-activated receptor-gamma. *J Exp Med* 2005; 201: 1205–15
143 Katz JA. Treatment of inflammatory bowel disease with corticosteroids. *Gastroenterol Clin North Am* 2004; 33: 171–89, vii
144 Karran P, Attard N. Thiopurines in current medical practice: molecular mechanisms and contributions to therapy-related cancer. *Nat Rev Cancer* 2008; 8: 24–36

The hygiene hypothesis and Type 1 diabetes

Anne Cooke

Department of Pathology, University of Cambridge, Tennis Court Rd, Cambridge, CB2 1QP, UK

Abstract

The incidence of some autoimmune diseases is increasing dramatically in the developed world. For example, the incidence of the autoimmune disease, Type 1 diabetes (T1D), is increasing in the UK at a rate of 4% per annum; faster than can be accounted for by genetic change. In the case of T1D, as for many autoimmune diseases, the development of the disease is known to have a genetic component with many genes playing a role in governing the development of disease [1]. However, the development of Type 1 diabetes is not wholly governed by genetics and a role for environmental factors is shown by the 40% concordance rate for development of T1D in identical twins. This lack of 100% concordance in identical twins which is indicative of environmental effects acting on a genetic background is also seen for some other autoimmune diseases such as multiple sclerosis (MS) and systemic lupus erythematosus (SLE). There has been considerable interest in analysing the basis for the dramatic rise in incidence of T1D in the developed world with particular emphasis being placed on the role that infection might play in exacerbating or preventing onset of this autoimmune condition. The evidence that infection may play a role in the prevention of T1D is discussed in this chapter.

What is the evidence for infection initiating autoimmune disease?

There have been considerable advances in the development of methodologies to analyse the genetic factors contributing to onset of autoimmune diseases such as T1D, rheumatoid arthritis (RA) and MS [2–5]. The establishment of large cohorts for detailed genetic analyses coupled with improved understanding of the biology of these diseases should lead to novel insights into the complex genetics underpinning these disorders.

In terms of environmental factors that might influence onset of autoimmune disorders these could be classed as agents that might precipitate disease and those that might prevent it. It is known that some autoimmune syndromes can be induced in some individuals following treatment with drugs such as procainamide or hydralazine [6]. The effects of such drugs has been attributed to epigenetic modification of chromatin, altered clearance of apoptotic bodies or the generation of neo-self epitopes leading to a breakdown in self tolerance and induction of autoimmunity [7, 8]. In terms of infection, a large number of viruses have been associated with onset of Type 1 diabetes and some have also been proposed as causal agents in the development of MS [9–16]. Initial studies suggesting that virus infections might play a role in precipitating diabetes onset relied on anecdotal patient reporting of infection prior to diabetes onset and the observed seasonal onset of this autoimmune disorder. A key role for a virally induced autoimmune pathology seemed to be conclusively established by the isolation of a virus from the pancreas of a patient that was able to transmit diabetes to animals in experimental studies [9, 17]. Viral mediated autoimmunity could arise following direct infection of the pancreatic beta cell [18, 19] or through cross reaction/mimicry between a virally encoded antigenic epitope and a self antigen expressed in the pancreatic beta cell [20–22]. However, despite many other attempts to isolate virus from the pancreata of patients with Type 1 diabetes the presence of such a virus infection in the pancreas does not seem to be a universal observation. With a greater understanding of the aetiology of Type 1 diabetes it became evident that there was a detectable autoimmune response against islet antigens several years before diabetes onset [23] thus dissociating the possibility of infection immediately preceding diabetes onset from playing a role as an initiator of autoimmunity. Several studies have been carried out to determine whether viral infections might play any role in the initiation of Type 1 diabetes. The observation of some sequence similarities between virally encoded antigens and self antigen raised the possibility that such a viral infection might break self tolerance and initiate a response against the pancreatic beta cell. Despite the large number of studies carried out such a link has not been conclusively shown. Other possible explanations for the dramatic rise in incidence of some autoimmune diseases have therefore been examined.

What is the evidence that diet plays a role in initiating Type 1 diabetes?

One of the environmental factors which have been studied in the context of the increasing incidence of T1D has been the role of dietary influences. Dietary modifiers that have been studied include cow's milk and food containing nitrites, nitrates or nitrosamine with no conclusive proof being found for such agents playing a role. The detection of antibodies to cow's milk proteins in the serum of individuals with T1D led to the proposition that early introduction of cow's milk protein to the infant diet may play a role in development of this autoimmune disease. It has also been postulated that beta-lactoglobulin which is present in cow's milk but not human milk may initiate an immune response in babies that is cross reactive with glycodelin. This response against glycodelin has been proposed to interfere with the ability of this molecule to modulate immune responses [24]. Several studies have been carried out to test the hypothesis of a link between cows milk products and diabetes without uniform conclusions being drawn [25–29].

Could infection inhibit onset of autoimmunity?

Type 1 diabetes is a disease of juvenile onset which arises through an autoimmune mediated destruction of the insulin secreting pancreatic beta cells. In humans, overt onset of T1D is manifest when 70% of the β cell mass has been destroyed and there is insufficient insulin to maintain glucose homeostasis [1]. Thereafter daily insulin injections are required for survival. In the 1920s the discovery of insulin by Banting and Best provided the means for replacing the hormone lost when the pancreatic beta cells were destroyed [30]. This, however, does not represent a cure for this autoimmune disease and there are also a range of complications, some of which are life threatening, which can arise following onset of diabetes including retinopathy, nephropathy and neuropathy. Therefore, considerable effort is being expended to develop strategies to prevent onset of this autoimmune disease as well as to reverse it once it has been initiated. It is interesting to note, that although this disease has been around for centuries and was lethal before the 1920s, the incidence of T1D is rising faster than can be accounted for by genetic change. This suggests that some environmental change has occurred that is influencing the onset of diabetes. Possible changes that could account for such a rise in autoimmune pathology include a change in exposure to infectious agents. Given the important role that genetic background plays in the predisposition to diabetes, this suggests that potentially lethal allelic variants of certain genes have been retained in the population. This could arise either because they have historically conferred a strong selective advantage or because they are in linkage disequilibrium with advantageous alleles. It is possible that such alleles might have historically provided increased resistance to infectious agents and improved survival and reproductive capacity for those carrying beneficial alleles. It has been postulated that certain infections that swept through Europe and Asia over the centuries might have contributed to the prevalence of certain alleles in particular population groups. For example, the high carrier frequency for the cystic fibrosis mutation (2%) in populations of European descent has been associated with resistance to cholera, typhoid and also tuberculosis [31]. Several different mutations in haemoglobin have been associated with increased resistance to malaria in the Mediterranean and in Africa providing another example of such a process [32].

The hygiene hypothesis

Our immune system has co-evolved with infectious agents with the innate and the adaptive immune systems having developed to enable the mammalian host to survive in the presence of evolving infectious agents. The complexity of lymphocyte subsets, the range of cytokines they are able to produce coupled with the integrated and tailored response ensured by effective interactions between innate and adaptive immune responses comprises the armoury enabling host survival. Infectious agents have also developed strategies to evade or even utilise the host immune system. This is exemplified by certain virus infections where the expression of virally encoded products to neutralise chemokines or expression of immunomodulatory cytokines enables the virus to evade the host immune system and establish latency. Such strategies not only facilitate the viral life cycle but also serve to limit host pathology. Induction of host derived immunomodulatory cytokines and harnessing of the host immune response is also seen in parasitic infections. For example, in the case of *Schistosoma mansoni* its complex life cycle in the mammalian definitive host is dependent on a host functioning immune system. Until relatively recently in evolutionary time humans would have been exposed to a range of infectious agents from an early age. These would have included not only bacteria and viruses but also parasitic worms. For example, hookworm was endemic in the southern states of America in the 1920s and most young children had been exposed to pinworm in the UK [33]. More recently, through vaccination strategies and increased public health measures such as improved sanitation and lower density living conditions in the developed world, the relationship between humans and infectious agents has dramatically altered. This has led to the proposal, sometimes called the 'hygiene hypothesis' that it is this lack of exposure to infections that is resulting in the dramatic rise of asthma, allergies and autoimmunity in the developed world [34–36]. This 'hygiene hypothesis' does not preclude the possibility that some cases of allergy or autoimmunity might be induced by exposure to agents such as certain infectious organisms or chemicals.

Some infections inhibit the development of autoimmune pathology

Animal models of allergy or autoimmunity have been used to examine whether infection might be able to inhibit or modulate the onset of pathology. These studies have been carried out using both experimentally induced models of disease as well as spontaneous models. The non obese diabetic (NOD) mouse spontaneously develops Type 1 diabetes and provides an excellent model of the human autoimmune disease [37]. In this murine model a mononuclear cell infiltration develops around the pancreatic islets from around 5 weeks of age. This infiltration comprising of $CD4^+$T cells, $CD8^+$ T cells, B cells and macrophages progressively moves into the islet and mediates selective beta cell destruction with diabetes developing predominantly in female mice from around 12 weeks of age. The NOD mouse has been used extensively to study the genetic basis of diabetes as well as its pathogenesis. Early studies in this model showed that exposure of young NOD mice to mycobacterial antigens inhibited the development of diabetes [38–42]. These studies were extended to other infectious agents and it was shown that infection with *Salmonella typhimurium* or with *S. mansoni* could also inhibit diabetes onset [35, 43, 44]. It has been shown furthermore that in many cases a live infection is not always required for diabetes prevention and equivalent results can be obtained using a bacterial or parasite derived product.

Analysis of the mechanisms by which such infectious agents or their products might inhibit autoimmune responses has revealed that there are multiple ways in which this might be mediated. Studies of diabetes prevention achieved by *Salmonella* infection have revealed that this organism induces changes in dendritic cells such that they are able to modulate trafficking of autoreactive T cells into the pancreas [45]. Several different mechanisms have been identified by which *Schistosome*-derived products modulate the immune response. Soluble extracts of schistosome eggs have been shown to induce IL-10 production from dendritic cells and additionally to induce the development of regulatory T cells [43, 46]. Soluble extracts of the Schistosome worm are able to induce iNKT cells [43] which have been shown to be capable of inhibiting diabetes onset in NOD mice [47–49]. Soluble worm extracts are also able to polarise the response against islet antigen from a diabetogenic Th1 response to a non pathogenic Th2 response [43, 50].

Schistosome products have also been shown to inhibit the development of an animal model of MS, experimental allergic encephalomyelitis (EAE) [51, 52]. As EAE is known to be controlled by both T regs and iNKT cells it may be that helminth mediated EAE prevention arises through the influence of the helminth on these cell types. Th17 cells have recently been shown to play pivotal roles in EAE development and it is possible that infections may influence the generation and development of this pathogenic cell type.

There are now many examples where either infectious agents themselves or their products have been shown to modulate inflammation [53–63]. There are additionally some data suggesting that the effects seen in animal models of human inflammatory disorders can be observed in humans. In a study of MS patients it was observed that those carrying parasitic infections had lower indices of disease progression as measured by disability score and MRI changes than non infected individuals. This improved outcome was associated with a switch to greater production of anti-inflammatory cytokines such as TGF-β and IL-10 as well as an increased proportional representation of regulatory T cells [64]. Furthermore, on the basis of studies in animal models [56, 65] patients with inflammatory bowel disease have been infected with the pig whipworm, *Trichuris suis*, and shown evidence of improvement in their clinical condition [66–68].

Concluding comments

The human immune system has been shaped by the infectious agents with which it has co-evolved. It is known from the study of prehistoric mummies that some parasites and mycobacteria found a niche in human at least 10,000–20,000 years ago. Historically, individuals would have been commonly infected or exposed to several agents that give rise to chronic infections, and they would have been selected on the basis of their ability to survive these infections. Key to survival would be a balanced, non-self-damaging, but effective response to infectious agents and immune regulation would play a part in this. Some infectious agents are able to elicit anti-inflammatory responses that minimise inflammation and it is this capacity that may hold in check allergic or autoreactive responses. By understanding the various ways, and by identifying the biomodulators used by different infectious agents to elicit such immune regulation there is the potential to develop novel therapeutic approaches to inflammation based pathologies in humans.

References

1 Redondo MJ, Fain PR, Eisenbarth GS (2001) Genetics of type 1A diabetes. *Recent Prog Horm Res* 56: 69–89

2 Hermanowski J, Bouzigon E, Forabosco P, Ng MY, Fisher SA, Lewis CM (2007) Meta-analysis of genome-wide linkage studies for multiple sclerosis, using an extended GSMA method. *Eur J Hum Genet* 15: 703–710

3 Hakonarson H, Grant SF, Bradfield JP, Marchand L, Kim CE, Glessner JT, Grabs R, Casalunovo T, Taback SP, Frackelton EC et al (2007) A genome-wide association study identifies KIAA0350 as a type 1 diabetes gene. *Nature* 448: 591–594

4 Hafler DA, Compston A, Sawcer S, Lander ES, Daly MJ, De Jager PL, de Bakker PI, Gabriel SB, Mirel DB, Ivinson AJ et al (2007) Risk alleles for multiple sclerosis identified by a genomewide study. *N Engl J Med* 357: 851–862

5 (2007) Genome-wide association study of 14,000 cases of seven common diseases and 3,000 shared controls. *Nature* 447: 661–678

6 Yung R, Chang S, Hemati N, Johnson K, Richardson B (1997) Mechanisms of drug-induced lupus. IV. Comparison of procainamide and hydralazine with analogs *in vitro* and *in vivo*. *Arthritis Rheum* 40: 1436–1443

7 Cooke A, Lydyard PM (1981) The role of T cells in autoimmune diseases. *Pathol Res Pract* 171: 173–196

8 Strickland FM, Richardson BC (2008) Epigenetics in human autoimmunity. Epigenetics in autoimmunity – DNA methylation in systemic lupus erythematosus and beyond. *Autoimmunity* 41: 278–286

9 Yoon JW, Austin M, Onodera T, Notkins AL (1979) Isolation of a virus from the pancreas of a child with diabetic ketoacidosis. *N Engl J Med* 300: 1173–1179

10 Sarchielli P, Trequattrini A, Usai F, Murasecco D, Gallai V (1993) Role of viruses in the etiopathogenesis of multiple sclerosis. *Acta Neurol (Napoli)* 15: 363–381

11 Clark D (2004) Human herpesvirus type 6 and multiple sclerosis. *Herpes* 11 (Suppl 2): 112A-119A

12 Green J, Casabonne D, Newton R (2004) Coxsackie B virus serology and Type 1 diabetes mellitus: a systematic review of published case-control studies. *Diabet Med* 21: 507–514
13 Christensen T (2006) The role of EBV in MS pathogenesis. *Int MS J* 13: 52–57
14 Drescher KM, Tracy SM (2008) The CVB and etiology of type 1 diabetes. *Curr Top Microbiol Immunol* 323: 259–274
15 Filippi CM, von Herrath MG (2008) Viral trigger for type 1 diabetes: pros and cons. *Diabetes* 57: 2863–2871
16 Lincoln JA, Hankiewicz K, Cook SD (2008) Could Epstein-Barr virus or canine distemper virus cause multiple sclerosis? *Neurol Clin* 26: 699–715, viii
17 Toniolo A, Onodera T, Jordan G, Yoon JW, Notkins AL (1982) Virus-induced diabetes mellitus. Glucose abnormalities produced in mice by the six members of the Coxsackie B virus group. *Diabetes* 31: 496–499
18 Coulson BS, Witterick PD, Tan Y, Hewish MJ, Mountford JN, Harrison LC, Honeyman MC (2002) Growth of rotaviruses in primary pancreatic cells. *J Virol* 76: 9537–9544
19 Onodera T, Jenson AB, Yoon JW, Notkins AL (1978) Virus-induced diabetes mellitus: reovirus infection of pancreatic beta cells in mice. *Science* 201: 529–531
20 Serreze DV, Leiter EH, Kuff EL, Jardieu P, Ishizaka K (1988) Molecular mimicry between insulin and retroviral antigen p73. Development of cross-reactive autoantibodies in sera of NOD and C57BL/KsJ db/db mice. *Diabetes* 37: 351–358
21 Honeyman MC, Stone NL, Harrison LC (1998) T-cell epitopes in type 1 diabetes autoantigen tyrosine phosphatase IA-2: potential for mimicry with rotavirus and other environmental agents. *Mol Med* 4: 231–239
22 Vreugdenhil GR, Geluk A, Ottenhoff TH, Melchers WJ, Roep BO, Galama JM (1998) Molecular mimicry in diabetes mellitus: the homologous domain in coxsackie B virus protein 2C and islet autoantigen GAD65 is highly conserved in the coxsackie B-like enteroviruses and binds to the diabetes associated HLA-DR3 molecule. *Diabetologia* 41: 40–46
23 Knip M, Siljander H (2008) Autoimmune mechanisms in type 1 diabetes. *Autoimmun Rev* 7: 550–557
24 Goldfarb MF (2008) Relation of time of introduction of cow milk protein to an infant and risk of type-1 diabetes mellitus. *J Proteome Res* 7: 2165–2167
25 Fort P, Lanes R, Dahlem S, Recker B, Weyman-Daum M, Pugliese M, Lifshitz F (1986) Breast feeding and insulin-dependent diabetes mellitus in children. *J Am Coll Nutr* 5: 439–441
26 Martin JM, Trink B, Daneman D, Dosch HM, Robinson B (1991) Milk proteins in the etiology of insulin-dependent diabetes mellitus (IDDM). *Ann Med* 23: 447–452
27 Rosenbauer J, Herzig P, Giani G (2008) Early infant feeding and risk of type 1 diabetes mellitus-a nationwide population-based case-control study in pre-school children. *Diabetes Metab Res Rev* 24: 211–222
28 Schrezenmeir J, Jagla A (2000) Milk and diabetes. *J Am Coll Nutr* 19: 176S-190S

29 Wasmuth HE, Kolb H (2000) Cow's milk and immune-mediated diabetes. *Proc Nutr Soc* 59: 573–579

30 Banting FG, Best CH (1922) The internal secretion of the pancreas. *Laboratory and Clinical Medicine* 7: 465–480

31 Poolman EM, Galvani AP (2007) Evaluating candidate agents of selective pressure for cystic fibrosis. *J R Soc Interface* 4: 91–98

32 Weatherall DJ, Clegg JB (2002) Genetic variability in response to infection: malaria and after. *Genes Immun* 3: 331–337

33 Gale EA (2002) The rise of childhood type 1 diabetes in the 20th century. *Diabetes* 51: 3353–3361

34 Strachan DP (1989) Hay fever, hygiene, and household size. *BMJ* 299: 1259–1260

35 Cooke A, Tonks P, Jones FM, O'Shea H, Hutchings P, Fulford AJ, Dunne DW (1999) Infection with *Schistosoma mansoni* prevents insulin dependent diabetes mellitus in non-obese diabetic mice. *Parasite Immunol* 21: 169–176

36 Raine T, Zaccone P, Dunne DW, Cooke A (2004) Can helminth antigens be exploited therapeutically to downregulate pathological Th1 responses? *Curr Opin Investig Drugs* 5: 1184–1191

37 Kikutani H, Makino S (1992) The murine autoimmune diabetes model: NOD and related strains. *Adv Immunol* 51: 285–322

38 Harada M, Kishimoto Y, Makino S (1990) Prevention of overt diabetes and insulitis in NOD mice by a single BCG vaccination. *Diabetes Res Clin Pract* 8: 85–89

39 Castro AP, Esaguy N, Aguas AP (1993) Effect of mycobacterial infection in the lupus-prone MRL/lpr mice: enhancement of life span of autoimmune mice, amelioration of kidney disease and transient decrease in host resistance. *Autoimmunity* 16: 159–166

40 Qin HY, Sadelain MW, Hitchon C, Lauzon J, Singh B (1993) Complete Freund's adjuvant-induced T cells prevent the development and adoptive transfer of diabetes in nonobese diabetic mice. *J Immunol* 150: 2072–2080

41 Baxter AG, Horsfall AC, Healey D, Ozegbe P, Day S, Williams DG, Cooke A (1994) Mycobacteria precipitate an SLE-like syndrome in diabetes-prone NOD mice. *Immunology* 83: 227–231

42 Bras A, Aguas AP (1996) Diabetes-prone NOD mice are resistant to *Mycobacterium avium* and the infection prevents autoimmune disease. *Immunology* 89: 20–25

43 Zaccone P, Fehervari Z, Jones FM, Sidobre S, Kronenberg M, Dunne DW, Cooke A (2003) *Schistosoma mansoni* antigens modulate the activity of the innate immune response and prevent onset of type 1 diabetes. *Eur J Immunol* 33: 1439–1449

44 Zaccone P, Raine T, Sidobre S, Kronenberg M, Mastroeni P, Cooke A (2004) *Salmonella typhimurium* infection halts development of type 1 diabetes in NOD mice. *Eur J Immunol* 34: 3246–3256

45 Raine T, Zaccone P, Mastroeni P, Cooke A (2006) *Salmonella typhimurium* infection in nonobese diabetic mice generates immunomodulatory dendritic cells able to prevent type 1 diabetes. *J Immunol* 177: 2224–2233

46 Kane CM, Cervi L, Sun J, McKee AS, Masek KS, Shapira S, Hunter CA, Pearce EJ

(2004) Helminth antigens modulate TLR-initiated dendritic cell activation. *J Immunol* 173: 7454–7461

47 Beaudoin L, Laloux V, Novak J, Lucas B, Lehuen A (2002) NKT cells inhibit the onset of diabetes by impairing the development of pathogenic T cells specific for pancreatic beta cells. *Immunity* 17: 725–736

48 Naumov YN, Bahjat KS, Gausling R, Abraham R, Exley MA, Koezuka Y, Balk SB, Strominger JL, Clare-Salzer M, Wilson SB (2001) Activation of CD1d-restricted T cells protects NOD mice from developing diabetes by regulating dendritic cell subsets. *Proc Natl Acad Sci USA* 98: 13838–13843

49 Sharif S, Arreaza GA, Zucker P, Mi QS, Sondhi J, Naidenko OV, Kronenberg M, Koezuka Y, Delovitch TL, Gombert JM et al (2001) Activation of natural killer T cells by alpha-galactosylceramide treatment prevents the onset and recurrence of autoimmune Type 1 diabetes. *Nat Med* 7: 1057–1062

50 Dunne DW, Cooke A (2005) A worm's eye view of the immune system: consequences for evolution of human autoimmune disease. *Nat Rev Immunol* 5: 420–426

51 La Flamme AC, Ruddenklau K, Backstrom BT (2003) Schistosomiasis decreases central nervous system inflammation and alters the progression of experimental autoimmune encephalomyelitis. *Infect Immun* 71: 4996–5004

52 Sewell D, Qing Z, Reinke E, Elliot D, Weinstock J, Sandor M, Fabry Z (2003) Immunomodulation of experimental autoimmune encephalomyelitis by helminth ova immunization. *Int Immunol* 15: 59–69

53 Mattsson L, Larsson P, Erlandsson-Harris H, Klareskog L, Harris RA (2000) Parasite-mediated down-regulation of collagen-induced arthritis (CIA) in DA rats. *Clin Exp Immunol* 122: 477–483

54 Costalonga M, Hodges JS, Herzberg MC (2002) Streptococcus sanguis modulates type II collagen-induced arthritis in DBA/1J mice. *J Immunol* 169: 2189–2195

55 Khan WI, Blennerhasset PA, Varghese AK, Chowdhury SK, Omsted P, Deng Y, Collins SM (2002) Intestinal nematode infection ameliorates experimental colitis in mice. *Infect Immun* 70: 5931–5937

56 Elliott DE, Li J, Blum A, Metwali A, Qadir K, Urban JF, Jr., Weinstock JV (2003) Exposure to schistosome eggs protects mice from TNBS-induced colitis. *Am J Physiol Gastrointest Liver Physiol* 284: G385–391

57 Maizels RM, Yazdanbakhsh M (2003) Immune regulation by helminth parasites: cellular and molecular mechanisms. *Nat Rev Immunol* 3: 733–744

58 McInnes IB, Leung BP, Harnett M, Gracie JA, Liew FY, Harnett W (2003) A novel therapeutic approach targeting articular inflammation using the filarial nematode-derived phosphorylcholine-containing glycoprotein ES-62. *J Immunol* 171: 2127–2133

59 Mangan NE, Fallon RE, Smith P, van Rooijen N, McKenzie AN, Fallon PG (2004) Helminth infection protects mice from anaphylaxis *via* IL-10-producing B cells. *J Immunol* 173: 6346–6356

60 Wohlleben G, Trujillo C, Muller J, Ritze Y, Grunewald S, Tatsch U, Erb KJ (2004) Hel-

minth infection modulates the development of allergen-induced airway inflammation. *Int Immunol* 16: 585–596
61 Harnett W, Harnett MM (2006) Filarial nematode secreted product ES-62 is an anti-inflammatory agent: therapeutic potential of small molecule derivatives and ES-62 peptide mimetics. *Clin Exp Pharmacol Physiol* 33: 511–518
62 Saunders KA, Raine T, Cooke A, Lawrence CE (2007) Inhibition of autoimmune type 1 diabetes by gastrointestinal helminth infection. *Infect Immun* 75: 397–407
63 Smith P, Mangan NE, Walsh CM, Fallon RE, McKenzie AN, van Rooijen N, Fallon PG (2007) Infection with a helminth parasite prevents experimental colitis *via* a macrophage-mediated mechanism. *J Immunol* 178: 4557–4566
64 Correale J, Farez M (2007) Association between parasite infection and immune responses in multiple sclerosis. *Ann Neurol* 61: 97–108
65 Elliott DE, Setiawan T, Metwali A, Blum A, Urban JF, Jr., Weinstock JV (2004) *Heligmosomoides polygyrus* inhibits established colitis in IL-10–deficient mice. *Eur J Immunol* 34: 2690–2698
66 Summers RW, Elliott DE, Qadir K, Urban JF, Jr., Thompson R, Weinstock JV (2003) *Trichuris suis* seems to be safe and possibly effective in the treatment of inflammatory bowel disease. *Am J Gastroenterol* 98: 2034–2041
67 Summers RW, Elliott DE, Urban JF, Jr., Thompson R, Weinstock JV (2005) *Trichuris suis* therapy in Crohn's disease. *Gut* 54: 87–90
68 Summers RW, Elliott DE, Urban JF, Jr., Thompson RA, Weinstock JV (2005) *Trichuris suis* therapy for active ulcerative colitis: a randomized controlled trial. *Gastroenterology* 128: 825–832

The hygiene hypothesis and affective and anxiety disorders

Graham A.W. Rook[1] *and Christopher A. Lowry*[2]

[1]Centre for Infectious Diseases and International Health, Windeyer Institute of Medical Sciences, Royal Free and University College Medical School, London, UK
[2]Department of Integrative Physiology, University of Colorado, Boulder, CO, USA

Abstract

Chronic inflammatory disorders are increasing in prevalence in the developed countries. The hygiene hypothesis proposes that our changing microbial environment has resulted in a deficit in immunoregulatory circuits so that there is a failure to terminate inappropriate inflammation. Several stress-related psychiatric disorders, particularly depression and anxiety disorders, are associated with raised levels of proinflammatory cytokines and of other markers of ongoing inflammation, even in the absence of any obvious inflammatory lesion. Moreover proinflammatory cytokines are known to induce depression, which is frequently seen when patients are treated with interleukin-2 (IL-2) or interferon-α (IFN-α). Therefore the occurrence of these psychiatric disorders in developed countries might be partly attributable to a failure of immunoregulation. We review the evidence that inflammation is associated with several patterns of psychiatric disturbance, and that regulatory cytokines such as IL-10 and transforming growth factor-β (TGF-β) can oppose these effects, and that anti-depressants might work in part via effects on inflammation in the periphery.

Introduction

The chapter is organised so that we first provide a brief description of the hygiene hypothesis, followed by 1) evidence supporting the hypothesis that proinflammatory cytokines are elevated in depressed patients, 2) evidence that proinflammatory cytokines can induce a depression-like syndrome in animals and humans, 3) the association of inflammatory gut diseases and vascular disease with stress-related psychiatric disorders, 4) the potential for treating depression using anti-inflammatory treatments or immunoregulatory compounds, and 5) future directions.

Hygiene hypothesis

The incidences of several chronic inflammatory disorders have been increasing strikingly in the developed countries. These include allergic disorders (asthma, hay fever), some autoimmune disease (for example Type 1 diabetes and multiple scle-

The Hygiene Hypothesis and Darwinian Medicine, edited by Graham A.W. Rook

rosis) [1] and inflammatory bowel diseases (IBD; ulcerative colitis and Crohn's disease) [2]. The 'hygiene', or 'old friends' hypothesis attributes some of these increases to a failure of immunoregulation [3]. Defective immunoregulation has been shown to play an important role in the pathogenesis of these disorders [4–8], and we know that a failure of immunoregulatory mechanisms can lead to simultaneous increases in all these diverse types of pathology, because genetic defects of Foxp3, a transcription factor that plays a crucial role in the development and function of Treg, leads to a syndrome known as X-linked autoimmunity–allergic dysregulation syndrome (XLAAD) that includes aspects of all of them [9].

The 'old friends' hypothesis suggests that the lack of appropriate levels of immunoregulatory pathways in rich northern countries is a consequence of diminished exposure to three categories of organism. Firstly, commensals in the gut; the nature and balance of these are changing in the developed world. Secondly, harmless organisms associated with mud, untreated water and fermenting vegetable matter, and thirdly helminth infections that are still common in developing countries but almost completely absent from rich ones [3]. (This is discussed in detail in the introductory chapter of this volume.) The former, namely commensals in the gut and harmless organisms normally present in the environment, need to be tolerated because they are harmless and were constantly present in the gut, food and water throughout mammalian evolution. The helminthic parasites need to be tolerated because although not always harmless, once they are established in the host any effort by the immune system to eliminate them is futile, and merely causes tissue damage [10].

A major mechanism by which these organisms prime immunoregulation and mediate protection from allergies, autoimmunity and IBD is explained in detail in the chapters by Graham Rook, and by Rick Maizels and Ursula Wiedermann in this volume. Briefly, it is postulated that rather than provoking aggressive immune responses, these organisms cause a pattern of maturation of dendritic cells (DC) such that these drive Treg rather than T helper cell 1 (Th1) or Th2 effector cells [11–13]. This in turn leads to two mechanisms that help to control inappropriate inflammation. First, continuing throughput of the 'old friends' causes continuous background activation of the DCreg and of Treg specific for the old friends themselves. The result is a constant background of bystander suppression of inflammatory responses, including background release of anti-inflammatory IL-10 and TGF-β. Secondly, these DCreg inevitably sample self, gut contents and allergens, and so induce Treg specific for the illicit target antigens of the three groups of chronic inflammatory disorder. These inhibitory mechanisms are aborted when there are legitimate 'danger' signals [14].

The validity of this hypothetical model is supported by clinical trials and experimental models in which exposure to microorganisms that were ubiquitous during mammalian evolutionary history, but are currently 'missing' from the environment in rich countries will treat allergy [15–17], autoimmunity [18] or intestinal inflammation [19–21].

We do not understand all of the ways in which DCreg and Treg block or terminate inflammatory responses. However we do know that the release of the anti-inflammatory cytokines, IL-10 and TGF-β is involved [15, 16, 22]. In the next sections we summarise the evidence that chronic inflammation is able to trigger or exacerbate certain psychiatric disorders, and that these anti-inflammatory mediators can oppose these effects.

Cytokines and inflammation in psychiatric disorders

Both Th2- and Th1-mediated inflammation have been associated with stress-related psychiatric syndromes, in particular anxiety and depression. Most studies have focussed on proinflammatory cytokines (IL-1, IL-6, tumour necrosis factor (TNF), IFN-α) or those released by cells associated with Th1 responses (IFN-γ, IL-2, IL-12).

Th1 and proinflammatory cytokines

The data showing that depression is associated with increased circulating levels of proinflammatory cytokines come from a range of direct and indirect sources, many of which are reviewed elsewhere [23–25], so we will give only a few brief examples.

First, patients presenting with major depression but no other obvious disorder often have raised levels of these cytokines [26]. Similarly in a community-based study of 3,024 well-functioning older persons 70–79 years of age, depression was found to correlate with higher circulating levels of IL-6 and TNF [27].

Secondly the occurrence of depression in chronic illnesses is related to the serum levels of IL-1 and IL-6 rather than to the severity of the symptoms. For instance cancer patients may have raised cytokine levels, especially if undergoing treatment with radiation or chemotherapy, which are associated with tissue damage. Plasma concentrations of IL-6 were found to be significantly increased in those with depression compared to those in healthy control subjects or in cancer patients without depression [28, 29]. Similarly, depression in heart failure patients correlated with serum TNF levels, and was independent of physical symptoms [30].

Other studies, rather than measuring cytokines in serum or plasma, have measured cytokines released from the peripheral blood cells *in vitro* following stimulation with specific antigens or non-specific mitogens such as endotoxin (lipopolysaccharide, LPS) or phytohaemagglutinin. Again, increased release of proinflammatory cytokines was related to psychiatric symptoms [31–33].

Since the proinflammatory cytokines, particularly IL-6, drive the acute phase response, there has also been an effort to document changes in negative and positive acute phase reactants in depressed individuals. Increased plasma concentrations of haptoglobin, C-reactive protein (CRP), α-1-antitrypsin, and ceruloplasmin, and lower zinc and albumin, have been reported in depression [27, 34, 35]. Others have reported increased neutrophil counts and increased levels of complement components C3 and C4 [36]. Changes in acute phase reactants were seen in bipolar disorder as well as in major depression [37, 38], but not in schizophrenics [38]. Together, these findings support the hypothesis that proinflammatory cytokines and associated acute phase responses are biomarkers of depression in at least a subset of patients.

Th2-mediated inflammation

There is little direct information on Th2 cytokine levels in psychiatric disorders. The psycho-neuro-immunology literature frequently refers to IL-10 as a Th2 cytokine, but this mediator should be considered regulatory, and is produced by many cell types, including some Th1-like IFN-γ-secreting cells, DC, macrophages and Treg cells, particularly Th3 cells. The 'signature' Th2 cytokines, IL-4, IL-5, and IL-13, are those that drive Th2-mediated inflammatory disorders, such as the allergic conditions. A few studies have tried to assay IL-4 in psychiatric disorders. Unfortunately measurement of IL-4 by ELISA is unreliable, because its bioactive levels are below the threshold for immunoassays, and because there is a splice variant, IL-4δ2, which is thought to be an antagonist [39, 40]. Commercially available antibodies are not known to distinguish between the two forms. Nevertheless, two recent studies suggest that the ratio of IFN-γ to IL-4 in plasma [41] or in culture supernatants of mitogen-stimulated blood cells [42] is higher in depressed or bipolar patients, than in controls. This argues against a role for Th2 cytokines in depression. However the inflammation in human allergic disorders is mixed, with components of Th1, and relatively high levels of proinflammatory cytokines such as TNF [43]. This is particularly true of obese asthmatics. Adipose tissue is a potent source of proinflammatory cytokines (IL-1, IL-6, TNF) and chemokines (IL-8, macrophage inflammatory protein-1 alpha (MIP-1a), eotaxin, chemokine (C-C motif) ligand 2 (CCL2; also known as monocyte chemotactic protein-1, MCP-1)), and of leptin, which is also proinflammatory [44]. Obesity increases the risk of asthma, and alters asthma toward a more difficult-to-control phenotype [45], and obese people are 2–3 times more likely to be depressed [46].

Thus, although the evidence for elevation of Th2 cytokines in depression is not convincing, allergic disorders involve significant Th1 and proinflammatory components. It is argued that nearly 50% of asthma patients may suffer from clinically significant depressive symptoms [47]. Results from a large multi-site epidemiological study of childhood psychiatric disorders in the U.S.A. indicated that childhood asthma was specifically associated with anxiety disorder [48], and those results were supported by studies in France and Puerto Rico [49, 50]. Interestingly there is a consensus that having a history of asthma is more often associated with anxiety disorder than with affective disorder [51]. This is in sharp contrast to the known association between several other chronic illnesses and depression [30], and poses an interesting dilemma. Is the anxiety a direct consequence of the psychological effects of the asthma, or is the anxiety exacerbated by the Th2 cytokines or lack of regulatory cytokines that are a feature of allergic disorders, or, finally, are both asthma and anxiety the result of a third, possibly genetic or environmental factor? Evidence for a genetic component has emerged from twin studies [52, 53].

Cytokines and depression in animal models

So in many depressed individuals there is evidence for increased levels of proinflammatory cytokines. Why does this matter? Evidence from human studies indicates that proinflammatory cytokines can induce depression, while evidence from animal studies indicates that proinflammatory cytokines can induce a depression-like syndrome. During an infection, or following injection of proinflammatory cytokines such as TNF or IL-1, or of substances such as LPS that cause systemic release of these cytokines, there is a coordinated set of adaptive behavioural changes caused by direct or indirect effects of proinflammatory mediators in the brain [54]. These behavioural changes, collectively known as 'sickness behaviour', overlap with the symptoms of depression, and many authors have suggested that depression is an aberrant form of chronic or dysregulated sickness behaviour [54]. More recently this has been explored with an elegant transgenic and gene knockout system. Mice subjected to chronic mild stress (CMS) for 5 weeks developed depression-like symptoms, reduced hippocampal neurogenesis, thought to be a biomarker of depression [55], and increased IL-1β levels in the hippocampus. In contrast, IL-1 receptor knockout mice, or mice with transgenic, brain-restricted overexpression of IL-1 receptor *antagonist* (IL-1Ra) did not show these changes. The effects of CMS could be mimicked by *subcutaneous* administration of IL-1β *via* osmotic minipumps for 4 weeks. Therefore elevation in brain IL-1 levels is necessary for producing depression-like symptoms induced by CMS, and moreover, the effects can be mimicked by *peripheral* administration of IL-1 [56]. Interestingly, transplantation of transgenic neural precursor cells overexpressing IL-Ra blocked the effects on mice of being isolated for 4 weeks (high hippocampal IL-1β levels, weight loss, impairment in hippocampal-dependent memory, and decreased hippocampal neurogenesis), further reinforcing the importance of IL-1 in a wide range of depression-associated phenomena [57].

Induction of depression by clinically administered cytokines

Direct proof that cytokines can modulate the emotional state in man too is provided by the occurrence of anxiety and depressive symptoms in patients treated with IL-2 or IFN-α (cytokines that are used to treat some forms of hepatitis and cancer [25]), or IFN-β, which is used to treat multiple sclerosis [58]. It is important to note that their effects are very complex. More than one syndrome induced by these cytokines has been recognised. Anorexia, fatigue, altered sleep, and pain developed within 2 weeks of IFN-α therapy in a large proportion of patients. Later during the treatment with IFN-α there was depressed mood, anxiety and cognitive dysfunction in the patients who developed DSM-IV criteria for major depression. There were also differences in the timing and nature of symptoms that depended on whether IL-2 or IFN-α or both were used [59]. This might imply that the different cytokines themselves exert different effects on the CNS, or that the additional cytokines released in response to their administration are doing so. Patients treated with IL-2, or a combination of IL-2 and IFN-α, show secondary rises in other cytokines such as IL-6 and the anti-inflammatory cytokines IL-10 and IL-1Ra [60].

What matters here is that different patterns of chronic cytokine release, due to failure of termination of Th1- or Th2-dominated inflammation, might give rise to different psychiatric syndromes.

It is significant that cytokine-induced depression can be treated by paroxetine, a serotonin re-uptake inhibitor antidepressant, suggesting that, by this criterion at least, it resembles spontaneously occurring depression [61]. In fact the same is probably true in animals with respect to depressive-like behavioural responses. In several animal models in which behavioural changes are evoked by LPS or cytokine administration, the symptoms can be attenuated with antidepressant drugs [62–64].

One important observation is that patients who had raised background levels of inflammatory cytokines *before the cytokine treatment was initiated*, were more likely to develop depression during treatment with IFN-α or IL-2 [65]. This implies that unexplained background inflammation was a predisposing factor, rather than a consequence of depression.

Several mechanisms involved in this induction of depression by cytokines are discussed in the following paragraphs.

So peripherally administered cytokines can trigger depression. There are many routes *via* which this CNS effect might occur, including passage of cytokines through regions where there is no blood–brain barrier (circumventricular organs), active transport, and transmission of cytokine signals *via* afferent fibers in the vagus nerve [66], or spinal cord [67].

Induction of indoleamine 2,3-dioxygenase (IDO) activity during inflammatory responses

During cytokine therapy (IFN-α, IFN-β, IL-2) there is increased activation of the enzyme indoleamine 2,3-dioxygenase (IDO), resulting in depletion of tryptophan [58, 68]. Therefore, since tryptophan levels are a limiting factor for synthesis of serotonin in the brain, inflammation can lead to serotonin deficiency. Such deficiency is known to be able to provoke depression in susceptible individuals (reviewed in [58]). Meanwhile the inflammatory response will increase expression of other inflammatory mediators such as IL-6, TNF and prostaglandin E_2 (PGE_2), which act synergistically with IFN-γ to induce IDO activity [58].

Pregnancy and post-partum depression

A similar mechanism may contribute to post-partum depression. During pregnancy there is a bias towards Th2 and regulatory mechanisms, often leading to remission of chronic inflammatory disorders [69]. Post-partum there is a Th1-rebound that can be accompanied by clinically obvious exacerbation of inflammatory disorders and depression [69]. This is associated with increased metabolism of tryptophan and with raised circulating IL-8 [70].

Glutamatergic pathways (N-methyl-D-aspartate (NMDA-) receptor-mediated neurotransmission)

Moreover there is a second relevant consequence of inflammation-mediated changes in tryptophan metabolism. Both IDO (expressed predominantly in macrophages, DCs, astrocytes, and microglia) and TDO (predominantly expressed in the liver) metabolise tryptophan to kynurenine [71]. This compound is then further metabolised to kynurenic acid (KYNA), which is an antagonist for N-methyl-D-aspartate (NMDA-) receptors, or to quinolinic acid and 3-hydroxykynurenine, both of which are agonists. In major depression there are reduced levels of KYNA relative to levels of quinolinic acid, suggesting increased signalling *via* NMDA receptors [58]. Both activated macrophages and microglial cells are able to produce quinolinic acid [72]. Moreover, significant microgliosis (an increase in density of microglia per unit area) was observed in the dorsolateral prefrontal cortex, anterior cingulate cortex, mediodorsal thalamus and hippocampus of suicide patients, whether suffering from schizophrenia or depression [73]. The glutamatergic system is known to influence serotonergic and noradrenergic neurotransmission, and there is some evidence for excessive glutamatergic activity in depression, including evidence that NMDA receptor antagonists have antidepressant properties (reviewed in [58, 74]).

Activation of serotonergic systems by proinflammatory cytokines

In addition to altering tryptophan metabolism and availability as described above, it is also clear that acute elevation of proinflammatory cytokines activates serotonergic systems in the brain. Thus, administration of LPS or IL-1β increases serotonin release in limbic forebrain structures including the hippocampus and prefrontal cortex [75, 76], structures implicated in the pathophysiology of depression [77]. We have recently demonstrated that this increase in limbic serotonergic activity is associated with activation of a small subpopulation of serotonergic neurons in a structure called the interfascicular part of the dorsal raphe nucleus (DRI). DRI serotonergic neurons project specifically to a distributed system implicated in the regulation of mood and in the pathophysiology of depression, including the medial prefrontal cortex (mPFC), frontal, cingulate, and entorhinal cortices, hippocampus and midline thalamus [78, 79]. It is reasonable that chronic elevation of proinflammatory cytokines, as described in depressed patients, would lead to dysregulation of this serotonergic system.

Significance of small changes in cytokine levels

It is possible then that chronic exposure to either the Th1-induced proinflammatory cytokine pattern, or to the Th2 or mixed proinflammatory pattern seen in allergic disorders, is able to influence psychiatric symptoms. But the changes in cytokine levels seen are usually very modest typically representing a doubling of control background levels [80]. Are such small increases biologically relevant? There is good evidence that very modest changes in levels of proinflammatory cytokines have biological effects. For instance diurnal variations in levels of TNF and IL-1 in blood and CSF play a role in the regulation of the spontaneous sleep–wake cycle [81]. Direct experimental studies also support this view. When very low doses of endotoxin (LPS; 0.2 ng/kg) were administered to healthy volunteers there was a doubling of circulating levels of cytokines and soluble cytokine receptors, and significantly increased deep non-rapid eye movement (NREM; slow-wave) sleep, despite the fact that this dose of LPS was too low to alter rectal temperature, heart rate, or cortisol levels [82]. Of particular interest is the observation that negative changes in mood following injection of a vaccine (*Salmonnella typhi*) that caused no subjective symptoms in normal volunteers, correlated significantly with a doubling in serum IL-6 [83].

The gut and depression

The fact that chronic inflammatory disorders are accompanied by release of proinflammatory cytokines provides a second reason, in addition to the stress of the disorder itself, for the raised incidence of depression that accompanies such disorders. However when the gastrointestinal tract is the site of the inflammation further mechanisms may be involved.

Gut permeability and systemic cytokine release

It has been postulated that the raised markers of chronic inflammation that are often seen in depression are attributable to leakiness of the gut epithelium. Levels of IgM and IgA directed against the LPS of several Gram-negative enterobacteria (*Hafnia alvei*, *Pseudomonas aeruginosa*, *Morganella morganii*, *Pseudomonas putida*, *Citrobacter koseri*, and *Klebsiella pneumoniae*) were raised in the serum of individuals with major depression [84]. Increased leakiness of the epithelium, leading to increased intake of gut contents, including powerfully inflammation-inducing endotoxins, would be expected to increase the levels of circulating proinflammatory cytokines and so lead to the symptoms of depression.

In fact this suggestion turns out to be of fundamental interest in the context of the 'old friends' hypothesis. A number of the other conditions that are considered in this volume are also thought by some authors to be associated with increased intestinal epithelial permeability [85].

There are reports that in clinically asymptomatic Crohn's disease patients, increased intestinal epithelial permeability precedes clinical relapse by as much as 1 year [86, 87]. Asymptomatic first-degree relatives of Crohn's disease patients can also have increased intestinal permeability [88]. Similarly, increased intestinal permeability has been observed in some human autoimmune diseases, including Type 1 diabetes (T1D). Subjects with islet autoimmunity showed an increase in intestinal permeability before clinical onset of the disease [89]. There was increased permeability to the disaccharide lactulose, suggesting a damaged barrier, but normal permeability to the monosaccharide mannitol suggesting that the mucosal surface was intact. Similar findings were reported from studies of the Bio-Breeding/Worcester (BB) diabetes-prone rat. Intestinal intraluminal zonulin levels (zonulin is a protein that regulates intercellular tight junctions) were elevated 35-fold compared to control BB diabetic-resistant rats. Upregulation of zonulin coincided with decreased small intestinal transepithelial electrical resistance, and was followed by the production of autoantibodies against pancreatic β cells. Approximately 25 days later T1D became clinically evident. Blocking the zonulin receptor reduced the cumulative incidence of T1D by 70%, and inhibited development of islet cell antibodies [90].

Similarly there is a single report of increased intestinal permeability in multiple sclerosis [91], and also sporadic reports of increased intestinal permeability in atopic eczema [92] and asthma [93].

These observations might provide an interesting link between the brain, the immune system, depression and the chronic inflammatory disorders. Increased expression of the stress- and anxiety-related neuropeptide corticotropin-releasing hormone (CRH) is found in cerebrospinal fluid (CSF) and in limbic brain regions in depression [94, 95], but CRH is also involved in the control of gut permeability [96, 97]. For example chronic administration of CRH *via* minipumps caused colonic barrier dysfunction in rats [97]. Stress-related release of CRH, arising from the autonomic component of the paraventricular nucleus of the hypothalamus or from Barrington's complex in the pons, is thought to alter gut function *via* increased parasympathetic outflow from both vagal and sacral pathways, leading to cholinergic recruitment of peripheral serotonin systems, either enterochromaffin cells or enteric neurons, and also by alterations in sympathetic outlflow [98]. Moreover CRH can be released in the periphery by T cells, cells within the submucosa and muscle layers of the gut, myenteric neurons, serotonin-containing enterochromaffin cells, and lamina propria cells of the mucosa in stomach and colon [98]. CRH is not only a regulator of intestinal permeability [96, 97], but also a potent proinflammatory cytokine [99]. In many cell types CRH activates the master regulator of inflammation nuclear factor (NF)-κB, and stimulates expression of IL-1β, IL-6, and TNF-α mRNAs [100], so some of the effects of CRH on permeability are secondary to the release of proinflammatory cytokines. Paracellular permeability is controlled by tight junctions, intermediate junctions (zonula adherens) and desmosomes, which together constitute a size- and cation-selective filter for small molecules. However, TNF, IL-17, IFN-γ and nitric oxide (NO) increase permeability. IFN-γ disrupts the tight junctions, and can modify para-epithelial traffic of inflammatory cells. By contrast, the immunoregulatory cytokine TGF-β decreases permeability [96, 97].

We are left with a chicken-and-egg situation (see below). Does depression trigger increased permeability, or does a dysregulated immune system cause increased permeability, leading secondarily to depression, and perhaps to the other chronic inflammatory disorders considered above? It is likely that both mechanisms occur (see Fig. 1).

The gut and anxiety disorders

Inflammatory conditions of the gastrointestinal tract have also been linked with certain anxiety disorders, particularly panic disorder [101]. As experimentally-induced panic attacks do not appear to increase proinflammatory cytokines in healthy volunteers, the association in panic patients would appear to be due to a shared vulnerability to gut inflammation and panic disorder [102].

Vascular disease and depression

Depression is very common in patients with coronary artery disease (17–27% in hospitalised patients) [103]. Similarly the metabolic syndrome, often linked with atherosclerosis, is associated with depression and accompanied by chronic activation of the immune system. A recent study confirmed that in patients with the metabolic syndrome, depression is increased and associated with raised CRP and IL-6 [104].

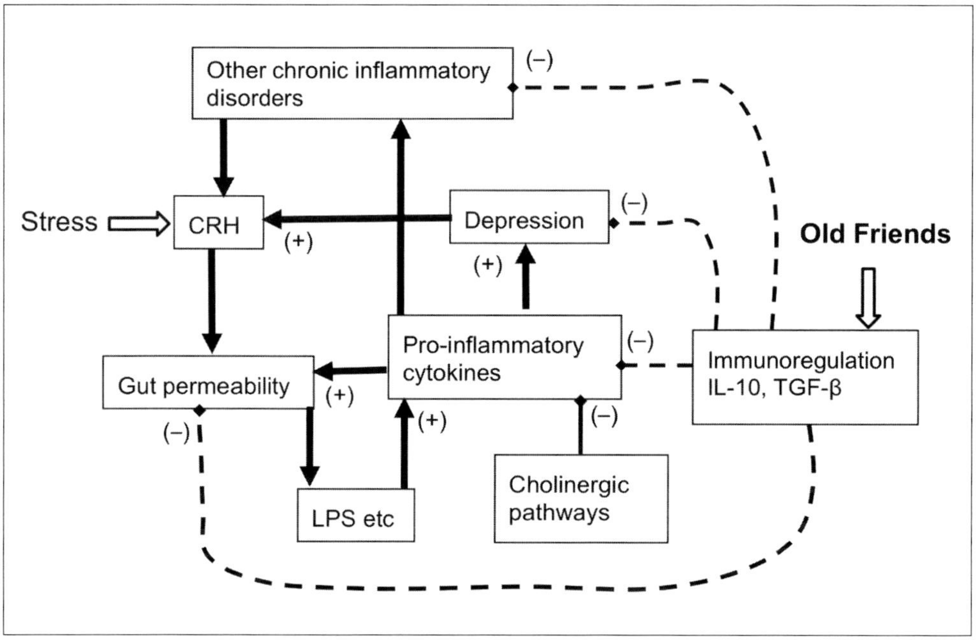

Figure 1
This cartoon attempts to show how inflammation, gut permeability and depression are interrelated, and shows ways in which increased immunoregulation driven by the 'old friends' can impact on these circuits at several points. LPS, lipopolysaccharide; CRH; corticotropin-releasing hormone

Brain-derived neurotrophic factor (BDNF), vascular endothelial growth factor (VEGF), and fibroblast growth factor (FGF)

Some work has implicated deficiency of brain-derived neurotrophic factor (BDNF) in both depression and atherosclerosis, and it has even been suggested that both might involve endothelial dysfunction [74]. Endothelial cells can make BDNF, and play a crucial role in brain neurogenesis, which may be diminished in depression. Serum levels of BDNF are low in depressed individuals, and levels rise during treatment [105]. Plasma levels of BDNF were also reduced in patients with acute coronary syndrome suggesting that BDNF might be implicated in the pathogenesis of human coronary atherosclerosis [106]. Similarly, significantly decreased expression of BDNF protein was also found in atopic dermatitis lesions compared with controls [107], and low BDNF might be implicated in MS, where treatment with IFN-β causes its increased expression [108].

A recent study has suggested that another endothelial growth factor might also be important [109]. Vascular endothelial growth factor (VEGF) was initially characterised for its role in angiogenesis, but it also exerts direct mitogenic effects on neural progenitors *in vitro*. Several classes of antidepressant drugs were shown to enhance production of VEGF [109].

Finally, microarray studies in human patients have highlighted deficiencies in the fibroblast growth factor (FGF) family of growth factors [110]. As with BDNF and VEGF, FGF plays an important role in angiogenesis, but also in cardioprotection [111] .

Could endothelial changes underlie some of the changes in gut permeability, risk of vascular lesions, and inflammation in all these inflammatory disorders, and if so, are the endothelial changes causes or consequences?

The causality dilemma

In the previous sections we have established that depression is often associated with multiple markers of inflammation, even when no cause for chronic inflammatory activity is apparent. We have also raised the possibility that anxiety disorders might be associated with long-term inflammation associated with Th2-biased conditions. Although the changes, such as the increased levels of circulating cytokines, are modest, they may well fall within the range expected to have effects on emotional states and behaviour. The next major issue we have to confront is whether the cytokines cause the depression (or anxiety), or *vice versa*.

It is well-established that exposure to psychological stress can lead to increased production of proinflammatory cytokines [112, 113]. Moreover it has been reported that this ability of stress to drive inflammatory responses, measured as serum IL-6 and DNA binding of NF-κB, is exaggerated in depressed individuals [114]. Compatible observations have been made by documenting the effects of taking a stressful exam, on production of cytokines by peripheral blood cells *in vitro*. Larger increases in release of IFN-γ, IL-1Rα, TNF and IL-6 were seen in cultures of cells from the most stressed students [31].

Insofar as depression is thought to be a form of psychological stress, or at least associated with enhanced stress-reactivity, these points seem to suggest that the depression might drive the cytokine release [115]. However many other observations, outlined below, support the reverse conclusion.

Direct induction of depression by cytokines in animals and man

The most powerful arguments in support of a causal role for the raised proinflammatory cytokines in affective and anxiety disorders have been outlined earlier, and will not be repeated here. In brief, cytokines cause depression-like symptoms in animals [54], and blocking the actions of IL-1 in the CNS stops this from occurring [56, 57]. Similarly, treatment of humans with proinflammatory cytokines induces depression that resembles spontaneous depression on brain activity scans, and can be treated with the standard antidepressant drugs [25].

Genetics

Links between polymorphisms of cytokine genes (IL-1 and TNF) and depression imply direct involvement rather than a bystander role [116, 117]. A single nucleotide polymorphism of the TNF gene (the *TNF2* allele) that leads to higher production of TNF was previously linked to autoimmune and inflammatory diseases [118], but it is also significantly associated with major depression in a Korean population [116]. However a recent study failed to find any associations between cytokine polymorphisms and childhood-onset mood disorders [119].

Vulnerability to depressive symptoms induced by IFN-$\alpha 2\beta$ has been linked to a polymorphism in the 5-HT_{1A} receptor, suggesting that effects of immune activation on serotonergic systems, as described above, may contribute to vulnerability to depression [120]. Interferon alpha downregulates 5-HT_{1A} receptors *in vitro* [121], and patients homozygous for the G allele of the C(-1019)G variant of the 5-HT_{1A} receptor have significantly increased incidence and severity of interferon-induced depression. Interestingly, panic patients [122], social anxiety disorder patients [123], and depressed patients have decreased 5-HT_{1A} receptor binding in brain [124, 125], while the G allele of the C(-1019)G variant of the 5-HT_{1A} receptor has been associated with depression, panic disorder, neuroticism, and reduced response to antidepressant drug treatment [126, 127]. Patients with panic disorder show subsensitivity of these receptors in laboratory challenge tests [128, 129]. Thus, proinflammatory cytokines may increase vulnerability to stress-related psychiatric disorders by downregulating 5-HT_{1A} receptors, presumably *via* indirect mechanisms.

A recent study has demonstrated high potency inhibition of TNF-α-induced inflammation by activation of 5-HT_{2A} receptors [130]. Although a direct link between vulnerability to depressive symptoms induced by proinflammatory cytokines and 5-HT_{2A} receptor function has not been demonstrated, microarray studies have identified altered expression levels of 5-HT_{2A} receptor mRNA in depressed patients [131]. Although less is known about associations between 5-HT_{2A} receptor function, anxiety disorders, and depression, compared to 5-HT_{1A} receptors, at least one study has reported association of polymorphism in the 5-HT_{2A} receptor gene and panic disorder [132], while recent independent studies have found that polymorphisms in the 5-HT_{2A} receptor are associated with symptom severity in PD [133, 134]. Thus, genetic studies support a role for both 5-HT_{1A} and 5-HT_{2A} receptors in the pathophysiology and symptom severity of anxiety disorders. Both

5-HT_{1A} and 5-HT_{2A} receptors have been identified as targets for development of novel antidepressant therapies [135].

Glucocorticoid resistance

Depression is commonly associated with hypercortisolaemia and glucocorticoid resistance [136]. This might imply that the raised levels of inflammatory cytokines seen in depression are secondary to a failure of glucocorticoid-mediated feedback leading to elevated cortisol, subsequent glucocorticoid resistance, and a decrease in glucocorticoid-mediated immunosuppression. However, recent analysis revealed that the raised cytokines cause the glucocorticoid resistance by impairing the function of glucocorticoid receptors [136–138]. Thus glucocorticoid resistance might constitute further evidence that failure of immunoregulation is a primary factor.

Treating depression with anti-inflammatory drugs

If chronic release of cytokines plays a role in driving the depression that accompanies chronic inflammation, then neutralising these cytokines *in vivo* should relieve the depression. At least two examples of this are known. First, administration of neutralising anti-TNF antibody to patients with Crohn's disease alleviated depressive symptoms in a way that could not be secondary to improvements in comfort and lifestyle [139]. Treatment with anti-TNF was also said to relieve symptoms of depression in patients with psoriasis [140]. Similarly the cyclooxygenase-2 inhibitor celecoxib has therapeutic effects in major depression [141]. Anti-inflammatory drugs of this class can both inhibit inflammation-induced increases in proinflammatory cytokines, and oppose their effects in the CNS (reviewed in [58]).

Anti-inflammatory effects of antidepressant drugs

A rather more tenuous kind of evidence that the increased serum cytokines might be contributing to depression, rather than merely a consequence of it, comes from the observation that the drugs used to treat depression modulate cytokine production. Clearly this would only constitute strong evidence if we could prove that this reduction in cytokines mediates some of the antidepressive effect. This might well be the case, and is discussed later, but it cannot be assumed.

The phenomenon has been demonstrated by stimulating blood cells *in vitro* with phytohaemagglutinin and endotoxin in the presence or absence of tricyclic antidepressants (TCA; imipramine, clomipramine, trimipramine), noradrenaline re-uptake inhibitors (NARI; reboxetine, desipramine), or selective serotonin re-uptake inhibitors (SSRI; citalopram, clomipramine, or sertraline). These antidepressants significantly reduced secretion of IFN-γ, and most also increased secretion of the anti-inflammatory cytokine IL-10 [142–145]. In similar experiments lithium was also reported to increase production of IL-10 [146, 147]. It has therefore been suggested that in general all these antidepressants reduce the IFN-γ/IL-10 ratio, and so exert an overall anti-inflammatory effect [143].

These findings are supported by the observation that when depressed patients were treated with amitriptyline, a tricyclic antidepressant, there was a fall in plasma CRP levels, and a fall in spontaneous release of TNF from whole blood during the first 3 h in culture [148]. Animal studies also support the view that antidepressants are anti-inflammatory [112]. For example, prolonged treatment of C57BL/6 mice subjected to the chronic mild stress model of depression with desipramine, another tricyclic antidepressant, led to increased production of IL-10 by their T cells when subsequently stimulated *in vitro* [149].

Peripheral *versus* central targets of antidepressants

It is not known how antidepressants exert their effects on proinflammatory cytokines. A full review of this topic is beyond the scope of this review, but it is of particular interest that a large part of a dose of an SSRI is taken up in peripheral sites such as the lung [150], and several possible molecular targets, present in the periphery as well as in the brain, have been identified. First, we described earlier the possibility that antidepressants exert direct effects on endothelial cells, increasing expression of BDNF [105] and VEGF [109], both of which are involved in neurogenesis. Secondly, the serotonin transporter (SERT), which is the target of SSRIs, is expressed in macrophages [151] and in mature DC, where it is involved in the uptake of 5-HT from mature T cells [152]. The 5-HT is then transferred to naïve T cells, and enhances their activation [152]. Clearly the role of SERT in DC provides a possible site of action for SSRIs in the culture systems described above. Interestingly, the expression of SERT in some cell types is up-regulated by IL-1 [153] and TNF but not by interleukin-6 [154]. IL-4 appears to cause a dose-dependent reduction of 5-HT uptake [155].

A functional polymorphism in the promoter region of SERT was found to influence the likelihood that stressful life events would trigger depression [156, 157]. The assumption that this is due entirely to effects in the CNS might need to be revised.

Finally, antidepressant drugs inhibit the activity of TDO, the enzyme responsible for metabolism of most of the tryptophan in the body, apart from that used for protein synthesis [158]. This action directly opposes the tendency, discussed in an earlier section, of mediators released by inflammation (LPS, IL-1, and IFN-γ), to increase catabolism of tryptophan by IDO, and of glucocorticoid hormones to increase catabolism of tryptophan by TDO [71]. Several antidepressant drugs have been shown to inhibit TDO, including imipramine, tranylcypramine [159], lofepramine [160], desmethylimipramine [160], paroxetine [161], fluoxetine [162, 163] and many others [164]. Therefore these drugs cause increased plasma and brain tryptophan concentrations and an increased rate of serotonin biosynthesis in the brain [165–167]. Moreover, in view of the immunoregulatory roles of both serotonin [152] and tryptophan [168], secondary effects on release of proinflammatory cytokines are probable. These studies establish a fundamental antagonism between stress hormones and proinflammatory cytokines on the one hand, and antidepressant drugs on the other.

Other treatments that are anti-inflammatory

Electroconvulsive therapy (ECT) was found to downregulate increased levels of the proinflammatory cytokine TNF in patients with major depression [169]. Another experimental treatment involves surgically implanting a pulse generator that transmits impulses to the left vagus nerve *via* an electrode [170]. This is of interest in the current context because the parasympathetic system modulates inflammation in the gut *via* tonic vagal inhibition of proinflammatory macrophages [171]. Moreover, tricyclic antidepressants restore vagal function and reduce intestinal inflammation [172]. Thus stimulating the vagal system might work in part *via* anti-inflammatory effects. Indeed, vagal nerve stimulation alters systemic cytokine concentration in human patients [173].

There is some speculation that treatment with probiotics might also be effective. Probiotics have been shown to exert anti-inflammatory effects in a variety of animal models of chronic inflammatory disorders, usually by inducing Treg cells [174–177]. However there have been few attempts to explore their efficacy in depression. The relationship between probiotics and antidepressant-like behavioural effects, however, has been explored in animal models. In one such study, Sprague-Dawley rats were treated for 14 days with *Bifidobacterium infantis* by the oral route [178]. Probiotic administration had no effect on performance in the forced swim test, a commonly used screen for antidepressant-like properties of drugs. However, there was a significantly reduced release of IFN-γ, TNF and IL-6 following mitogen stimulation of peripheral blood cells ($p < 0.05$) in probiotic-treated rats relative to controls. Furthermore, there was a marked increase in plasma concentrations of tryptophan ($p < 0.005$) and kynurenic acid ($p < 0.05$), and a reduced concentration of 5-hydroxyindoleacetic acid (5-HIAA; the major metabolite of serotonin) in the frontal cortex, suggesting reduced serotonin metabolism in this area. Thus oral *B. infantis* attenuated proinflammatory responses, and caused increased levels of the serotonin precursor, tryptophan. Interestingly, oral probiotic *Lactobacillus* strains also protected against the long-term changes in gut paracellular permeability that are caused by subjecting mice to maternal separation [179].

Regulatory cytokines (IL-10, TGF-β)

If we are to apply the hygiene hypothesis to depression and anxiety, we need to show that there is a deficit in immunoregulation in this disorder. We are not aware of any studies of Treg function in depression. However there is some evidence that circulating levels of IL-10 and TGF-β, the two most important regulatory cytokines, are associated with alleviation of symptoms.

IL-10

Levels of IL-10 are frequently raised in patients with depression [180]. However, this can be seen as secondary to the increases in proinflammatory cytokines such as IFN-γ, IL-1, IL-6 and TNF. Hence the crucial factor may be the *ratio* of IL-10 to the proinflammatory mediators. For instance, when the ratio of IL-10 is calculated relative to levels of proinflammatory cytokines, it is found that a high TNF/IL-10 ratio correlates with depression. In patients with chronic heart failure, those who were depressed had higher levels of TNF but lower levels of IL-10 than those who were not depressed [181]. Moreover the TNF/IL-10 ratio correlated significantly with the severity of depressive symptoms [181].

There are many other types of evidence that IL-10 may have antidepressant actions. First, we know that IL-10 opposes a number of the CNS effects of inflammatory agents.

1. Inhibits spontaneous sleep in rabbits [182]
2. Inhibits sickness behaviour induced by LPS [183]
3. Reduces hyperalgesia caused by LPS [184]
4. Reduces LPS-induced TNF release in the brain [185]
5. Reduces LPS-induced fever in mice [186]
6. IL-10 KO mice have prolonged fever to very low dose LPS i.p. [187]

Secondly, as outlined earlier, antidepressants reduce secretion of IFN-γ, but increase secretion of IL-10 [142–144, 146, 147].

When recall responses to influenza vaccine were elicited in peripheral blood cells *in vitro*, perceived stress by the donors was related to IL-2, whereas optimism was associated with a greater IL-10 response [32]. Similarly anger and fatigue were associated with significantly lower release of IL-10 [33].

IL-10 might also oppose anxiety symptoms. When cells from stressed students were stimulated *in vitro*, cells from students scoring high for anxiety showed large increases in release of IFN-γ, IL-1Ra, TNF and IL-6, whereas cells from the least anxious students released more IL-10 [31].

Finally, it is interesting that whereas TNF, IL-1β, and chemokine (C-X-C motif) ligand 1 (CXCL1) mediate hypernociception (exaggerated perception of pain), the anti-inflammatory cytokine IL-10 has the reverse effect, decreasing pain perception [188]. Increased awareness of pain is common in depression.

TGF-β

The second major regulatory cytokine, TGF-β, has received little attention in this context. In a microarray comparison of post-mortem tissue from the prefrontal cortex of patients with bipolar disorder with tissue from matched controls, it was found that expression of mRNA encoding TGF-β was significantly lower in the patients' samples. This was confirmed in an RT-PCR study of material from the same individuals [189]. It is difficult to relate this finding to the activity of the peripheral immune system. However, studies in Korea and the Netherlands and Turkey have looked at TGF-β in the periphery and shown that depression is accompanied by a low ratio of TGF-β to proinflammatory cytokines (IL-12 or IFN-γ), and that TGF-β levels showed a significant negative correlation with depression [41, 190]. Moreover TGF-β levels rose significantly after treatment [41, 190, 191]. Interestingly, decreased levels of TGF-β were also described in depressed bulimic patients [192].

The *Mycobacterium vaccae* animal model

There is then considerable evidence that anti-inflammatory cytokines can block sickness behaviour/depression-like behaviour and that absolute or relative levels of these cytokines tend to be low in patients with depression and anxiety disorders, whereas levels of proinflammatory cytokines tend to be high. Taken together this suggests that the hygiene hypothesis might be relevant to certain types of anxiety disorders and depression. The study of a saprophytic environmental mycobacterium, *M. vaccae*, has recently provided direct experimental links between the hygiene hypothesis and events in the CNS. *M. vaccae* induces Treg that downregulate chronic inflammatory states *via* a mechanism that depends on IL-10 and TGF-β [15]. It has undergone clinical trials for allergic disorders, psoriatic arthritis and some cancers. In several studies the patients who had received one or more intradermal injections of a heat-killed preparation of this organism showed unexpected improvements in quality of life scores [193–195]. This led to investigation of the properties of *M. vaccae* in a mouse model, and to the discovery that intratracheal or subcutaneous administration of heat-killed *M. vaccae* activated a specific subset of serotonergic neurons in the interfascicular part of the dorsal raphe nucleus (DRI) of mice [78]. *M. vaccae*-induced activation of DRI serotonergic neurons was associated with increases in serotonin metabolism within the medial prefrontal cortex (mPFC), consistent with an effect of immune activation on mesolimbocortical serotonergic systems. These effects were temporally associated with reductions in immobility in the forced swim test, which is a standard test for antidepressant activity [78].

The hygiene hypothesis and psychiatric disorders

Studies have documented three primary factors involved in determining vulnerability to depression. These include genetic influences, adverse early life experiences, and major stressful life events [157, 196]. Interactions among these vulnerability factors appear to be particularly important [156]. Many authors have suggested previously that persistently raised proinflammatory cytokines might be a further aetiological factor [23–25]. In this review, and elsewhere [197], we add to that hypothesis the suggestion that in developed countries the immunoregulatory failure that leads to persistently high proinflammatory cytokine levels might be the deficit in regulatory T cell activity that has already been implicated in the increased incidences of chronic inflammatory disorders such as allergies and autoimmunity (hygiene hypothesis). Treatments that induce regulatory immune responses, such as *M. vaccae*, could have therapeutic value in patients with anxiety disorders or depression associated with chronic proinflammatory states, whether accompanied by a clinically apparent inflammatory disorder (allergic disease, autoimmunity, inflammatory bowel disease, cardiovascular disease), or not. Indeed, as others have suggested, an anti-inflammatory effect might be an integral part of the activity of several currently used antidepressants [142–145].

References

1 Bach JF (2002) The effect of infections on susceptibility to autoimmune and allergic diseases. *N Engl J Med* 347: 911–920

2 Sawczenko A, Sandhu BK, Logan RF, Jenkins H, Taylor CJ, Mian S, Lynn R (2001) Prospective survey of childhood inflammatory bowel disease in the British Isles. *Lancet* 357: 1093–1094

3 Rook GA, Adams V, Hunt J, Palmer R, Martinelli R, Brunet LR (2004) Mycobacteria and other environmental organisms as immunomodulators for immunoregulatory disorders. *Springer Semin Immunopathol* 25: 237–255

4 Akdis M, Verhagen J, Taylor A, Karamloo F, Karagiannidis C, Crameri R, Thunberg S, Deniz G, Valenta R, Fiebig H et al (2004) Immune responses in healthy and allergic individuals are characterized by a fine balance between allergen-specific T regulatory 1 and T helper 2 cells. *J Exp Med* 199: 1567–1575

5 Viglietta V, Baecher-Allan C, Weiner HL, Hafler DA (2004) Loss of functional suppression by $CD4^+CD25^+$ regulatory T cells in patients with multiple sclerosis. *J Exp Med* 199: 971–979

6 Kriegel MA, Lohmann T, Gabler C, Blank N, Kalden JR, Lorenz HM (2004) Defective suppressor function of human $CD4^+$ $CD25^+$ regulatory T cells in autoimmune polyglandular syndrome type II. *J Exp Med* 199: 1285–1291

7 Duchmann R, Kaiser I, Hermann E, Mayet W, Ewe K, Meyer zum Buschenfelde KH (1995) Tolerance exists towards resident intestinal flora but is broken in active inflammatory bowel disease (IBD) *Clin Exp Immunol* 102: 448–455

8 Kraus TA, Toy L, Chan L, Childs J, Mayer L (2004) Failure to induce oral tolerance to a soluble protein in patients with inflammatory bowel disease. *Gastroenterology* 126: 1771–1778

9 Wildin RS, Smyk-Pearson S, Filipovich AH (2002) Clinical and molecular features of the immunodysregulation, polyendocrinopathy, enteropathy, X linked (IPEX) syndrome. *J Med Genet* 39: 537–545

10 Babu S, Blauvelt CP, Kumaraswami V, Nutman TB (2006) Regulatory networks induced by live parasites impair both Th1 and Th2 pathways in patent lymphatic filariasis: implications for parasite persistence. *J Immunol* 176: 3248–3256

11 van der Kleij D, Latz E, Brouwers JF, Kruize YC, Schmitz M, Kurt-Jones EA, Espevik T, de Jong EC, Kapsenberg ML, Golenbock DT et al (2002) A novel host-parasite lipid cross-talk. Schistosomal lyso-phosphatidylserine activates Toll-like receptor 2 and affects immune polarization. *J Biol Chem* 277: 48122–48129

12 Adams VC, Hunt JR, Martinelli R, Palmer R, Rook GA, Brunet LR (2004) *Mycobacterium vaccae* induces a population of pulmonary $CD11c^+$ cells with regulatory potential in allergic mice. *Eur J Immunol* 34: 631–638

13 Smits HH, Engering A, van der Kleij D, de Jong EC, Schipper K, van Capel TM, Zaat BA, Yazdanbakhsh M, Wierenga EA, van Kooyk Y et al (2005) Selective probiotic bacteria induce IL-10-producing regulatory T cells *in vitro* by modulating dendritic cell function through dendritic cell-specific intercellular adhesion molecule 3–grabbing nonintegrin. *J Allergy Clin Immunol* 115: 1260–1267

14 Pasare C, Medzhitov R (2003) Toll pathway-dependent blockade of $CD4^+CD25^+$ T cell-mediated suppression by dendritic cells. *Science* 299: 1033–1036

15 Zuany-Amorim C, Sawicka E, Manlius C, Le Moine A, Brunet LR, Kemeny DM, Bowen G, Rook G, Walker C (2002) Suppression of airway eosinophilia by killed *Mycobacterium vaccae*-induced allergen-specific regulatory T-cells. *Nat Med* 8: 625–629

16 Wilson MS, Taylor MD, Balic A, Finney CA, Lamb JR, Maizels RM (2005) Suppression of allergic airway inflammation by helminth-induced regulatory T cells. *J Exp Med* 202: 1199–1212

17 Ricklin-Gutzwiller ME, Reist M, Peel JE, Seewald W, Brunet LR, Roosje PJ (2007) Intradermal injection of heat-killed *Mycobacterium vaccae* in dogs with atopic dermatitis: a multicentre pilot study. *Vet Dermatol* 18: 87–93

18 Zaccone P, Fehervari Z, Jones FM, Sidobre S, Kronenberg M, Dunne DW, Cooke A (2003) *Schistosoma mansoni* antigens modulate the activity of the innate immune response and prevent onset of type 1 diabetes. *Eur J Immunol* 33: 1439–1449

19 Di Giacinto C, Marinaro M, Sanchez M, Strober W, Boirivant M (2005) Probiotics ame-

liorate recurrent Th1–mediated murine colitis by inducing IL-10 and IL-10-dependent TGF-{beta}-bearing regulatory cells. *J Immunol* 174: 3237–3246
20 Summers RW, Elliott DE, Urban JF, Jr., Thompson RA, Weinstock JV (2005) *Trichuris suis* therapy for active ulcerative colitis: a randomized controlled trial. *Gastroenterology* 128: 825–832
21 Summers RW, Elliott DE, Urban JF, Jr., Thompson R, Weinstock JV (2005) *Trichuris suis* therapy in Crohn's disease. *Gut* 54: 87–90
22 Taylor A, Verhagen J, Blaser K, Akdis M, Akdis CA (2006) Mechanisms of immune suppression by interleukin-10 and transforming growth factor-beta: the role of T regulatory cells. *Immunology* 117: 433–442
23 O'Brien SM, Scott LV, Dinan TG (2004) Cytokines: abnormalities in major depression and implications for pharmacological treatment. *Hum Psychopharmacol* 19: 397–403
24 Schiepers OJ, Wichers MC, Maes M (2005) Cytokines and major depression. Prog. Neuropsychopharmacol. *Biol Psychiatry* 29: 201–217
25 Capuron L, Miller AH (2004) Cytokines and psychopathology: lessons from interferon-alpha. *Biol Psychiatry* 56: 819–824
26 Maes M (1999) Major depression and activation of the inflammatory response system. *Adv Exp Med Biol* 461: 25–46
27 Penninx BW, Kritchevsky SB, Yaffe K, Newman AB, Simonsick EM, Rubin S, Ferrucci L, Harris T, Pahor M (2003) Inflammatory markers and depressed mood in older persons: results from the health, aging and body domposition study. *Biol Psychiatry* 54: 566–572
28 Musselman DL, Miller AH, Porter MR, Manatunga A, Gao F, Penna S, Pearce BD, Landry J, Glover S, McDaniel JS et al (2001) Higher than normal plasma interleukin-6 concentrations in cancer patients with depression: preliminary findings. *Am J Psychiatry* 158: 1252–1257
29 Raison CL, Miller AH (2003) Depression in cancer: new developments regarding diagnosis and treatment. *Biol Psychiatry* 54: 283–294
30 Ferketich AK, Ferguson JP, Binkley PF (2005) Depressive symptoms and inflammation among heart failure patients. *Am Heart J* 150: 132–136
31 Maes M, Song C, Lin A, De Jongh R, Van Gastel A, Kenis G, Bosmans E, De Meester I, Benoy I, Neels H et al (1998) The effects of psychological stress on humans: increased production of pro-inflammatory cytokines and a Th1-like response in stress-induced anxiety. *Cytokine* 10: 313–318
32 Kohut ML, Cooper MM, Nickolaus MS, Russell DR, Cunnick JE (2002) Exercise and psychosocial factors modulate immunity to influenza vaccine in elderly individuals. *J Gerontol A Biol Sci Med Sci* 57: M557–562
33 Costanzo ES, Lutgendorf SK, Kohut ML, Nisly N, Rozeboom K, Spooner S, Benda J, McElhaney JE (2004) Mood and cytokine response to influenza virus in older adults. *J Gerontol A Biol Sci Med Sci* 59: 1328–1333
34 Maes M, Scharpe S, Van Grootel L, Uyttenbroeck W, Cooreman W, Cosyns P, Suy E (1992) Higher alpha 1-antitrypsin, haptoglobin, ceruloplasmin and lower retinol bind-

ing protein plasma levels during depression: further evidence for the existence of an inflammatory response during that illness. *J Affect Disord* 24: 183–192

35 Maes M, De Vos N, Demedts P, Wauters A, Neels H (1999) Lower serum zinc in major depression in relation to changes in serum acute phase proteins. *J Affect Disord* 56: 189–194

36 Kronfol Z, House JD (1989) Lymphocyte mitogenesis, immunoglobulin and complement levels in depressed patients and normal controls. *Acta Psychiatr Scand* 80: 142–147

37 Huang TL, Lin FC (2007) High-sensitivity C-reactive protein levels in patients with major depressive disorder and bipolar mania. *Prog Neuropsychopharmacol Biol Psychiatry* 31: 370–372

38 Kronfol Z, Turner R, House JD, Winokur G (1986) Elevated blood neutrophil concentration in mania. *J Clin Psychiatry* 47: 63–65

39 Atamas SP, Choi J, Yurovsky VV, White B (1996) An alternative splice variant of human IL-4, IL-4 delta 2, inhibits IL-4-stimulated T cell proliferation. *J Immunol* 156: 435

40 Dheda K, Chang JS, Breen RA, Kim LU, Haddock JA, Huggett JF, Johnson MA, Rook GA, Zumla A (2005) *In vivo* and *in vitro* studies of a novel cytokine, interleukin-4delta2, in Pulmonary Tuberculosis. *Am J Respir Crit Care Med* 172: 501–508

41 Myint AM, Leonard BE, Steinbusch HW, Kim YK (2005) Th1, Th2, and Th3 cytokine alterations in major depression. *J Affect Disord* 88: 167–173

42 Kim YK, Jung HG, Myint AM, Kim H, Park SH (2007) Imbalance between pro-inflammatory and anti-inflammatory cytokines in bipolar disorder. *J Affect Disord* 104: 91–95

43 Berry M, Brightling C, Pavord I, Wardlaw A (2007) TNF-alpha in asthma. *Curr Opin Pharmacol* 7: 279–282

44 Shore SA (2007) Obesity and asthma: implications for treatment. *Curr Opin Pulm Med* 13: 56–62

45 Beuther DA, Weiss ST, Sutherland ER (2006) Obesity and asthma. *Am J Respir Crit Care Med* 174: 112–119

46 Wyatt SB, Winters KP, Dubbert PM (2006) Overweight and obesity: prevalence, consequences, and causes of a growing public health problem. *Am J Med Sci* 331: 166–174

47 Chapman DP, Perry GS, Strine TW (2005) The vital link between chronic disease and depressive disorders. *Prev Chronic Dis* 2: A14

48 Ortega AN, Huertas SE, Canino G, Ramirez R, Rubio-Stipec M (2002) Childhood asthma, chronic illness, and psychiatric disorders. *J Nerv Ment Dis* 190: 275–281

49 Vila G, Nollet-Clemencon C, de Blic J, Mouren-Simeoni MC, Scheinmann P (2000) Prevalence of DSM IV anxiety and affective disorders in a pediatric population of asthmatic children and adolescents. *J Affect Disord* 58: 223–231

50 Ortega AN, McQuaid EL, Canino G, Goodwin RD, Fritz GK (2004) Comorbidity of asthma and anxiety and depression in Puerto Rican children. *Psychosomatics* 45: 93–99

51 Goodwin RD, Messineo K, Bregante A, Hoven CW, Kairam R (2005) Prevalence of probable mental disorders among pediatric asthma patients in an inner-city clinic. *J Asthma* 42: 643–647

52 Wamboldt MZ, Hewitt JK, Schmitz S, Wamboldt FS, Rasanen M, Koskenvuo M, Romanov K, Varjonen J, Kaprio J (2000) Familial association between allergic disorders and depression in adult Finnish twins. *Am J Med Genet* 96: 146–153
53 Wamboldt MZ, Schmitz S, Mrazek D (1998) Genetic association between atopy and behavioral symptoms in middle childhood. *J Child Psychol Psychiatry* 39: 1007–1016
54 Dantzer R (2001) Cytokine-induced sickness behavior: where do we stand? *Brain Behav Immun* 15: 7–24
55 Banasr M, Duman RS (2007) Regulation of neurogenesis and gliogenesis by stress and antidepressant treatment. *CNS Neurol Disord Drug Targets* 6: 311–320
56 Goshen I, Kreisel T, Ben-Menachem-Zidon O, Licht T, Weidenfeld J, Ben-Hur T, Yirmiya R (2008) Brain interleukin-1 mediates chronic stress-induced depression in mice *via* adrenocortical activation and hippocampal neurogenesis suppression. *Mol Psychiatry* 13: 717–728
57 Ben Menachem-Zidon O, Goshen I, Kreisel T, Ben Menahem Y, Reinhartz E, Ben Hur T, Yirmiya R (2008) Intrahippocampal transplantation of transgenic neural precursor cells overexpressing interleukin-1 receptor antagonist blocks chronic isolation-induced impairment in memory and neurogenesis. *Neuropsychopharmacology* 33: 2251–2262
58 Muller N, Schwarz MJ (2008) COX-2 inhibition in schizophrenia and major depression. *Curr Pharm Des* 14: 1452–1465
59 Capuron L, Ravaud A, Dantzer R (2000) Early depressive symptoms in cancer patients receiving interleukin 2 and/or interferon alfa-2b therapy. *J Clin Oncol* 18: 2143–2151
60 Capuron L, Ravaud A, Gualde N, Bosmans E, Dantzer R, Maes M, Neveu PJ (2001) Association between immune activation and early depressive symptoms in cancer patients treated with interleukin-2-based therapy. *Psychoneuroendocrinology* 26: 797–808
61 Musselman DL, Lawson DH, Gumnick JF, Manatunga AK, Penna S, Goodkin RS, Greiner K, Nemeroff CB, Miller AH (2001) Paroxetine for the prevention of depression induced by high-dose interferon alfa. *N Engl J Med* 344: 961–966
62 Yirmiya R, Weidenfeld J, Polak Y, Morag M, Morag A, Avitsur R, Barak O, Reichenberg A, Cohen E, Shavit Y et al (1999) Cytokines, depression due to a generalised medical condition, and antidepressant drugs. *Adv Exp Med Biol* 461: 283–316
63 Yirmiya R (1996) Endotoxin produces a depressive-like episode in rats. *Brain Res* 711: 163–174
64 Castanon N, Bluthé RM, Dantzer R (2001) Chronic treatment with the atypical atidepressant tianeptine attenuates sickness behaviour induced by peripheral but not central liopolysaccharide and interleukin-1β in the rat. *Psychopharmacology* 154: 50–60
65 Wichers MC, Kenis G, Leue C, Koek G, Robaeys G, Maes M (2006) Baseline immune activation as a risk factor for the onset of depression during interferon-alpha treatment. *Biol Psychiatry* 60: 77–79
66 Konsman JP, Parnet P, Dantzer R (2002) Cytokine-induced sickness behaviour: mechanisms and implications. *Trends Neurosci* 25: 154–159
67 Lowry CA, Lightman SL, Nutt DL (2009) That warm fuzzy feeling: brain serotonergic neurons and the regulation of emotion. *J Psychopharmacol* 23: 392–400

68 Capuron L, Neurauter G, Musselman DL, Lawson DH, Nemeroff CB, Fuchs D, Miller AH (2003) Interferon-alpha-induced changes in tryptophan metabolism; relationship to depression and paroxetine treatment. *Biol Psychiatry* 54: 906–914

69 Ostensen M, Forger F, Nelson JL, Schuhmacher A, Hebisch G, Villiger PM (2005) Pregnancy in patients with rheumatic disease: anti-inflammatory cytokines increase in pregnancy and decrease post partum. *Ann Rheum Dis* 64: 839–844

70 Maes M, Verkerk R, Bonaccorso S, Ombelet W, Bosmans E, Scharpe S (2002) Depressive and anxiety symptoms in the early puerperium are related to increased degradation of tryptophan into kynurenine, a phenomenon which is related to immune activation. *Life Sci* 71: 1837–1848

71 Ruddick JP, Evans AK, Nutt DJ, Lightman SL, Rook GA, Lowry CA (2006) Tryptophan metabolism in the central nervous system: medical implications. *Expert Rev Mol Med* 8: 1–27

72 Saito K, Crowley JS, Markey SP, Heyes MP (1993) A mechanism for increased quinolinic acid formation following acute systemic immune stimulation. *J Biol Chem* 268: 15496–15503

73 Steiner J, Bielau H, Brisch R, Danos P, Ullrich O, Mawrin C, Bernstein HG, Bogerts B (2008) Immunological aspects in the neurobiology of suicide: elevated microglial density in schizophrenia and depression is associated with suicide. *J Psychiatr Res* 42: 151–157

74 Belmaker RH, Agam G (2008) Major depressive disorder. *N Engl J Med* 358: 55–68

75 Linthorst AC, Reul JM (1998) Brain neurotransmission during peripheral inflammation. *Ann NY Acad Sci* 840: 139–152

76 Lavicky J, Dunn AJ (1995) Endotoxin administration stimulates cerebral catecholamine release in freely moving rats as assessed by microdialysis. *J Neurosci Res* 40: 407–413

77 Lowry CA, Evans AK, Gasser PJ, Hale MW, Staub DR, Shekhar A (2008) Topographical organization and chemoarchitecture of the dorsal raphe nucleus and the median raphe nucleus. In: JM Monti, BL Pandi-Perumal, BL Jacobs, DL Nutt (eds): *Serotonin and Sleep: Molecular, Functional and Clinical Aspects*. Birkhäuser, Basel, 25–68

78 Lowry CA, Hollis JH, de Vries A, Pan B, Brunet LR, Hunt JR, Paton JF, van Kampen E, Knight DM, Evans AK et al (2007) Identification of an immune-responsive mesolimbocortical serotonergic system: Potential role in regulation of emotional behavior. *Neuroscience* 146: 756–772

79 Drevets WC, Price JL, Furey ML (2008) Brain structural and functional abnormalities in mood disorders: implications for neurocircuitry models of depression. *Brain Struct Funct* 213: 93–118

80 Pollmacher T, Haack M, Schuld A, Reichenberg A, Yirmiya R (2002) Low levels of circulating inflammatory cytokines – do they affect human brain functions? *Brain Behav Immun* 16: 525–532

81 Kapsimalis F, Richardson G, Opp MR, Kryger M (2005) Cytokines and normal sleep. *Curr Opin Pulm Med* 11: 481–484

82 Mullington J, Korth C, Hermann DM, Orth A, Galanos C, Holsboer F, Pollmacher T

(2000) Dose-dependent effects of endotoxin on human sleep. *Am J Physiol Regul Integr Comp Physiol* 278: R947–955

83 Wright CE, Strike PC, Brydon L, Steptoe A (2005) Acute inflammation and negative mood: mediation by cytokine activation. *Brain Behav Immun* 19: 345–350

84 Maes M, Kubera M, Leunis JC (2008) The gut-brain barrier in major depression: intestinal mucosal dysfunction with an increased translocation of LPS from gram negative enterobacteria (leaky gut) plays a role in the inflammatory pathophysiology of depression. *Neuroendocrinol Lett* 29: 117–124

85 Meddings J (2008) The significance of the gut barrier in disease. *Gut* 57: 438–440

86 D'Inca R, Di Leo V, Corrao G, Martines D, D'Odorico A, Mestriner C, Venturi C, Longo G, Sturniolo GC (1999) Intestinal permeability test as a predictor of clinical course in Crohn's disease. *Am J Gastroenterol* 94: 2956–2960

87 Wyatt J, Vogelsang H, Hubl W, Waldhoer T, Lochs H (1993) Intestinal permeability and the prediction of relapse in Crohn's disease. *Lancet* 341: 1437–1439

88 Fasano A, Shea-Donohue T (2005) Mechanisms of disease: the role of intestinal barrier function in the pathogenesis of gastrointestinal autoimmune diseases. *Nat Clin Pract Gastroenterol Hepatol* 2: 416–422

89 Bosi E, Molteni L, Radaelli MG, Folini L, Fermo I, Bazzigaluppi E, Piemonti L, Pastore MR, Paroni R (2006) Increased intestinal permeability precedes clinical onset of type 1 diabetes. *Diabetologia* 49: 2824–2827

90 Watts T, Berti I, Sapone A, Gerarduzzi T, Not T, Zielke R, Fasano A (2005) Role of the intestinal tight junction modulator zonulin in the pathogenesis of type I diabetes in BB diabetic-prone rats. *Proc Natl Acad Sci USA* 102: 2916–2921

91 Yacyshyn B, Meddings J, Sadowski D, Bowen-Yacyshyn MB (1996) Multiple sclerosis patients have peripheral blood $CD45RO^+$ B cells and increased intestinal permeability. *Dig Dis Sci* 41: 2493–2498

92 Hijazi Z, Molla AM, Al-Habashi H, Muawad WM, Molla AM, Sharma PN (2004) Intestinal permeability is increased in bronchial asthma. *Arch Dis Child* 89: 227–229

93 Benard A, Desreumeaux P, Huglo D, Hoorelbeke A, Tonnel AB, Wallaert B (1996) Increased intestinal permeability in bronchial asthma. *J Allergy Clin Immunol* 97: 1173–1178

94 Merali Z, Du L, Hrdina P, Palkovits M, Faludi G, Poulter MO, Anisman H (2004) Dysregulation in the suicide brain: mRNA expression of corticotropin-releasing hormone receptors and GABA(A) receptor subunits in frontal cortical brain region. *J Neurosci* 24: 1478–1485

95 Lee R, Geracioti TD, Jr., Kasckow JW, Coccaro EF (2005) Childhood trauma and personality disorder: positive correlation with adult CSF corticotropin-releasing factor concentrations. *Am J Psychiatry* 162: 995–997

96 Gareau MG, Silva MA, Perdue MH (2008) Pathophysiological mechanisms of stress-induced intestinal damage. *Curr Mol Med* 8: 274–281

97 Teitelbaum AA, Gareau MG, Jury J, Yang PC, Perdue MH (2008) Chronic peripheral administration of corticotropin-releasing factor causes colonic barrier dysfunction similar to psychological stress. *Am J Physiol Gastrointest Liver Physiol* 295: G452–459

98 Stengel A, Tache Y (2009) Neuroendocrine control of the gut during stress: corticotropin-releasing factor signaling pathways in the spotlight. *Annu Rev Physiol* 71: 219–239
99 Calcagni E, Elenkov I (2006) Stress system activity, innate and T helper cytokines, and susceptibility to immune-related diseases. *Ann NY Acad Sci* 1069: 62–76
100 Zbytek B, Slominski AT (2007) CRH mediates inflammation induced by lipopolysaccharide in human adult epidermal keratinocytes. *J Invest Dermatol* 127: 730–732
101 Lydiard RB (2005) Increased prevalence of functional gastrointestinal disorders in panic disorder: clinical and theoretical implications. *CNS Spectr* 10: 899–908
102 de la Fontaine L, Schwarz MJ, Eser. D, Muller N, Rupprecht R, Zwanzger P (2008) Effects of experimentally induced panic attacks on neuroimmunological markers. *J Neural Transm* Published online DOI 10.1007/s00702-00008-00140-00706
103 Lesperance F, Frasure-Smith N (2007) Depression and heart disease. *Cleve Clin J Med* 74 (Suppl 1): S63–66
104 Capuron L, Su S, Miller AH, Bremner JD, Goldberg J, Vogt GJ, Maisano C, Jones L, Murrah NV, Vaccarino V (2008) Depressive symptoms and metabolic syndrome: is inflammation the underlying link? *Biol Psychiatry* 64: 896–900
105 Gervasoni N, Aubry JM, Bondolfi G, Osiek C, Schwald M, Bertschy G, Karege F (2005) Partial normalization of serum brain-derived neurotrophic factor in remitted patients after a major depressive episode. *Neuropsychobiology* 51: 234–238
106 Manni L, Nikolova V, Vyagova D, Chaldakov GN, Aloe L (2005) Reduced plasma levels of NGF and BDNF in patients with acute coronary syndromes. *Int J Cardiol* 102: 169–171
107 Groneberg DA, Fischer TC, Peckenschneider N, Noga O, Dinh QT, Welte T, Welker P (2007) Cell type-specific regulation of brain-derived neurotrophic factor in states of allergic inflammation. *Clin Exp Allergy* 37: 1386–1391
108 Lalive PH, Kantengwa S, Benkhoucha M, Juillard C, Chofflon M (2008) Interferon-beta induces brain-derived neurotrophic factor in peripheral blood mononuclear cells of multiple sclerosis patients. *J Neuroimmunol* 197: 147–151
109 Warner-Schmidt JL, Duman RS (2007) VEGF is an essential mediator of the neurogenic and behavioral actions of antidepressants. *Proc Natl Acad Sci USA* 104: 4647–4652
110 Akil H, Evans SJ, Turner CA, Perez J, Myers RM, Bunney WE, Jones EG, Watson SJ (2008) The fibroblast growth factor family and mood disorders. *Novartis Found Symp* 289: 94–96; discussion 97–100, 193–105
111 Kardami E, Detillieux K, Ma X, Jiang Z, Santiago JJ, Jimenez SK, Cattini PA (2007) Fibroblast growth factor-2 and cardioprotection. *Heart Fail Rev* 12: 267–277
112 Leonard BE, Song C (1999) Stress, depression, and the role of cytokines. *Adv Exp Med Biol* 461: 251–265
113 Watkins LR, Maier SF (1999) Implications of immune-to-brain communication for sickness and pain. *Proc Natl Acad Sci USA* 96: 7710–7713
114 Pace TW, Mletzko TC, Alagbe O, Musselman DL, Nemeroff CB, Miller AH, Heim CM (2006) Increased stress-induced inflammatory responses in male patients with major depression and increased early life stress. *Am J Psychiatry* 163: 1630–1633

115 Ising M, Kunzel HE, Binder EB, Nickel T, Modell S, Holsboer F (2005) The combined dexamethasone/CRH test as a potential surrogate marker in depression. *Prog Neuropsychopharmacol Biol Psychiatry* 29: 1085–1093
116 Jun TY, Pae CU, Hoon H, Chae JH, Bahk WM, Kim KS, Serretti A (2003) Possible association between –G308A tumour necrosis factor-alpha gene polymorphism and major depressive disorder in the Korean population. *Psychiatr Genet* 13: 179–181
117 Fertuzinhos SM, Oliveira JR, Nishimura AL, Pontual D, Carvalho DR, Sougey EB, Otto PA, Zatz M (2004) Analysis of IL-1alpha, IL-1beta, and IL-1RA [correction of IL-RA] polymorphisms in dysthymia. *J Mol Neurosci* 22: 251–256
118 Wilson AG, Gordon C, di Giovine FS, de Vries N, van de Putte LB, Emery P, Duff GW (1994) A genetic association between systemic lupus erythematosus and tumor necrosis factor alpha. *Eur J Immunol* 24: 191–195
119 Misener VL, Gomez L, Wigg KG, Luca P, King N, Kiss E, Daroczi G, Kapornai K, Tamas Z, Mayer L et al (2008) Cytokine Genes TNF, IL1A, IL1B, IL6, IL1RN and IL10, and childhood-onset mood disorders. *Neuropsychobiology* 58: 71–80
120 Kraus MR, Al-Taie O, Schafer A, Pfersdorff M, Lesch KP, Scheurlen M (2007) Serotonin-1A receptor gene HTR1A variation predicts interferon-induced depression in chronic hepatitis C. *Gastroenterology* 132: 1279–1286
121 Cai W, Khaoustov VI, Xie Q, Pan T, Le W, Yoffe B (2005) Interferon-alpha-induced modulation of glucocorticoid and serotonin receptors as a mechanism of depression. *J Hepatol* 42: 880–887
122 Neumeister A, Bain E, Nugent AC, Carson RE, Bonne O, Luckenbaugh DA, Eckelman W, Herscovitch P, Charney DS, Drevets WC (2004) Reduced serotonin type 1A receptor binding in panic disorder. *J Neurosci* 24: 589–591
123 Lanzenberger RR, Mitterhauser M, Spindelegger C, Wadsak W, Klein N, Mien LK, Holik A, Attarbaschi T, Mossaheb N, Sacher J et al (2007) Reduced serotonin-1A receptor binding in social anxiety disorder. *Biol Psychiatry* 61: 1081–1089
124 Hirvonen J, Karlsson H, Kajander J, Lepola A, Markkula J, Rasi-Hakala H, Nagren K, Salminen JK, Hietala J (2008) Decreased brain serotonin 5-HT1A receptor availability in medication-naive patients with major depressive disorder: an *in vivo* imaging study using PET and [carbonyl-11C]WAY-100635. *Int J Neuropsychopharmacol* 11: 465–476
125 Nash JR, Sargent PA, Rabiner EA, Hood SD, Argyropoulos SV, Potokar JP, Grasby PM, Nutt DJ (2008) Serotonin 5-HT1A receptor binding in people with panic disorder: positron emission tomography study. *Br J Psychiatry* 193: 229–234
126 Drago A, Ronchi DD, Serretti A (2008) 5-HT1A gene variants and psychiatric disorders: a review of current literature and selection of SNPs for future studies. *Int J Neuropsychopharmacol* 11: 701–721
127 Le Francois B, Czesak M, Steubl D, Albert PR (2008) Transcriptional regulation at a HTR1A polymorphism associated with mental illness. *Neuropharmacology* 55: 977–985

128 Lesch KP, Wiesmann M, Hoh A, Muller T, Disselkamp-Tietze J, Osterheider M, Schulte HM (1992) 5-HT1A receptor-effector system responsivity in panic disorder. *Psychopharmacology (Berl)* 106: 111–117

129 Hennig J, Becker H, Netter P (1996) 5–HT agonist-induced changes in peripheral immune cells in healthy volunteers: the impact of personality. *Behav Brain Res* 73: 359–363

130 Yu B, Becnel J, Zerfaoui M, Rohatgi R, Boulares AH, Nichols CD (2008) Serotonin 5–hydroxytryptamine(2A) receptor activation suppresses tumor necrosis factor-alpha-induced inflammation with extraordinary potency. *J Pharmacol Exp Ther* 327: 316–323

131 Evans SJ, Choudary PV, Vawter MP, Li J, Meador-Woodruff JH, Lopez JF, Burke SM, Thompson RC, Myers RM, Jones EG et al (2003) DNA microarray analysis of functionally discrete human brain regions reveals divergent transcriptional profiles. *Neurobiol Dis* 14: 240–250

132 Inada Y, Yoneda H, Koh J, Sakai J, Himei A, Kinoshita Y, Akabame K, Hiraoka Y, Sakai T (2003) Positive association between panic disorder and polymorphism of the serotonin 2A receptor gene. *Psychiatry Res* 118: 25–31

133 Unschuld PG, Ising M, Erhardt A, Lucae S, Kloiber S, Kohli M, Salyakina D, Welt T, Kern N, Lieb R et al (2007) Polymorphisms in the serotonin receptor gene HTR2A are associated with quantitative traits in panic disorder. *Am J Med Genet B Neuropsychiatr Genet* 144B: 424–429

134 Yoon HK, Yang JC, Lee HJ, Kim YK (2008) The association between serotonin-related gene polymorphisms and panic disorder. *J Anxiety Disord* 22: 1529–1534

135 Celada P, Puig M, Amargos-Bosch M, Adell A, Artigas F (2004) The therapeutic role of 5-HT1A and 5-HT2A receptors in depression. *J Psychiatry Neurosci* 29: 252–265

136 Pace TW, Hu F, Miller AH (2007) Cytokine-effects on glucocorticoid receptor function: relevance to glucocorticoid resistance and the pathophysiology and treatment of major depression. *Brain Behav Immun* 21: 9–19

137 Miller AH, Pariante CM, Pearce BD (1999) Effects of cytokines on glucocorticoid receptor expression and function. Glucocorticoid resistance and relevance to depression. *Adv Exp Med Biol* 461: 107–116

138 Pariante CM, Pearce BD, Pisell TL, Sanchez CI, Po C, Su C, Miller AH (1999) The proinflammatory cytokine, interleukin-1alpha, reduces glucocorticoid receptor translocation and function. *Endocrinology* 140: 4359–4366

139 Lichtenstein GR, Bala M, Han C, DeWoody K, Schaible T (2002) Infliximab improves quality of life in patients with Crohn's disease. *Inflamm Bowel Dis* 8: 237–243

140 Tyring S, Gottlieb A, Papp K, Gordon K, Leonardi C, Wang A, Lalla D, Woolley M, Jahreis A, Zitnik R et al (2006) Etanercept and clinical outcomes, fatigue, and depression in psoriasis: double-blind placebo-controlled randomised phase III trial. *Lancet* 367: 29–35

141 Müller N, Schwarz MJ, Dehning S, Douhe A, Cerovecki A, Goldstein-Muller B, Spellmann I, Hetzel G, Maino K, Kleindienst N et al (2006) The cyclooxygenase-2 inhibitor

celecoxib has therapeutic effects in major depression: results of a double-blind, randomized, placebo controlled, add-on pilot study to reboxetine. *Mol Psychiatry* 11: 680–684
142 Maes M, Song C, Lin AH, Bonaccorso S, Kenis G, De Jongh R, Bosmans E, Scharpe S (1999) Negative immunoregulatory effects of antidepressants: inhibition of interferon-gamma and stimulation of interleukin-10 secretion. *Neuropsychopharmacology* 20: 370–379
143 Kubera M, Lin AH, Kenis G, Bosmans E, van Bockstaele D, Maes M (2001) Anti-Inflammatory effects of antidepressants through suppression of the interferon-gamma/interleukin-10 production ratio. *J Clin Psychopharmacol* 21: 199–206
144 Leonard BE (2001) The immune system, depression and the action of antidepressants. *Prog Neuropsychopharmacol Biol Psychiatry* 25: 767–780
145 Diamond M, Kelly JP, Connor TJ (2006) Antidepressants suppress production of the Th1 cytokine interferon-gamma, independent of monoamine transporter blockade. *Eur Neuropsychopharmacol* 16: 481–490
146 Rapaport MH, Manji HK (2001) The effects of lithium on *ex vivo* cytokine production. *Biol Psychiatry* 50: 217–224
147 Maes M, Song C, Lin AH, Pioli R, Kenis G, Kubera M, Bosmans E (1999) In vitro immunoregulatory effects of lithium in healthy volunteers. *Psychopharmacology (Berl)* 143: 401–407
148 Lanquillon S, Krieg JC, Bening-Abu-Shach U, Vedder H (2000) Cytokine production and treatment response in major depressive disorder. *Neuropsychopharmacology* 22: 370–379
149 Kubera M, Maes M, Holan V, Basta-Kaim A, Roman A, Shani J (2001) Prolonged desipramine treatment increases the production of interleukin-10, an anti-inflammatory cytokine, in C57BL/6 mice subjected to the chronic mild stress model of depression. *J Affect Disord* 63: 171–178
150 Suhara T, Sudo Y, Yoshida K, Okubo Y, Fukuda H, Obata T, Yoshikawa K, Suzuki K, Sasaki Y (1998) Lung as reservoir for antidepressants in pharmacokinetic drug interactions. *Lancet* 351: 332–335
151 Rudd ML, Nicolas AN, Brown BL, Fischer-Stenger K, Stewart JK (2005) Peritoneal macrophages express the serotonin transporter. *J Neuroimmunol* 159: 113–118
152 O'Connell PJ, Wang X, Leon-Ponte M, Griffiths C, Pingle SC, Ahern GP (2006) A novel form of immune signaling revealed by transmission of the inflammatory mediator serotonin between dendritic cells and T cells. *Blood* 107: 1010–1017
153 Ramamoorthy S, Ramamoorthy JD, Prasad PD, Bhat GK, Mahesh VB, Leibach FH, Ganapathy V (1995) Regulation of the human serotonin transporter by interleukin-1 beta. *Biochem Biophys Res Commun* 216: 560–567
154 Mossner R, Heils A, Stober G, Okladnova O, Daniel S, Lesch KP (1998) Enhancement of serotonin transporter function by tumor necrosis factor alpha but not by interleukin-6. *Neurochem Int* 33: 251–254
155 Mossner R, Daniel S, Schmitt A, Albert D, Lesch KP (2001) Modulation of serotonin transporter function by interleukin-4. *Life Sci* 68: 873–880

156 Caspi A, Sugden K, Moffitt TE, Taylor A, Craig IW, Harrington H, McClay J, Mill J, Martin J, Braithwaite A et al (2003) Influence of life stress on depression: moderation by a polymorphism in the 5–HTT gene. *Science* 301: 386–389
157 Hariri AR, Holmes A (2006) Genetics of emotional regulation: the role of the serotonin transporter in neural function. *Trends Cogn Sci* 10: 182–191
158 Botting NP (1995) Chemistry and neurochemistry of the kynurenine pathway of tryptophan metabolism. *Chem Soc Rev* 24: 401–412
159 Mangoni A (1974) The "kynurenine shunt" and depression. *Adv Biochem Psychopharmacol* 11: 293–298
160 Badawy AA, Morgan CJ, Dacey A, Stoppard T (1991) The effects of lofepramine and desmethylimipramine on tryptophan metabolism and disposition in the rat. *Biochem Pharmacol* 42: 921–929
161 Badawy AA, Morgan CJ (1991) Effects of acute paroxetine administration on tryptophan metabolism and disposition in the rat. *Brit J Pharmacol* 102: 429–433
162 Bano S, Sherkheli MA (2003) Inhibition of tryptophan – pyrrolase activity and elevation of brain tryptophan concentration by fluoxetine in rats. *J Coll Physicians Surg Pak* 13: 5–10
163 Badawy AAB, Morgan JC, Bano S, Buckland P, McGuffin P (1996) Mechanism of enhancement of rat brain serotonin synthesis by acute fluoxetine administration. *J Neurochem* 66: 436–437
164 Badawy AA, Evans M (1982) Inhibition of rat liver tryptophan pyrrolase activity and elevation of brain tryptophan concentration by acute administration of small doses of antidepressants. *Brit J Pharmacol* 77: 59–67
165 Salter M, Hazelwood R, Pogson CI, Iyer R, Madge DJ (1995) The effects of a novel and selective inhibitor of tryptophan 2,3-dioxygenase on tryptophan and serotonin metabolism in the rat. *Biochem Pharmacol* 49: 1435–1442
166 Salter M, Hazelwood R, Pogson CI, Iyer R, Madge DJ, Jones HT, Cooper BR, Cox RF, Wang CM, Wiard RP (1995) The effects of an inhibitor of tryptophan 2,3-dioxygenase and a combined inhibitor of tryptophan 2,3-dioxygenase and 5-HT reuptake in the rat. *Neuropharmacology* 34: 217–227
167 Reinhard JF, Jr., Flanagan EM, Madge DJ, Iyer R, Salter M (1996) Effects of 540C91 [(E)-3-[2-(4'-pyridyl)-vinyl]-1H-indole], an inhibitor of hepatic tryptophan dioxygenase, on brain quinolinic acid in mice. *Biochem Pharmacol* 51: 159–163
168 Mellor AL, Munn DH (2003) Tryptophan catabolism and regulation of adaptive immunity. *J Immunol* 170: 5809–5813
169 Hestad KA, Tonseth S, Stoen CD, Ueland T, Aukrust P (2003) Raised plasma levels of tumor necrosis factor alpha in patients with depression: normalization during electroconvulsive therapy. *J Ect* 19: 183–188
170 Shafique S, Dalsing MC (2006) Vagus nerve stimulation therapy for treatment of drug-resistant epilepsy and depression. *Perspect Vasc Surg Endovasc Ther* 18: 323–327
171 Borovikova LV, Ivanova S, Zhang M, Yang H, Botchkina GL, Watkins LR, Wang H,

Abumrad N, Eaton JW, Tracey KJ (2000) Vagus nerve stimulation attenuates the systemic inflammatory response to endotoxin. *Nature* 405: 458–462
172 Ghia JE, Blennerhassett P, Collins SM (2008) Impaired parasympathetic function increases susceptibility to inflammatory bowel disease in a mouse model of depression. *J Clin Invest* 118: 2209–2218
173 Corcoran C, Connor TJ, O'Keane V, Garland MR (2005) The effects of vagus nerve stimulation on pro- and anti-inflammatory cytokines in humans: a preliminary report. *Neuroimmunomodulation* 12: 307–309
174 Feleszko W, Jaworska J, Rha RD, Steinhausen S, Avagyan A, Jaudszus A, Ahrens B, Groneberg DA, Wahn U, Hamelmann E (2007) Probiotic-induced suppression of allergic sensitization and airway inflammation is associated with an increase of T regulatory-dependent mechanisms in a murine model of asthma. *Clin Exp Allergy* 37: 498–505
175 Sheil B, McCarthy J, O'Mahony L, Bennett MW, Ryan P, Fitzgibbon JJ, Kiely B, Collins JK, Shanahan F (2004) Is the mucosal route of administration essential for probiotic function? Subcutaneous administration is associated with attenuation of murine colitis and arthritis. *Gut* 53: 694–700
176 Mazmanian SK, Round JL, Kasper DL (2008) A microbial symbiosis factor prevents intestinal inflammatory disease. *Nature* 453: 620–625
177 Calcinaro F, Dionisi S, Marinaro M, Candeloro P, Bonato V, Marzotti S, Corneli RB, Ferretti E, Gulino A, Grasso F et al (2005) Oral probiotic administration induces interleukin-10 production and prevents spontaneous autoimmune diabetes in the non-obese diabetic mouse. *Diabetologia* 48: 1565–1575
178 Desbonnet L, Garrett L, Clarke G, Bienenstock J, Dinan TG (2008) The probiotic Bifidobacteria infantis: An assessment of potential antidepressant properties in the rat. *J Psychiatr Res* 43: 164–174
179 Varghese AK, Verdu EF, Bercik P, Khan WI, Blennerhassett PA, Szechtman H, Collins SM (2006) Antidepressants attenuate increased susceptibility to colitis in a murine model of depression. *Gastroenterology* 130: 1743–1753
180 Kubera M, Kenis G, Bosmans E, Zieba A, Dudek D, Nowak G, Maes M (2000) Plasma levels of interleukin-6, interleukin-10, and interleukin-1 receptor antagonist in depression: comparison between the acute state and after remission. *Pol J Pharmacol* 52: 237–241
181 Parissis JT, Adamopoulos S, Rigas A, Kostakis G, Karatzas D, Venetsanou K, Kremastinos DT (2004) Comparison of circulating proinflammatory cytokines and soluble apoptosis mediators in patients with chronic heart failure with versus without symptoms of depression. *Am J Cardiol* 94: 1326–1328
182 Kushikata T, Fang J, Krueger JM (1999) Interleukin-10 inhibits spontaneous sleep in rabbits. *J Interferon Cytokine Res* 19: 1025–1030
183 Bluthé RM, Castanon N, Pousset F, Bristow A, Ball C, Lestage J, Michaud B, Kelley KW, Dantzer R (1999) Central injection of IL-10 antagonizes the behavioural effects of lipopolysaccharide in rats. *Psychoneuroendocrinology* 24: 301–311

184 Kanaan SA, Poole S, Saade NE, Jabbur S, Safieh-Garabedian B (1998) Interleukin-10 reduces the endotoxin-induced hyperalgesia in mice. *J Neuroimmunol* 86: 142–150

185 Di Santo E, Sironi M, Pozzi P, Gnocchi P, Isetta AM, Delvaux A, Goldman M, Marchant A, Ghezzi P (1995) Interleukin-10 inhibits lipopolysaccharide-induced tumor necrosis factor and interleukin-1 beta production in the brain without affecting the activation of the hypothalamus-pituitary-adrenal axis. *Neuroimmunomodulation* 2: 149–154

186 Kim WK, Ganea D, Jonakait GM (2002) Inhibition of microglial CD40 expression by pituitary adenylate cyclase-activating polypeptide is mediated by interleukin-10. *J Neuroimmunol* 126: 16–24

187 Leon LR, Kozak W, Kluger MJ (1998) Role of IL-10 in inflammation. Studies using cytokine knockout mice. *Ann NY Acad Sci* 856: 69–75

188 Amaral FA, Sachs D, Costa VV, Fagundes CT, Cisalpino D, Cunha TM, Ferreira SH, Cunha FQ, Silva TA, Nicoli JR et al (2008) Commensal microbiota is fundamental for the development of inflammatory pain. *Proc Natl Acad Sci USA* 105: 2193–2197

189 Bezchlibnyk YB, Wang JF, McQueen GM, Young LT (2001) Gene expression differences in bipolar disorder revealed by cDNA array analysis of post-mortem frontal cortex. *J Neurochem* 79: 826–834

190 Sutcigil L, Oktenli C, Musabak U, Bozkurt A, Cansever A, Uzun O, Sanisoglu SY, Yesilova Z, Ozmenler N, Ozsahin A et al (2007) Pro- and anti-inflammatory cytokine balance in major depression: effect of sertraline therapy. *Clin Dev Immunol* 2007: 76396

191 Lee KM, Kim YK (2006) The role of IL-12 and TGF-beta1 in the pathophysiology of major depressive disorder. *Int Immunopharmacol* 6: 1298–1304

192 Raymond NC, Dysken M, Bettin K, Eckert ED, Crow SJ, Markus K, Pomeroy C (2000) Cytokine production in patients with anorexia nervosa, bulimia nervosa, and obesity. *Int J Eat Disord* 28: 293–302

193 Dalbeth N, Yeoman S, Dockerty JL, Highton J, Robinson E, Tan PL, Herman D, McQueen FM (2004) A randomised placebo controlled trial of delipidated, deglycolipidated *Mycobacterium vaccae* as immunotherapy for psoriatic arthritis. *Ann Rheum Dis* 63: 718–722

194 O'Brien ME, Anderson H, Kaukel E, O'Byrne K, Pawlicki M, Von Pawel J, Reck M (2004) SRL172 (killed *Mycobacterium vaccae*) in addition to standard chemotherapy improves quality of life without affecting survival, in patients with advanced non-small-cell lung cancer: phase III results. *Ann Oncol* 15: 906–914

195 O'Brien ME, Saini A, Smith IE, Webb A, Gregory K, Mendes R, Ryan C, Priest K, Bromelow KV, Palmer RD et al (2000) A randomized phase II study of SRL172 (*Mycobacterium vaccae*) combined with chemotherapy in patients with advanced inoperable non-small-cell lung cancer and mesothelioma. *Br J Cancer* 83: 853–857

196 Heim C, Nemeroff CB (1999) The impact of early adverse experiences on brain systems involved in the pathophysiology of anxiety and affective disorders. *Biol Psychiatry* 46: 1509–1522

197 Rook GAW, Lowry CA (2008) The hygiene hypothesis and psychiatric disorders. *Trends Immunol* 29: 150–158

Immune regulation in atherosclerosis and the hygiene hypothesis

Hafid Ait-Oufella, Alain Tedgui and Ziad Mallat

Paris Cardiovascular Research Center, INSERM and Assistance Publique-Hôpitaux de Paris, Hôpital Européen Georges Pompidou, 75015 Paris, France

Abstract

Atherosclerosis is a chronic inflammatory disease of the arterial wall where both innate and adaptive immune responses contribute to disease initiation and progression. The hygiene hypothesis implies that dysregulation of the immune response has led to increased susceptibility to immuno-inflammatory diseases. Recent studies established that subtypes of T cells, regulatory T cells, actively involved in the maintenance of immunological tolerance, inhibit the development and progression of atherosclerosis. Here, we review the immune regulatory pathways of atherosclerosis and discuss the potential implication of pathogens and their associated molecular patterns in the regulation of the immuno-inflammatory response of atherosclerosis.

Atherosclerosis is an immune regulated disease

Atherosclerosis is a pathological condition of the arterial wall that underlies adverse vascular events including coronary artery disease, stroke, abdominal aortic aneurysms and ischemic gangrene, responsible for most of the cardiovascular morbidity and mortality in the Western world. Epidemiological studies also indicate that the prevalence of atherosclerosis is increasing due to the adoption of the Western life-style in developing countries and the accumulation of metabolic risk factors [1, 2].

Role of endothelial cell activation

Experimental and clinical studies have provided strong evidence supporting a crucial role for inflammation in the development and progression of atherosclerosis. Atherosclerosis is initiated by focal endothelial activation in large- and medium-size arteries induced by biochemical and physical stimulation, including hypercholesterolemia and hypertension. Initially, lipoproteins infiltrate the artery wall to an extent that exceeds the capacity for elimination, and are retained in the intimal space [3].

The Hygiene Hypothesis and Darwinian Medicine, edited by Graham A. W. Rook

LDL modifications, through enzymatic attack or non-enzymatic oxidation, release bioactive phospholipids that can activate endothelial cells. Activated endothelial cells express leukocyte adhesion molecules and release chemokines, which promote leukocyte recruitment (monocytes, lymphocytes) into the intima [4].

Innate immunity and atherosclerosis

Cytokines and growth factors produced in the inflamed intima, such as MCP-1 and M-CSF [5], induce monocyte diapedesis into the plaque where they differentiate into macrophages. This step is critical for the development of atherosclerosis and is associated with upregulation of pattern-recognition receptors for innate immunity, including scavenger receptors (SRs) and toll-like receptors (TLRs). SRs lead to phagocytosis of bacterial components, apoptotic debris and oxLDL particles, and to foam cell transformation [6]. TLRs also bind molecules with pathogen-like molecular patterns and induce a signal cascade through Myd88 that provokes cell activation [7]. The activated macrophages produce proinflammatory cytokines, metalloproteinases and cytotoxic and nitrogen radical molecules. Lipid-laden macrophages participate in the progression of atherosclerosis and to growth of necrotic cores through molecule secretion (cytokines, chemokines, proteases) and cellular interaction with vascular cells (endothelial and smooth muscle) and inflammatory actors such as T cells.

Pathogenic Th1-driven responses in atherosclerosis

Experimental studies have clearly shown that the adaptive immune system affects the development of atherosclerosis [8]. In humans, most of the T cells in atherosclerotic plaques express $\alpha\beta$-T cell receptor (TCR) and are of the T helper type 1 (Th1). They are responsible for cell-mediated immunity and secrete interferon (IFN)-γ, and IL-2, in contrast to Th2 cells, which secrete IL-4, IL-5, IL-10 and IL-13, and provide help for antibody production by B cells. In mice, deficiency in both T and B cells, as occurs in apoE$^{-/-}$ or LDLr$^{-/-}$ mice on a recombination-activating gene (Rag)-deficient background, is associated with a significant reduction in early atherosclerotic lesion development [9, 10]. Moreover, atherosclerosis is enhanced after transfer of CD4^{+} T cells from apoE$^{-/-}$ mice into apoE$^{-/-}$xSCID immunodeficient mice, indicating a proatherogenic role of T cells [11]. The transplanted cells produced high levels of IFN-γ, suggesting a Th1-related proatherogenic effect. Subsequent studies have shown that cytokines or transcription factors involved in the differentiation and/or activation of Th1 cells also contribute to the atherosclerotic process. Differentiation of Th1 cells requires TCR activation by DCs. Distinct subsets of DCs elicit distinct T helper responses [12]. IL-12 production by DCs plays a critical role in Th1 differentiation as DCs from IL-12$^{-/-}$ mice fail to induce Th1 responses [13]. IL-12 activates the transcription factor signal transducer and activator of transcription 4 (STAT4) and a unique Th1 transcription factor, T-box expressed in T cells (T-bet), leading to upregulation of IFN-γ, the prototypic Th1 cytokine. Interleukin-12 synergises with IL-18 for full induction of IFN-γ. In addition, proinflammatory mediators such as tumour necrosis factor (TNF) and IL-1, as well as co-stimulatory signals, including CD40/CD40L interaction, contribute to DC maturation and induction of Th1 cells. All these proinflammatory mediators, co-stimulatory molecules and transcription factors involved in Th1 differentiation and activation are expressed in atherosclerotic plaques of mice and humans, and are required for initial plaque development [14–16], as well as for the perpetuation of plaque inflammation and 'instability' [15–18] in mouse models of atherosclerosis. Overall, these results provide convincing elements to incriminate Th1-related responses in the promotion of plaque development and progression.

Th2-driven responses do not invariably prevent atherosclerosis

Differentiation toward Th2 cells requires specific factors. IL-6, IL-13 and OX40-L (CD134; OX40 ligand) may play a role in DC-induced Th2 differentiation [19]. Particularly, interleukin-4 activates the Th2 transcription factor Gata3 through STAT6, induces interleukin-5 and downregulates IFN-γ. Counter-regulation between Th1 and Th2 may result from a balance between T-bet and Gata3 [20]. It has therefore been proposed that Th2-biased responses antagonise proatherogenic Th1 effects and thereby should confer atheroprotection. In fact, a number of experiments, especially those exploring the role of humoral immunity in atherosclerosis, suggest that Th2-driven humoral immune responses may be atheroprotective. A switch toward a Th2 cytokine profile in mouse models of atherosclerosis is associated with increased production of 'protective' anti-oxidised LDL (oxLDL) antibodies [21]. Furthermore, splenectomy in cholesterol-fed apoE$^{-/-}$ mice, which is associated with reduced levels of IgM and Th2-related IgG anti-oxLDL antibodies, increases atherosclerosis [22]. Production of high titers of IgM-type anti-oxLDL antibodies, as observed following immunization of apoE$^{-/-}$ mice with malondialdehyde-LDL, is associated with reduced lesion size [23–25]. These antibodies arise from B1 cells and appear to be under the control of IL-5 produced by modified LDL-specific Th2 cells [26, 27]. A switch toward the production of Th2-related IgG1 antibodies has been reported in mice overexpressing IL-10, which was associated with a reduction in lesion size [28]. In addition, promoting Th2 responses in mice with mild hypercholesterolemia resulted in a reduction of early fatty-streak formation [29]. However, other data indicate that Th2 responses may be proatherogenic. Deficiency in IL-4, the prototypic Th2-related cytokine, is associated with a decrease in atherosclerotic lesion formation [30], particularly at the advanced stages of lesion progression [31].

Thus, even though initial lesion development in mice is mostly under the control of Th1-related immunity and could be counter-regulated by the promotion of a Th2 response, this may occur at the risk of favouring plaque progression as the lesion progresses in a hypercholesterolemic context. Therefore, the attractive concept of the Th1 and Th2 yin and yang controlling the development of atherosclerosis may not stand at all stages of plaque development. It would therefore be risky to promote Th2 responses as a strategy to modulate atherosclerosis.

Regulatory T cells and immunoregulatory cytokines TGF-β and IL-10 are potent inhibitors of atherosclerosis

Since both Th1- and Th2-mediated responses may potentially promote atherogenesis, we hypothesised that this disease is probably under the control of regulatory cell subsets, capable of suppression of both Th1 and Th2 pathogenic responses. This hypothesis was supported by our initial studies showing that IL-10 and TGF-β, two major mediators of regulatory T cell (Treg) function, are potent inhibitors of atherosclerosis in experimental models of the disease [32, 33].

IL-10, Treg cell function and atherosclerosis

Experiments using specific deletion of IL-10 in lymphocytes have revealed the importance of this cytokine in the protection against inflammatory processes. Mice with deficiency of IL-10 are susceptible to inflammatory diseases such as colitis [34]. Furthermore, adoptive transfer of IL-10-deficient $CD4^{+}$ T cells into lymphopenic mice induces severe colitis despite the ability of the recipient's innate immune system to produce IL-10 [35]. In lymphocytes, the production of IL-10 has been associated with the Th2 subset and Treg cells. Among the regulatory T cells, both natural nTreg and induced Tr1 cells have the capacity to produce IL-10. Tr1 cells exhibit their suppressor function by a cell contact-independent, cytokine-dependent mechanism that involves both IL-10 and TGF-β. Several experiments have revealed the requirement for IL-10 to modulate the activation of DCs that prime Tr1 development. Specifically, it has been shown that culture of bone marrow cells in the presence of IL-10 induces the differentiation of tolerogenic DCs expressing $CD11c^{low}CD45RB^{high}$, which have the capacity to induce the Tr1 phenotype *in vitro* and *in vivo* [36].

The role of endogenous IL-10 has been clearly established in mouse models of atherosclerosis. We and others have shown that IL-10 deficiency in C57BL/6 mice fed an atherogenic cholate-containing diet promotes early atherosclerotic lesion formation, characterised by increased infiltration of inflammatory cells, particularly activated T cells, and by increased production of proinflammatory cytokines [33].

Similar results have been reported in IL-10$^{-/-}$/apoE$^{-/-}$ mice fed a chow diet [37]. More recently, using a model of chimeric LDLr$^{-/-}$ mice in which bone marrow cells were deficient in IL-10, we showed that the absence of IL-10 induced a clear switch toward a Th1 immune response, associated with enhanced accumulation of T cells and macrophages within the lesions [38]. These results provided evidence that leukocyte-derived IL-10 is instrumental in the prevention of atherosclerotic lesion development and in the modulation of cellular and collagen plaque composition, at least in part, through a systemic immune response modulation [38]. The effect of IL-10 disruption in specific cell subtypes (macrophages, DCs or T cells) on lesion development and progression is still unknown. Consistent with a protective role of IL-10 in atherosclerosis, systemic or local overexpression of IL-10 by adenoviral gene transfer in collar-induced carotid atherosclerosis of LDLr$^{-/-}$ mice was found to be highly efficient in preventing atherosclerosis [39]. It is noteworthy that overexpression of IL-10 by activated T lymphocytes reduced atherosclerosis in LDLr$^{-/-}$ mice [28]. The authors attributed these effects to a switch towards a Th2-like phenotype but failed to report on IL-4 production. In fact, the mouse strain used in that study has been shown to be unable to generate Th2 responses [40], leading us to suggest that the protective effect on atherosclerosis was associated with a Tr1-like phenotype. This is consistent with studies showing that transfer of clones of Tr1 cells reduces lesion development in apoE$^{-/-}$ mice [41] and that promotion of the endogenous adaptive Tr1 cell response plays a significant role in limiting disease development during the natural course of atherosclerosis [42].

TGF-β, Treg cell function and atherosclerosis

The importance of TGF-β in the immune system was highlighted by the discovery that TGF-β-deficient mice develop multiple inflammatory diseases [43, 44]. These were associated with enhanced T cell proliferation, activation, and a switch of T cell differentiation toward both Th1 and Th2 profiles. The activation of T cells in this setting results from the fact that TGF-β inhibits the proliferation, activation and differentiation of T cells towards Th1 and Th2 [45]. In addition, TGF-β1 has been shown to maintain Treg cells in the periphery by acting as a co-stimulatory factor for expression of Foxp3 [46]. This dual effect on effector T cells and Treg cells is likely to contribute to regulation of peripheral T cell tolerance by TGF-β.

Studies using either TGF-β-neutralising antibodies [32] soluble TGF-β receptors [47] or genetic deficiency in TGF-β [48], demonstrated an anti-atherosclerotic effect of TGF-β in apoE$^{-/-}$ mice. In these murine models, accelerated development of atherosclerosis was observed, with increased infiltration of inflammatory cells within lesions, together with reduced collagen content [32, 47]. Therefore, TGF-β has anti-inflammatory effects in addition to its stabilising effects within the lesions through the induction of extracellular matrix synthesis.

The availability of mice with specific deletion of TGF-β signalling in T cells facilitated the study of the specific role of TGF-β in T cell-induced atherosclerosis [49, 50]. The transplantation of bone marrow from T cell dominant-negative TGF-β receptor type II mice into irradiated LDLr$^{-/-}$ mice revealed accelerated atherogenesis [50] and increased differentiation of T cells toward both Th1 and Th2 phenotypes [49, 50]. These studies clearly showed a protective role of T cell-specific TGF-β activity against atherogenesis by inhibiting activation of both Th1 and Th2 pathways. These experimental studies were in agreement with data in humans showing that patients with acute coronary syndromes displayed reduced circulating IL-10 levels [51] and that low levels of serum TGF-β activity were associated with more widespread atherosclerosis [52].

The cellular source of TGF-β within atherosclerotic lesions is multiple since all atheroma-associated cells have the capacity to produce this cytokine. Treg cells, which can be both source and target of TGF-β may, contribute to its production and/or protective activity. Strategies using mouse models with genetic deficiency of Treg cells or strategies using CD25-neutralising antibodies clearly demonstrated a protective role of Treg cells against atherogenesis [53]. Moreover, Treg depletion did not influence lesion size or inflammatory phenotype when the host T cells did not respond to TGF-β, suggesting that this factor is required for the atheroprotective effect of Treg cells. Furthermore, reduction in atherosclerosis in ApoE$^{-/-}$ mice has also been achieved through adoptive transfer of CD4^{+}CD25^{+} regulatory T cells [53, 54]. Other studies have been published supporting an anti-atherogenic role for this Treg cell subtype. Recent studies highlighted the role of inducible co-stimulatory molecule (ICOS) on regulatory T cell responses in atherosclerosis [55]. LDLr$^{-/-}$ mice transplanted with ICOS-deficient bone marrow showed accelerated atherosclerosis and enhanced infiltration of CD4^{+} T cells, as well as increased macrophage content. This was associated with decreased numbers of Foxp3^{+} Treg cells and impaired *in vitro* Treg-suppressive function in ICOS-deficient mice compared with control mice, suggesting that ICOS modulates atherosclerosis through its effect on Treg cell responses [55]. Compound deficient apoE$^{-/-}$/Cxcl10$^{-/-}$ mice fed a Western-style diet demonstrated significant reductions in atherogenesis as compared with *apoE*$^{-/-}$ controls, and this was associated with increased Foxp3 expression, as well as IL-10 and TGF-β1 immunostaining [56]. These studies in mice seem to be relevant to the clinical situation in humans where defective Treg cell number/function has been associated with the presence of advanced stable or unstable coronary artery disease [57, 58].

The hygiene hypothesis and atherosclerosis

The hygiene hypothesis stipulates that reduced exposure to microorganisms has led to immune dysregulation, which has promoted an increase in susceptibility to immuno-inflammatory disorders. If by immune dysregulation, one means decreased Th1 and increased Th2 responses, we would argue against any role of the hygiene hypothesis in the development and progression of atherosclerosis. In contrast, if immune dysregulation means an alteration in regulatory T cell function, which is probably the preferred current explanation linking hygiene status to the susceptibility to immuno-inflammatory disorders, we would support a significant role for the hygiene hypothesis in modulating the susceptibility to atherosclerosis. However, in order to address the relation between the hygiene hypothesis and atherosclerosis, it will be important to examine 1) the evidence linking pathogens and their associated molecular patterns to atherosclerosis, 2) and whether modulation of immune responses by pathogens and related products may alter the susceptibility to atherosclerosis.

> *"Pathogens are probably not necessary for the development or complications of atherosclerosis unless there is a defect in the ability to mount an anti-inflammatory response."*

Certain common bacteria such as *C. pneumoniae*, *Mycoplasma pneumoniae*, and *Helicobacter pylori*, which cause chronic infections, as well as the herpes virus family members cytomegalovirus (CMV), Epstein-Barr virus, and herpes-simplex virus type I have all been implicated in atherosclerosis. However, a meta-analysis supports a possible pathogenic role in coronary artery diseases only for *C. pneumoniae* and CMV [59].

C. pneumoniae and CMV are ubiquitous, with 50–100% of the human population infected. CMV seropositivity correlates with atherosclerosis, restenosis after coronary angioplasty, and transplant vascular sclerosis (TVS, also known as transplant arteriosclerosis or chronic vascular rejection of organ grafts) [60]. CMV antigens and DNA have been detected both in TVS and atherosclerotic vessels [61, 62]. CMV infection in cardiac-transplant recipients doubles the rate of long-term graft failure because of accelerated TVS. CMV infection also accelerates experimental TVS [63]. CMV targets vascular endothelial and smooth muscle cells and causes endothelial expression of intercellular-adhesion molecule 1 (ICAM-1) and several chemokines [64, 65]. In animal studies, an atherosclerosis-like process is induced by infecting chickens with an avian herpesvirus, Marek's disease virus [66]. CMV infection of apoE$^{-/-}$ mice increases lesion size, lesion T cell content, and systemic levels of IFN-γ and TNF [67]. Although this result suggests that an immune response to CMV may aggravate early atherosclerosis, an inflammatory micro-environment in the atherosclerotic artery may also result in reactivation of virus. Viral reactivation could conceivably contribute to plaque activation and acute coronary syndromes. Indeed, seropositivity to CMV increases the risk of death in patients with coronary artery disease [68], and patients with acute myocardial infarction frequently develop CMV antigenemia [69].

Initial sero-epidemiological evidence for an association between *C. pneumoniae* and atherosclerosis [70] was supported by detection of this intracellular pathogen in (occasional cells of) atherosclerotic lesions [71]. However, the most recent clinical trials, including Weekly Intervention with Zithromax for Atherosclerosis and its Related Disorders (WIZARD) [72], Azithromycin in Acute Coronary Syndrome (AZACS) [73], Antibiotic Therapy After Acute Myocardial Infarction (ANTIBIO), Pravastatin or Atorvastatin Evaluation and Infection Therapy (PROVE-IT) [74], and Azithromycin and Coronary Events Study (ACES) [75], assessing the potential benefits of antibiotic therapy with the goal of targeting *Chlamydia pneumoniae* showed no effect of treatment in patients with CAD. Several reports indicate that inoculation of atherosclerosis-prone mice with high doses of *C. pneumoniae* fosters atherosclerosis [76, 77]. Yet, others have detected no aggravating effects [78] and if present, the atherogenic effect of *C. pneumoniae* requires elevated serum cholesterol levels [77]. Moreover, experimental studies showed that infection is not necessary for initiation or progression of atherosclerosis in apoE-deficient mice. Atherosclerosis develops identically in germ-free animals and in animals raised with ambient levels of microbial challenge [79]. However, we have clearly shown that under IL-10 deficiency, mice housed in conventional conditions develop greater atherosclerosis than mice housed in SPF conditions [33], clearly suggesting that ambient microorganisms may induce both pro- and anti-atherogenic pathways and that their impact on atherogenesis may depend on the ability to mount pro- or anti-inflammatory responses. One must therefore conclude that pathogens do not serve as etiologic agents for atherosclerosis, even though one cannot rule out a role in disease exacerbation in the presence of a defective anti-inflammatory response.

Pro-atherogenic pathways of pathogens and their associated molecular patterns

A refined version of the infectious hypothesis has been formulated a few years ago and evoked an atherogenic role of microorganisms-derived proteins, called pathogen associated molecular patterns (PAMPs). PAMPs are attractive as they could stimulate TLRs on vascular and inflammatory cells and could modulate endothelial cell activation, monocyte adhesion, foam cell formation and cytokine production. For instance, LPS-associated TLR-4 signalling promotes a proinflammatory phenotype in vascular SMC, inducing the release of MCP-1 and IL-6, and enhancing IL-1 expression [80]. Dendritic cells, after TLR activation, secrete high levels of IL-12, TNF-α, and IFN-γ, but no IL-10 [81]. In animal models, administration of purified PAMPs or heat-killed bacteria stimulates atherogenesis [82, 83]. PAMPs such as peptidoglycan [84], LPS and bacterial DNA [85] have been identified as constituents of human atheroma. Thus it is possible that PAMPs, rather than viable organisms, potentiate atherosclerosis. Recent genetic studies in atherosclerosis-prone apoE$^{-/-}$ or LDLr$^{-/-}$ mice have revealed a central role for TLR-signalling in the development of atherosclerosis. Genetic deletion of Myd88 results in a 60% reduction in plaque burden [86] and deletion of TLR-4 or TLR-2 alone induce a significant reduction of aortic lesion area (30% to 60%) [82, 86]. Conversely, recurrent administration of the TLR-4 agonist LPS, or the synthetic BLP mimetic and TLR2 agonist PAM3CSK4 [83], accelerates atherosclerosis development. In human atheroma, some TLRs (1,2,4,5 and 6) are upregulated when compared to healthy arteries [87]. Furthermore, a tight link exists between TLR-signalling and macrophage cholesterol metabolism. Stimulation of TLR2- or TLR4-dependent signalling in macrophages potentiates foam cell formation *in vitro* while TLR-deficient macrophages are resistant to foam cell differentiation [88].

A number of pathways exist by which arteries may become exposed to bacterial PAMPs. The most often suggested explanation is that circulating monocytes may phagocytose bacteria at sites of chronic infections and thereafter deliver their (killed and partially degraded) cargo to the developing lesion. PAMPs can translocate into the circulation even in relatively healthy subjects. For example, peptidoglycan is present in the circulation of healthy subjects after, but not prior to, bacterial colonisation of the gut [89], and many recent studies have shown that endotoxin is present, albeit at very low concentrations, in the circulation of all healthy human subjects [90, 91]. Moreover, epidemiologic studies have reported a positive correlation between circulating endotoxin concentrations and the incidence of carotid atherosclerosis, suggesting that subclinical endotoxemia may exert atherogenic effects on the vasculature over the longer term [90]. Theses points could constitute an explanation for the epidemiologic association between periodontal diseases (mucosal site for chronic microorganism infections) and cardiovascular diseases [92].

Modulation of immune responses by pathogens and related products leading to anti-atherogenic effects

Microorganisms could also modulate adaptive immune responses and, through this route, could influence atherosclerosis. Cholera bacterial toxins and molecules from helminth pathogens appear to be capable of driving DCs into a phenotype that selectively enhances Th2 induction [93, 94]. In the lung, *Bordetella pertussis* infection stimulates DCs to produce large amounts of IL-10, inhibits IL-12 production and consecutively induces expansion of Tr-1 like regulatory cell population, known to be atheroprotective [95].

Bacterial DNA contains CpG motifs that can trigger an immune response through TLR-9. A recent study has reported that administration of CpG induced a deviation of the inflammatory response toward a Th2 profile, inhibited MCP-1 release, and VCAM expression and reduced atherosclerotic lesion size in apoE$^{-/-}$ mice [96].

Others have reproducibly shown that administration of complete and incomplete Freund's adjuvant clearly reduces the development of atherosclerosis in several mouse models [97–99]. The mechanisms involved in this protection are currently unknown. We speculate an important role for Treg cells in Freund's adjuvant is anti-atherogenic effects. In support of this hypothesis, one study showed that the anti-atherogenic effects of Freund's adjuvant was lost in mice with CD4 deficiency [97], suggesting a role for CD4 cells in mediating the protective effect of Freund's adjuvant. We believe that these are Treg cells since several studies have now reported increased Treg cell number and function following administration of Freund's adjuvant [100].

Binder et al. have shown molecular mimicry between epitopes of oxLDL and *S. pneumoniae* and indicated that these immune responses can have beneficial effects in the context of atherosclerosis [101]. Immunization of $Ldlr^{-/-}$ mice with *Streptococcus pneumoniae* induced high circulating levels of protective oxLDL-specific T15 IgM antibodies cross-reactive with pneumococcal determinants, culminating in a significant decrease in the extent of atherosclerosis.

Finally, viruses and related products may also modulate immune responses towards an anti-atherogenic effect. Measles virus (MV) is responsible for an acute childhood disease that each year infects over 40 million individuals and causes the death of more than 1 million, primarily in the developing world [102]. The high mortality is associated with transitional MV-induced immunosuppression, enabling secondary infections. *In vitro* and *in vivo*, MV-infected DCs produce less IL-12 and inhibit T cell proliferation [103]. We have recently shown that repetitive administration of MV nucleoprotein to $apoE^{-/-}$ mice modulates DC/T cell functions, induces a Tr-1 like phenotype and reduces atherosclerosis [42]. These findings identify a novel mechanism of immune modulation by measles virus nucleoprotein through the promotion of a regulatory T cell response and suggest that this property may be harnessed for treating atherosclerosis.

Conclusion

The last decade has witnessed major advances in our understanding of the pathophysiology of atherosclerosis. The discovery of endogenous counter-regulators of the pathogenic immune response in atherosclerosis led to the identification of an important role for Treg cells in the control of lesion development and/or progression. Pathogens, their components and their associated molecular patterns clearly impact atherogenesis through the modulation of natural and adaptive immune responses.

References

1 Bonow RO (2002) Primary prevention of cardiovascular disease: a call to action. *Circulation* 106: 3140–3141

2 Lopez AD, Murray CC (1998) The global burden of disease, 1990–2020. *Nat Med* 4: 1241–1243
3 Skalen K, Gustafsson M, Rydberg EK, Hulten LM, Wiklund O, Innerarity TL, Boren J (2002) Subendothelial retention of atherogenic lipoproteins in early atherosclerosis. *Nature* 417: 750–754
4 Eriksson EE, Xie X, Werr J, Thoren P, Lindbom L (2001) Importance of primary capture and L-selectin-dependent secondary capture in leukocyte accumulation in inflammation and atherosclerosis *in vivo*. *J Exp Med* 194: 205–218
5 Smith JD, Trogan E, Ginsberg M, Grigaux C, Tian J, Miyata M (1995) Decreased atherosclerosis in mice deficient in both macrophage colony-stimulating factor (op) and apolipoprotein E. *Proc Natl Acad Sci USA* 92: 8264–8268
6 Peiser L, Mukhopadhyay S, Gordon S (2002) Scavenger receptors in innate immunity. *Curr Opin Immunol* 14: 123–128
7 Janeway CA, Jr., Medzhitov R (2002) Innate immune recognition. *Annu Rev Immunol* 20: 197–216
8 Binder CJ, Chang MK, Shaw PX, Miller YI, Hartvigsen K, Dewan A, Witztum JL (2002) Innate and acquired immunity in atherogenesis. *Nat Med* 8: 1218–1226
9 Dansky HM, Charlton SA, Harper MM, Smith JD (1997) T and B lymphocytes play a minor role in atherosclerotic plaque formation in the apolipoprotein E-deficient mouse. *Proc Natl Acad Sci USA* 94: 4642–4646
10 Daugherty A, Pure E, Delfel-Butteiger D, Chen S, Leferovich J, Roselaar SE, Rader DJ (1997) The effects of total lymphocyte deficiency on the extent of atherosclerosis in apolipoprotein $E^{-/-}$ mice. *J Clin Invest* 100: 1575–1580
11 Zhou X, Nicoletti A, Elhage R, Hansson GK (2000) Transfer of CD4(+) T cells aggravates atherosclerosis in immunodeficient apolipoprotein E knockout mice. *Circulation* 102: 2919–2922
12 Banchereau J, Briere F, Caux C, Davoust J, Lebecque S, Liu YJ, Pulendran B, Palucka K (2000) Immunobiology of dendritic cells. *Annu Rev Immunol* 18: 767–811
13 Maldonado-Lopez R, De Smedt T, Michel P, Godfroid J, Pajak B, Heirman C, Thielemans K, Leo O, Urbain J, Moser M (1999) $CD8alpha^+$ and $CD8alpha^-$ subclasses of dendritic cells direct the development of distinct T helper cells *in vivo*. *J Exp Med* 189: 587–592
14 Mach F, Schönbeck U, Sukhova GK, Atkinson E, Libby P (1998) Reduction of atherosclerosis in mice by inhibition of CD40 signalling. *Nature* 394: 200–203
15 de Nooijer R, von der Thusen JH, Verkleij CJ, Kuiper J, Jukema JW, van der Wall EE, van Berkel JC, Biessen EA (2004) Overexpression of IL-18 decreases intimal collagen content and promotes a vulnerable plaque phenotype in apolipoprotein-E-deficient mice. *Arterioscler Thromb Vasc Biol* 24: 2313–2319
16 Buono C, Binder CJ, Stavrakis G, Witztum JL, Glimcher LH, Lichtman AH (2005) T-bet deficiency reduces atherosclerosis and alters plaque antigen-specific immune responses. *Proc Natl Acad Sci USA* 102: 1596–1601
17 Mallat Z, Corbaz A, Scoazec A, Graber P, Alouani S, Esposito B, Humbert Y, Chvatchko

Y, Tedgui A (2001) Interleukin-18/interleukin-18 binding protein signaling modulates atherosclerotic lesion development and stability. *Circ Res* 89: E41–45

18 Schonbeck U, Sukhova GK, Shimizu K, Mach F, Libby P (2000) Inhibition of CD40 signaling limits evolution of established atherosclerosis in mice. *Proc Natl Acad Sci USA* 97: 7458–7463

19 Banchereau J, Steinman RM (1998) Dendritic cells and the control of immunity. *Nature* 392: 245–252

20 Szabo SJ, Kim ST, Costa GL, Zhang X, Fathman CG, Glimcher LH (2000) A novel transcription factor, T-bet, directs Th1 lineage commitment. *Cell* 100: 655–669

21 Zhou X, Paulsson G, Stemme S, Hansson GK (1998) Hypercholesterolemia is associated with a T helper (Th) 1/Th2 switch of the autoimmune response in atherosclerotic apo E-knockout mice. *J Clin Invest* 101: 1717–1725

22 Caligiuri G, Nicoletti A, Poirier B, Hansson GK (2002) Protective immunity against atherosclerosis carried by B cells of hypercholesterolemic mice. *J Clin Invest* 109: 745–753

23 Palinski W, Miller E, Witztum JL (1995) Immunization of low density lipoprotein (LDL) receptor-deficient rabbits with homologous malondialdehyde-modified LDL reduces atherogenesis. *Proc Natl Acad Sci USA* 92: 821–825

24 Freigang S, Horkko S, Miller E, Witztum JL, Palinski W (1998) Immunization of LDL receptor-deficient mice with homologous malondialdehyde-modified and native LDL reduces progression of atherosclerosis by mechanisms other than induction of high titers of antibodies to oxidative neoepitopes. *Arterioscler Thromb Vasc Biol* 18: 1972–1982

25 George J, Afek A, Gilburd B, Levkovitz H, Shaish A, Goldberg I, Kopolovic Y, Wick G, Shoenfeld Y, Harats D (1998) Hyperimmunization of apo-E-deficient mice with homologous malondialdehyde low-density lipoprotein suppresses early atherogenesis. *Atherosclerosis* 138: 147–152

26 Binder CJ, Hartvigsen K, Chang MK, Miller M, Broide D, Palinski W, Curtiss LK, Corr M, Witztum JL (2004) IL-5 links adaptive and natural immunity specific for epitopes of oxidized LDL and protects from atherosclerosis. *J Clin Invest* 114: 427–437

27 Binder CJ, Shaw PX, Chang MK, Boullier A, Hartvigsen K, Horkko S, Miller YI, Woelkers DA, Corr M, Witztum JL (2005) The role of natural antibodies in atherogenesis. *J Lipid Res* 46: 1353–1363

28 Pinderski LJ, Fischbein MP, Subbanagounder G, Fishbein MC, Kubo N, Cheroutre H, Curtiss LK, Berliner JA, Boisvert WA (2002) Overexpression of interleukin-10 by activated T lymphocytes inhibits atherosclerosis in LDL receptor-deficient Mice by altering lymphocyte and macrophage phenotypes. *Circ Res* 90: 1064–1071

29 Huber SA, Sakkinen P, David C, Newell MK, Tracy RP (2001) T helper-cell phenotype regulates atherosclerosis in mice under conditions of mild hypercholesterolemia. *Circulation* 103: 2610–2616

30 King VL, Szilvassy SJ, Daugherty A (2002) Interleukin-4 deficiency decreases atherosclerotic lesion formation in a site-specific manner in female LDL receptor$^{-/-}$ mice. *Arterioscler Thromb Vasc Biol* 22: 456–461

31 Davenport P, Tipping PG (2003) The role of interleukin-4 and interleukin-12 in the progression of atherosclerosis in apolipoprotein E-deficient mice. *Am J Pathol* 163: 1117–1125
32 Mallat Z, Gojova A, Marchiol-Fournigault C, Esposito B, Kamate C, Merval R, Fradelizi D, Tedgui A (2001) Inhibition of transforming growth factor-beta signaling accelerates atherosclerosis and induces an unstable plaque phenotype in mice. *Circ Res* 89: 930–934
33 Mallat Z, Besnard S, Duriez M, Deleuze V, Emmanuel F, Bureau MF, Soubrier F, Esposito B, Duez H, Fievet C et al (1999) Protective role of interleukin-10 in atherosclerosis. *Circ Res* 85: e17–24
34 Davidson NJ, Leach MW, Fort MM, Thompson-Snipes L, Kuhn R, Muller W, Berg DJ, Rennick DM (1996) T helper cell 1-type $CD4^+$ T cells, but not B cells, mediate colitis in interleukin 10-deficient mice. *J Exp Med* 184: 241–251
35 Asseman C, Mauze S, Leach MW, Coffman RL, Powrie F (1999) An essential role for interleukin 10 in the function of regulatory T cells that inhibit intestinal inflammation. *J Exp Med* 190: 995–1004
36 Wakkach A, Fournier N, Brun V, Breittmayer JP, Cottrez F, Groux H (2003) Characterization of dendritic cells that induce tolerance and T regulatory 1 cell differentiation *in vivo*. *Immunity* 18: 605–617
37 Caligiuri G, Rudling M, Ollivier V, Jacob MP, Michel JB, Hansson GK, Nicoletti A (2003) Interleukin-10 deficiency increases atherosclerosis, thrombosis, and low-density lipoproteins in apolipoprotein E knockout mice. *Mol Med* 9: 10–17
38 Potteaux S, Esposito B, van Oostrom O, Brun V, Ardouin P, Groux H, Tedgui A, Mallat Z (2004) Leukocyte-derived interleukin 10 is required for protection against atherosclerosis in low-density lipoprotein receptor knockout mice. *Arterioscler Thromb Vasc Biol* 24: 1474–1478
39 Von Der Thusen JH, Kuiper J, Fekkes ML, De Vos P, Van Berkel TJ, Biessen EA (2001) Attenuation of atherogenesis by systemic and local adenovirus-mediated gene transfer of interleukin-10 in $LDLr^{-/-}$ mice. *Faseb J* 15: 2730–2732
40 Hagenbaugh A, Sharma S, Dubinett SM, Wei SH, Aranda R, Cheroutre H, Fowell DJ, Binder S, Tsao B, Locksley RM et al (1997) Altered immune responses in interleukin 10 transgenic mice. *J Exp Med* 185: 2101–2110
41 Mallat Z, Gojova A, Brun V, Esposito B, Fournier N, Cottrez F, Tedgui A, Groux H (2003) Induction of a regulatory T cell type 1 response reduces the development of atherosclerosis in apolipoprotein E-knockout mice. *Circulation* 108: 1232–1237
42 Ait-Oufella H, Horvat B, Kerdiles Y, Herbin O, Gourdy P, Khallou-Laschet J, Merval R, Esposito B, Tedgui A, Mallat Z (2007) Measles virus nucleoprotein induces a regulatory immune response and reduces atherosclerosis in mice. *Circulation* 116: 1707–1713
43 Kulkarni AB, Karlsson S (1993) Transforming growth factor-beta 1 knockout mice. A mutation in one cytokine gene causes a dramatic inflammatory disease. *Am J Pathol* 143: 3–9
44 Shull MM, Ormsby I, Kier AB, Pawlowski S, Diebold RJ, Yin M, Allen R, Sidman C,

Proetzel G, Calvin D et al (1992) Targeted disruption of the mouse transforming growth factor-beta 1 gene results in multifocal inflammatory disease. *Nature* 359: 693–699
45 Li MO, Wan YY, Sanjabi S, Robertson AK, Flavell RA (2006) Transforming growth factor-beta regulation of immune responses. *Annu Rev Immunol* 24: 99–146
46 Cobbold SP, Castejon R, Adams E, Zelenika D, Graca L, Humm S, Waldmann H (2004) Induction of foxP3+ regulatory T cells in the periphery of T cell receptor transgenic mice tolerized to transplants. *J Immunol* 172: 6003–6010
47 Lutgens E, Gijbels M, Smook M, Heeringa P, Gotwals P, Koteliansky VE, Daemen MJ (2002) Transforming growth factor-beta mediates balance between inflammation and fibrosis during plaque progression. *Arterioscler Thromb Vasc Biol* 22: 975–982
48 Grainger DJ, Mosedale DE, Metcalfe JC, Bottinger EP (2000) Dietary fat and reduced levels of TGFbeta1 act synergistically to promote activation of the vascular endothelium and formation of lipid lesions. *J Cell Sci* 113 (Pt 13): 2355–2361
49 Gojova A, Brun V, Esposito B, Cottrez F, Gourdy P, Ardouin P, Tedgui A, Mallat Z, Groux H (2003) Specific abrogation of transforming growth factor-beta signaling in T cells alters atherosclerotic lesion size and composition in mice. *Blood* 102: 4052–4058
50 Robertson AK, Rudling M, Zhou X, Gorelik L, Flavell RA, Hansson GK (2003) Disruption of TGF-beta signaling in T cells accelerates atherosclerosis. *J Clin Invest* 112: 1342–1350
51 Smith DA, Irving SD, Sheldon J, Cole D, Kaski JC (2001) Serum levels of the antiinflammatory cytokine interleukin-10 are decreased in patients with unstable angina. *Circulation* 104: 746–749
52 Grainger DJ, Kemp PR, Metcalfe JC, Liu AC, Lawn RM, Williams NR, Grace AA, Schofield PM, Chauhan A (1995) The serum concentration of active transforming growth factor-beta is severely depressed in advanced atherosclerosis. *Nat Med* 1: 74–79
53 Ait-Oufella H, Salomon BL, Potteaux S, Robertson AK, Gourdy P, Zoll J, Merval R, Esposito B, Cohen JL, Fisson S et al (2006) Natural regulatory T cells control the development of atherosclerosis in mice. *Nat Med* 12: 178–180
54 Mor A, Planer D, Luboshits G, Afek A, Metzger S, Chajek-Shaul T, Keren G, George J (2007) Role of naturally occurring CD4+ CD25+ regulatory T cells in experimental atherosclerosis. *Arterioscler Thromb Vasc Biol* 27: 893–900
55 Gotsman I, Grabie N, Gupta R, Dacosta R, MacConmara M, Lederer J, Sukhova G, Witztum JL, Sharpe AH, Lichtman AH (2006) Impaired regulatory T-cell response and enhanced atherosclerosis in the absence of inducible costimulatory molecule. *Circulation* 114: 2047–2055
56 Heller EA, Liu E, Tager AM, Yuan Q, Lin AY, Ahluwalia N, Jones K, Koehn SL, Lok VM, Aikawa E et al (2006) Chemokine CXCL10 promotes atherogenesis by modulating the local balance of effector and regulatory T cells. *Circulation* 113: 2301–2312
57 Mor A, Luboshits G, Planer D, Keren G, George J (2006) Altered status of CD4(+) CD25(+) regulatory T cells in patients with acute coronary syndromes. *Eur Heart J* 27: 2530–2537
58 de Boer OJ, van der Meer JJ, Teeling P, van der Loos CM, van der Wal AC (2007) Low

numbers of FOXP3 positive regulatory T cells are present in all developmental stages of human atherosclerotic lesions. *PLoS ONE* 2: e779

59 Danesh J (2005) Antibiotics in the prevention of heart attacks. *Lancet* 365: 365–367

60 Epstein SE, Zhou YF, Zhu J (1999) Infection and atherosclerosis: emerging mechanistic paradigms. *Circulation* 100: e20–28

61 Zhou YF, Leon MB, Waclawiw MA, Popma JJ, Yu ZX, Finkel T, Epstein SE (1996) Association between prior cytomegalovirus infection and the risk of restenosis after coronary atherectomy. *N Engl J Med* 335: 624–630

62 Hendrix MG, Salimans MM, van Boven CP, Bruggeman CA (1990) High prevalence of latently present cytomegalovirus in arterial walls of patients suffering from grade III atherosclerosis. *Am J Pathol* 136: 23–28

63 Lemstrom K, Sihvola R, Bruggeman C, Hayry P, Koskinen P (1997) Cytomegalovirus infection-enhanced cardiac allograft vasculopathy is abolished by DHPG prophylaxis in the rat. *Circulation* 95: 2614–2616

64 Streblow DN, Kreklywich C, Yin Q, De La Melena VT, Corless CL, Smith PA, Brakebill C, Cook JW, Vink C, Bruggeman CA et al (2003) Cytomegalovirus-mediated upregulation of chemokine expression correlates with the acceleration of chronic rejection in rat heart transplants. *J Virol* 77: 2182–2194

65 Sedmak DD, Knight DA, Vook NC, Waldman JW (1994) Divergent patterns of ELAM-1, ICAM-1, and VCAM-1 expression on cytomegalovirus-infected endothelial cells. *Transplantation* 58: 1379–1385

66 Fabricant CG, Fabricant J (1999) Atherosclerosis induced by infection with Marek's disease herpesvirus in chickens. *Am Heart J* 138: S465–468

67 Hsich E, Zhou YF, Paigen B, Johnson TM, Burnett MS, Epstein SE (2001) Cytomegalovirus infection increases development of atherosclerosis in Apolipoprotein-E knockout mice. *Atherosclerosis* 156: 23–28

68 Muhlestein JB, Horne BD, Carlquist JF, Madsen TE, Bair TL, Pearson RR, Anderson JL (2000) Cytomegalovirus seropositivity and C-reactive protein have independent and combined predictive value for mortality in patients with angiographically demonstrated coronary artery disease. *Circulation* 102: 1917–1923

69 Prosch S, Wendt CE, Reinke P, Priemer C, Oppert M, Kruger DH, Volk HD, Docke WD (2000) A novel link between stress and human cytomegalovirus (HCMV) infection: sympathetic hyperactivity stimulates HCMV activation. *Virology* 272: 357–365

70 Saikku P, Leinonen M, Mattila K, Ekman MR, Nieminen MS, Makela PH, Huttunen JK, Valtonen V (1988) Serological evidence of an association of a novel Chlamydia, TWAR, with chronic coronary heart disease and acute myocardial infarction. *Lancet* 2: 983–986

71 Kuo CC, Gown AM, Benditt EP, Grayston JT (1993) Detection of *Chlamydia pneumoniae* in aortic lesions of atherosclerosis by immunocytochemical stain. *Arterioscler Thromb* 13: 1501–1504

72 O'Connor CM, Dunne MW, Pfeffer MA, Muhlestein JB, Yao L, Gupta S, Benner RJ, Fisher MR, Cook TD (2003) Azithromycin for the secondary prevention of coronary

heart disease events: the WIZARD study: a randomized controlled trial. *Jama* 290: 1459–1466

73 Cercek B, Shah PK, Noc M, Zahger D, Zeymer U, Matetzky S, Maurer G, Mahrer P (2003) Effect of short-term treatment with azithromycin on recurrent ischaemic events in patients with acute coronary syndrome in the azithromycin in Acute Coronary Syndrome (AZACS) trial: a randomised controlled trial. *Lancet* 361: 809–813

74 Cannon CP, Braunwald E, McCabe CH, Grayston JT, Muhlestein B, Giugliano RP, Cairns R, Skene AM (2005) Antibiotic treatment of *Chlamydia pneumoniae* after acute coronary syndrome. *N Engl J Med* 352: 1646–1654

75 Grayston JT, Kronmal RA, Jackson LA, Parisi AF, Muhlestein JB, Cohen JD, Rogers WJ, Crouse JR, Borrowdale SL, Schron E et al (2005) Azithromycin for the secondary prevention of coronary events. *N Engl J Med* 352: 1637–1645

76 Moazed TC, Campbell LA, Rosenfeld ME, Grayston JT, Kuo CC (1999) *Chlamydia pneumoniae* infection accelerates the progression of atherosclerosis in apolipoprotein E-deficient mice. *J Infect Dis* 180: 238–241

77 Hu H, Pierce GN, Zhong G (1999) The atherogenic effects of chlamydia are dependent on serum cholesterol and specific to *Chlamydia pneumoniae*. *J Clin Invest* 103: 747–753

78 Caligiuri G, Rottenberg M, Nicoletti A, Wigzell H, Hansson GK (2001) *Chlamydia pneumoniae* infection does not induce or modify atherosclerosis in mice. *Circulation* 103: 2834–2838

79 Wright SD, Burton C, Hernandez M, Hassing H, Montenegro J, Mundt S, Patel S, Card DJ, Hermanowski-Vosatka A, Bergstrom JD et al (2000) Infectious agents are not necessary for murine atherogenesis. *J Exp Med* 191: 1437–1442

80 Yang X, Coriolan D, Murthy V, Schultz K, Golenbock DT, Beasley D (2005) Proinflammatory phenotype of vascular smooth muscle cells: role of efficient Toll-like receptor 4 signaling. *Am J Physiol Heart Circ Physiol* 289: H1069–1076

81 Krutzik SR, Tan B, Li H, Ochoa MT, Liu PT, Sharfstein SE, Graeber TG, Sieling PA, Liu YJ, Rea TH et al (2005) TLR activation triggers the rapid differentiation of monocytes into macrophages and dendritic cells. *Nat Med* 11: 653–660

82 Mullick AE, Tobias PS, Curtiss LK (2005) Modulation of atherosclerosis in mice by Toll-like receptor 2. *J Clin Invest* 115: 3149–3156

83 Schoneveld AH, Oude Nijhuis MM, van Middelaar B, Laman JD, de Kleijn DP, Pasterkamp G (2005) Toll-like receptor 2 stimulation induces intimal hyperplasia and atherosclerotic lesion development. *Cardiovasc Res* 66: 162–169

84 Laman JD, Schoneveld AH, Moll FL, van Meurs M, Pasterkamp G (2002) Significance of peptidoglycan, a proinflammatory bacterial antigen in atherosclerotic arteries and its association with vulnerable plaques. *Am J Cardiol* 90: 119–123

85 Ott SJ, El Mokhtari NE, Musfeldt M, Hellmig S, Freitag S, Rehman A, Kuhbacher T, Nikolaus S, Namsolleck P, Blaut M et al (2006) Detection of diverse bacterial signatures in atherosclerotic lesions of patients with coronary heart disease. *Circulation* 113: 929–937

86 Michelsen KS, Wong MH, Shah PK, Zhang W, Yano J, Doherty TM, Akira S, Rajavashisth TB, Arditi M (2004) Lack of Toll-like receptor 4 or myeloid differentiation factor 88 reduces atherosclerosis and alters plaque phenotype in mice deficient in apolipoprotein E. *Proc Natl Acad Sci USA* 101: 10679–10684
87 Edfeldt K, Swedenborg J, Hansson GK, Yan ZQ (2002) Expression of toll-like receptors in human atherosclerotic lesions: a possible pathway for plaque activation. *Circulation* 105: 1158–1161
88 Cao F, Castrillo A, Tontonoz P, Re F, Byrne GI (2007) *Chlamydia pneumoniae*-induced macrophage foam cell formation is mediated by Toll-like receptor 2. *Infect Immun* 75: 753–759
89 Lehtonen L, Eerola E, Oksman P, Toivanen P (1995) Muramic acid in peripheral blood leukocytes of healthy human subjects. *J Infect Dis* 171: 1060–1064
90 Wiedermann CJ, Kiechl S, Dunzendorfer S, Schratzberger P, Egger G, Oberhollenzer F, Willeit J (1999) Association of endotoxemia with carotid atherosclerosis and cardiovascular disease: prospective results from the Bruneck Study. *J Am Coll Cardiol* 34: 1975–1981
91 Erridge C, Spickett CM, Webb DJ (2007) Non-enterobacterial endotoxins stimulate human coronary artery but not venous endothelial cell activation *via* Toll-like receptor 2. *Cardiovasc Res* 73: 181–189
92 Demmer RT, Desvarieux M (2006) Periodontal infections and cardiovascular disease: the heart of the matter. *J Am Dent Assoc* (137) Suppl: 14S-20S; quiz 38S
93 Gagliardi MC, Sallusto F, Marinaro M, Langenkamp A, Lanzavecchia A, De Magistris MT (2000) Cholera toxin induces maturation of human dendritic cells and licences them for Th2 priming. *Eur J Immunol* 30: 2394–2403
94 Whelan M, Harnett MM, Houston KM, Patel V, Harnett W, Rigley KP (2000) A filarial nematode-secreted product signals dendritic cells to acquire a phenotype that drives development of Th2 cells. *J Immunol* 164: 6453–6460
95 McGuirk P, McCann C, Mills KH (2002) Pathogen-specific T regulatory 1 cells induced in the respiratory tract by a bacterial molecule that stimulates interleukin 10 production by dendritic cells: a novel strategy for evasion of protective T helper type 1 responses by Bordetella pertussis. *J Exp Med* 195: 221–231
96 Cheng X, Chen Y, Xie JJ, Yao R, Yu X, Liao MY, Ding YJ, Tang TT, Liao YH, Cheng Y (2008) Suppressive oligodeoxynucleotides inhibit atherosclerosis in ApoE(-/-) mice through modulation of Th1/Th2 balance. *J Mol Cell Cardiol* 45: 168–175
97 Zhou X, Robertson AK, Rudling M, Parini P, Hansson GK (2005) Lesion development and response to immunization reveal a complex role for CD4 in atherosclerosis. *Circ Res* 96: 427–434
98 Hansen PR, Chew M, Zhou J, Daugherty A, Heegaard N, Jensen P, Mouritsen S, Falk E (2001) Freunds adjuvant alone is antiatherogenic in apoE-deficient mice and specific immunization against TNFalpha confers no additional benefit. *Atherosclerosis* 158: 87–94
99 Khallou-Laschet J, Tupin E, Caligiuri G, Poirier B, Thieblemont N, Gaston AT, Vandaele

M, Bleton J, Tchapla A, Kaveri SV et al (2006) Atheroprotective effect of adjuvants in apolipoprotein E knockout mice. *Atherosclerosis* 184: 330–341

100 Tian B, Hao J, Zhang Y, Tian L, Yi H, O'Brien TD, Sutherland DE, Hering BJ, Guo Z (2009) Upregulating $CD4^{+}CD25^{+}FOXP3^{+}$ regulatory T cells in pancreatic lymph nodes in diabetic NOD mice by adjuvant immunotherapy. *Transplantation* 87: 198–206

101 Binder CJ, Horkko S, Dewan A, Chang MK, Kieu EP, Goodyear CS, Shaw PX, Palinski W, Witztum JL, Silverman GJ (2003) Pneumococcal vaccination decreases atherosclerotic lesion formation: molecular mimicry between *Streptococcus pneumoniae* and oxidized LDL. *Nat Med* 9: 736–743

102 Murray CJ, Lopez AD (1997) Global mortality, disability, and the contribution of risk factors: Global Burden of Disease Study. *Lancet* 349: 1436–1442

103 Marie JC, Kehren J, Trescol-Biemont MC, Evlashev A, Valentin H, Walzer T, Tedone R, Loveland B, Nicolas JF, Rabourdin-Combe C et al (2001) Mechanism of measles virus-induced suppression of inflammatory immune responses. *Immunity* 14: 69–79

The ‘delayed infection’ (aka ‘hygiene’) hypothesis for childhood leukaemia

Mel Greaves

Section of Haemato-Oncology, The Institute of Cancer Research, Brookes Lawley Building, 15 Cotswold Road, Sutton, Surrey SM2 5NG, UK

Abstract

The common variant of childhood acute lymphoblastic leukaemia (cALL) is the most frequent paediatric cancer subtype. Its incidence rate appears to have increased substantially in Western societies during the mid-20th century and continues to increase at ~1%/year. Worldwide cALL appears to track with affluence of societies. The ‘delayed infection’ hypothesis, first formulated in 1988, parallels the hygiene hypothesis and has an evolutionary foundation in the concept of a mismatch between prior genetic selection and programming (of the immune system) and contemporary social circumstances. In essence, the hypothesis predicts that ALL is triggered by an abnormal immune response to one or more common microbial infections and that the abnormality arises for two reasons: (i) infectious exposures being delayed beyond the immunologically anticipated period of infancy; (ii) some degree of inherited genetic susceptibility *via*, for example, allelic variation in genes involved in the MHC and/or immune response network. The hypothesis also has a framework in the underlying cell and molecular biology of ALL and its natural history. Epidemiological studies of social contacts in infancy (as a proxy for common infections) and risk of ALL provide indirect but strong support for the hypothesis. The idea still requires mechanistic and genetic endorsement and the appropriate studies are in progress.

Background: childhood leukaemia descriptive epidemiology

Leukaemia, in common with all cancers, develops *via* sequential mutation and clonal selection. The end product, in the absence of early and effective intervention, is a weed-like sub-species of cell that hijacks tissue ecosystems with lethal impact. This process is Darwinian natural selection in action – at the level of somatic cells [1].

Paediatric acute leukaemia comprises a group of distinctive cancers differing in cellular origins, phenotypes, genetic abnormalities and clinical response (Tab. 1). The most frequent or common (c) subtype is B cell precursor acute lymphoblastic leukaemia (cALL) which has a very distinctive age incidence peak at 2–5 years [2]. The overall annual incidence of ALL in USA, Europe, Australia/New Zealand

The Hygiene Hypothesis and Darwinian Medicine, edited by Graham A. W. Rook

Table 1 - Biological subtypes of blood cell cancer in children

A. Acute leukaemia		
1. Acute lymphoblastic leukaemia (ALL) ~80%[1]		
- B cell precursor/common	(80%)[2]	(2–5 yr peak)[3]
- T cell precursor	(15%)[2]	(2-15 yr)
- pro-B/monocyte precursor	(5%)[2]	(infants, <18 mths)
2. Acute myeloblastic leukaemia (AML) ~20%[1]		
(subtypes)		
B. Non-Hodgkin's lymphoma		
(subtypes)		

[1] *as percentage of total ALL.*
[2] *as percentage of total acute leukaemias.*
[3] *This subtype (and others) are heterogeneous in terms of underlying chromosomal and mutational drivers of cancer. For cALL for example, ~25% share a chimaeric fusion gene ETV6-RUNX1 (also known as TEL-AML1), ~35% a hyperdiploid karyotype and the remainder have diverse, less common genetic lesions. Most (~85%) infant ALL (pro-B/monocyte) have chimaeric gene fusions involving the important developmental gene MLL. T-ALL have a wide diversity of genetic alterations though many are chromosome translocations involving T cell receptors (γ, β).*
Other very rare subtypes exist, e.g., juvenile chronic myelomonocytic leukaemia, acute mixed (lympho/myeloid) lineage.

and Japan is in the range of 25–45 cases/10^6 equating to a risk (0–15 years) of ~1 in 2000 [3]. Epidemiological studies and cancer registry data suggest that the age peak of ALL at 2–5 years has an interesting history and geography distinct from other subtypes of childhood leukaemia. In the USA (whites) and UK, the peak first appeared during the period 1920–1940, later in Japan and in US blacks (1960s) and later still in China (1970s) [2]. On a worldwide basis, and where reliable registration data is available, the disease appears to track with affluence. In this respect, it parallels childhood Type 1 diabetes and allergies; indeed international incidence rates of ALL and Type 1 diabetes are significantly correlated, hinting at shared risk factors [4]. Currently in the UK, Scandinavia and USA, cALL is increasing at ~1% per year [5].

Clinicians treating childhood leukaemia in the first half of the 20th century favoured an infectious aetiology, primarily because of the correspondence of diagnosis in time with common childhood infections such as measles. However leukaemia was clearly not itself contagious and no evidence was available to support the contention. Some favoured an anonymous but specific virus, others an indirect mechanism that comes close to the hypothesis discussed in this chapter.

> "*We incline on our evidence to the belief that the solution of the problem of leukaemia lies rather in some peculiar reaction to infection than in the existence of some specific infective agent.*"
>
> J Poynton, H Thursfield and D Paterson, Great Ormond Street Hospital for Sick Children, London, 1922 [6]

As epidemiological science became established (after the 1950s), numerous relatively small-scale studies have, over several decades, sought to implicate many different types of environmental exposure (Tab. 2). The only unambiguous observation derives from the 1945 atomic bomb exposure which was associated with a significant increase in ALL [7]. Ionising radiation is therefore a cause of childhood ALL but it is unlikely to be *the*, or a, major causal factor. The problems with prior case/control epidemiological studies were compounded: under-powering, lack of appropriate control group selection, no distinction drawn between subtypes of leukaemia that might have distinctive aetiologies and prevailing ignorance of the

Table 2 - Postulated causal exposures

- Car exhaust fumes
- Pesticides
- Ionising radiation
- Non-ionising radiation
- Electric fields
- Vitamin K injection at birth
- Hot dogs or hamburgers (depending on whether the consumer (patient) was in California or Colorado)
- Domestic animals
- Organic dust from cotton, wool or synthetic fibres
- Natural light deprivation through melatonin disruption
- Artificial, fluorescent light exposure in hospital neonatal care units
- Parental cigarette smoking
- Maternal medicinal drug taking (during pregnancy)
- Maternal alcohol consumption (during pregnancy)
- Chemical contamination in drinking water
- Infections

natural history of the disease and therefore likely timing of key exposures. The only other established risk factors for childhood leukaemia are rare: inheritance of predisposing mutations (e.g., for Fanconi's anaemia or Bloom's syndrome), Down's syndrome or genotoxic therapy – collectively amounting for no more than 5% of cases [8]. The major causal mechanisms and exposures in ALL have therefore remained unidentified.

A biological framework for the natural history of childhood ALL

We have taken the stance that unravelling the aetiological, causal mechanisms for childhood leukaemia would benefit from taking into account the biological heterogeneity of disease and a clearer understanding of its natural (preclinical) history. We therefore focussed on identifying and validating the distinctive biological subtypes of ALL [2] and, subsequently, in identifying the timing of key events in the natural history of the major subtype of childhood leukaemia – cALL. Figure 1 illustrates the current picture we have of the sequential events that drive the clinical emergence of cALL, their developmental timing and incidence or probability rates (reviewed in [9, 10]). Reassuringly, the data endorses an entirely speculative 'two hit' model for cALL that we proposed in 1988 [11]. The key observations are:

1. cALL is usually initiated *in utero via* chromosome translocation or hyperdiploidy. These events generate a clinically silent and persistent (→ 15 years) preleukaemic clone which requires additional genetic abnormalities to convert to overt, clinically diagnosable ALL. This 'two hit' principle is endorsed by modelling with murine [12] and human cells [13].
2. The initiation of ALL prenatally occurs some 100× the frequency of the disease itself [14].
3. Common or recurrent genetic abnormalities involving gene (or region) copy number variation (CNV) or sequence mutation are identifiable in ALL samples [15]. CNV are postnatal in origin and are critical or essential 'secondary' mutations [16].
4. It is suggested (but currently unproven) that the 2–4 secondary CNV/mutations in ALL occur as a bolus or suite proximal to diagnosis (~a few months prior) and may promote or trigger clonal evolution leading to rapid proliferative exposures, marrow failure and diagnosis of ALL.

This model immediately identifies some previously entirely cryptic causal complexities: (i) two distinctive, developmental stages of mutation and clonal selection and therefore two potential windows of exposure; and (ii) rare disease prevalence is associated with relatively common initiation but low penetrance. This suggests a key bottleneck in the development of ALL is the postnatal trigger of secondary mutations [17].

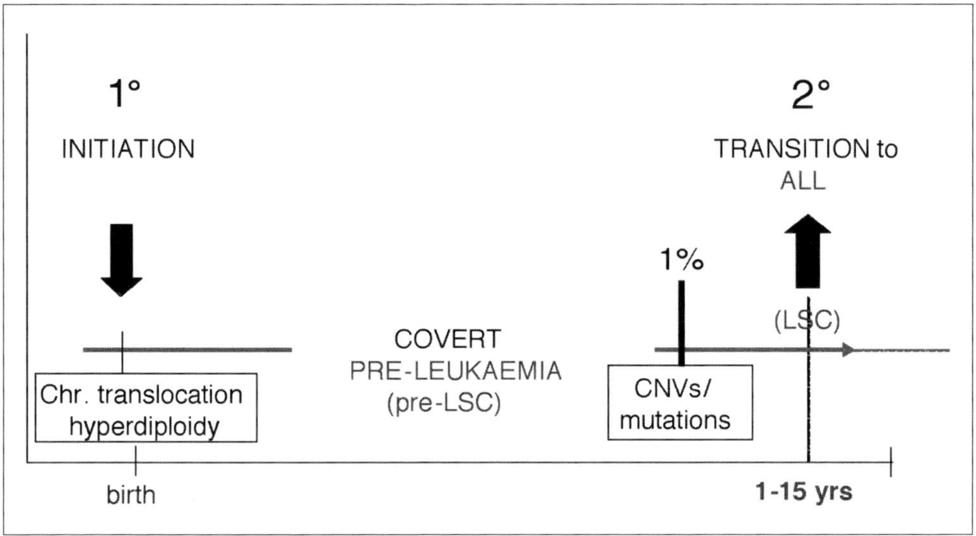

Figure 1
Natural history of paediatric acute leukaemias. Chr, chromosome; LSC, leukaemic stem cell; ALL, acute lymphoblastic leukaemia; CNVs, gene copy number variations.

The 'delayed infection' hypothesis for cALL

In the context of the two-hit, pre-/postnatal hypothesis, an entirely speculative explanation was proffered for the aetiology of ALL [11, 17]. It was suggested that the first hit, *in utero*, was largely independent of exogenous genotoxic exposure and a consequence of endogeneous, developmental stress during foetal lymphopoiesis, i.e., spontaneous mutation reflecting intrinsically accident-prone systems of DNA and cellular maintenance. Such a level of mistake could be tolerated in an evolutionary context if they only very rarely led to lethal disease outcomes. The second hit was proposed to have external origins in the form of common infections. Specifically, it was suggested that a deficit of common infections in infancy, followed by subsequent 'delayed' infection would provide, *via* immune system deregulation, the proliferative stress to the bone marrow that could trigger the critical secondary mutation. The logic of this explanation is entirely Darwinian (Box 1). George Williams' memorable phrase "*evolution has no eyes to the future*" aptly captures the problem. If environmental or, for humans, social circumstances rapidly change, then prior genetic programming which was contingent upon prevailing selective pressure, becomes mismatched with the new circumstances. There are several examples of this in medicine [18], and especially for cancer [19]. For childhood leukaemia, the evolutionary context is the organisation of the immune system. The 'delayed infec-

The 'delayed infection' hypothesis –
A genetic adaptation – lifestyle mismatch?

- **Evolutionary adaptation**
  1. The immune system has been evolutionarily *programmed* to anticipate infectious challenge after birth
  2. The neonatal immune network is not hardwired and *requires* modulation by infectious exposure
  3. Human genetic variants in immune response genes (strength of signal)
     - *adaptive selection by past plagues / epidemics*
- **The mismatched lifestyle/cultural factors**
  Affluent societies / families provide greatly reduced opportunities for 'natural' infectious exposure in infancy (- reduced family size, hygiene measures).
- **The consequences of mismatch:**
  1. Unprimed immune systems in infancy
  2. Later (inevitable) childhood infections precipitate highly dysregulated immune responses
  3. Proliferative / apoptotic stress to bone marrow
     - *selection of pre-leukaemic stem cells?*
- **Definition of those at risk:**
  1. Those with pre-existing pre-leukaemia (foetal) clone
     - *developmental accident / imperfect fidelity of DNA maintenance/repair?*
  2. Those with minimal infectious exposure in infancy
     - *social circumstances?*
  3. Those who have particular immune response gene alleles
     - *historical contingency / adaptive selection?*

Box 1

tion' hypothesis was proposed independently of Strachan's 'hygiene hypothesis' for childhood allergies [20] but clearly they are very similar.

Epidemiological testing

The 'delayed infection' hypothesis has not been easy to test by epidemiological methods. If one specific transforming virus, like, say, EBV or HTLV1 [21], had been involved, this would have changed matters. Extensive screening for viral footprints by all available PCR-based methods have failed to identify such a culprit [17]. In the absence of such evidence, the hypothesis has remained that the infectious trigger is one or, more probably, several common infections (bacterial and/or viral). In this situation, the appropriate tactic has to be to resort to reliable proxies of infectious exposure. A UK nationwide case/control study of childhood leukaemia was designed (around 1990-1992) to look at this aspect – alongside several other candidate exposures including natural ionising radiation and electromagnetic fields (EMF) and electric fields [22]. The UK Childhood Cancer Study (UKCCS) found no evidence for the latter exposures contributing anything other than a very small fraction of causal exposures [23, 24]. Our principle test or prediction of the 'delayed infection' hypothesis was that social interactions at playgroups in infancy should be protective. And it was (Tab. 3) [25]. The study has been replicated in California [26] and Scandinavia [27] with essentially the same positive association with lowered risk of ALL. The UKCCS, and particularly the California study [26], provided evidence for a significant trend or dose response with greater social contact levels providing more risk reduction. Playgroup or day care attendance (and levels or number of contacts) is a well established proxy for exposure to common infections that spread by person-person contact [28] and it is difficult to draw any conclusion from these data other than early infectious exposure is protective for ALL.

Two other subsidiary predictions were also made. It was anticipated that risk might be modified by birth order/parity and by early exposure to known or specific infections. Studies of these two parameters have produced mixed or conflicting results [29]. Two recent reports indicate that GP records of infections in infancy may tell a different story than parental recall as used in most case/control questionnaires [30, 31]. While this may be true, the failure of infection history as documented in GP records in these studies to reflect any protective impact [30, 31] does not necessarily negate the hypothesis. The authors correctly concluded that the GP record stud-

Table 3 - Social contact in first year of life and risk of acute lymphoblastic leukaemia [25]

Cases	**# 1277**	**OR (CI)**
Controls	**6268**	
None		1.00
Social activity but no day care		0.73 (0.62–0.87)
Informal day care		0.62 (0.51–0.75)
Formal day care		0.48 (0.37–0.62)
	p for trend = <0.001	

OR (CI): odds ratios (95% confidence intervals).

ies provided no evidence for the hypothesis. The problem with these studies is the unstated, underlying premise: that protective impact of infectious exposure in infancy is necessarily reflected in symptoms prompting a visit to GPs [32]. Clearly, many infections are asymptomatic or sub-clinical. In the context of the parallel 'hygiene hypothesis' for allergies, G Rook suggests *via* the 'old friends' variant hypothesis that the critical infectious exposure might well be innocuous, symbiont species.

Validation

Although case/control studies of ALL assessing the infectious hypothesis are ongoing, they may contribute little more. Validation may rest more on genetics and mechanistic insights. If the hypothesis is correct, then it is likely that inherited genetic variation should impact on risk and we would anticipate or predict that relevant genes should include some that encode products integral to immune network regulation, e.g., HLA or key regulators such as IL-12, IL-10, TGF-β, etc. There is some evidence for this (in the UKCCS) *via* association with particular HLA-DP alleles, particular alleles either decreasing or increasing risk [33]. Small-scale pilot genetic screening implicated IL-12 [34] but such studies are known to be generally faulty. We have therefore embarked on a genome-wide (SNP) association study (GWAS) in the UK involving some 900 cases (and similar number of controls from the Wellcome Consortium Study). Similar (though usually larger) GWAS for adult cancer have proven very productive though the picture they paint of genetic susceptibility is complex [35, 36].

Endorsement of the model also requires some mechanistic insight into how an abnormal or dysregulated immune response might help promote transition of pre-leukaemic to leukaemic stem cells (Fig. 1) *via* the acquisition of further genetic lesions. A lead here comes from screening candidate cytokine molecules that are known to be key immuno-modulatory components of the regulatory T

Table 4 - Model systems for exploring the molecular pathogenesis of childhood ALL initiated by ETV6-RUNX1 gene fusion

		Reference
1.	Hormone inducible ETV6-RUNX1 in a murine cell line (BaF3) *in vitro*	[37]
2.	Transgenic mice expressing human ETV6-RUNX1 selectivity in the B cell lineage under the control of the IGH enhancer mouse stem cells or human cord blood	[37]
3.	Mice repopulated with murine or human stem cells retrovirally infected with ETV6-RUNX1 fusion gene construct	[12, 13]

cell network. We first developed three model systems with either murine or human progenitor cells transfected to express the leukaemia initiating fusion gene ETV6-RUNX1 (Tab. 4). Screening of several candidate molecules – γ-interferon, TNF and TGF-β in each of these models revealed that the response to one of them – TGF-β, was strikingly altered by expression of the leukaemia gene [37]. B progenitor cells induced to express ETV6-RUNX1 proliferate more slowly than normal. TGF-β is a potent suppressor of proliferation in haemopoietic stem cells and progenitors. In normal B cell progenitors, this is signalled *via* SMAD3 activation and downstream transcriptional activation of the cell cycle inhibitor p27. In the presence of ETV6-RUNX1 protein, this response is blocked. ETV6-RUNX1 binds to SMAD3 and recruits transcriptional co-repressor molecules NCoR and Sin3A; SMAD complexes are then unable to activate the p27 promoter. The consequence of this is that normal progenitors cease dividing but the cells expressing ETV6-RUNX1 continues to divide. The selective effect appears to be most pronounced for modelled human (cord blood) pre-leukaemic stem cells that preferentially expand at the expense of normal B progenitors. This altered response to TGF-β therefore provides a possible route *via* which the relatively small size pre-leukaemic clone in patients (10^{-3}–10^{-4} of blood lymphoid cells) could expand both in bone marrow niches and in blood (Fig. 2). Increase in numbers of these cells by one or more orders of magnitude is a

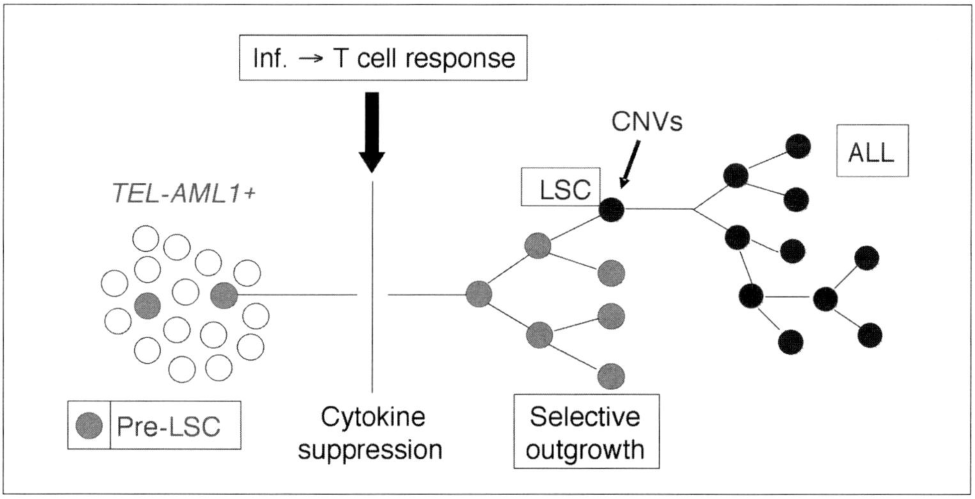

Figure 2
Infection, the immune response and 'selection' of pre-leukaemic clones
LSC, leukaemic stem cell; CNV, gene copy number variations (mostly deletions); TEL-AML1 (aka ETV6-RUNX1) gene fusion resulting from chromosome translocation; A common (pre-natal) initiating event in childhood ALL. Inf, common (but delayed) infection.

prerequisite for the acquisition of additional mutations (in the absence of facilitation by genetic instability). How the mutations might actively be induced *via* immune response-derived TGF-β is unclear. It is unlikely to be simply stochastic. Oxidative damage *via* free radicals in the context of a persistent inflammatory response is a possibility, as in the activation of intrinsic mutagenic enzymes in B cell progenitors – RAG1/2 or AID [38, 39]. Further *in vivo* mouse and human cell (in NOD/SCID) modelling may further endorse the involvement of these signalling pathways in the infection-driven pathogenesis of ALL.

'Natural' experiments

Alongside continued modelling experiments, further support for the aetiological model might come from 'experiments of nature' in which changes of social circumstances appear to be followed by changes in the incidence rates of ALL. Three socio-political changes during the latter half of the 20th century fit with this expectation. Costa Rica has a very modest GDP; it is not a wealthy country. Yet it is top of the league in childhood leukaemia rates at ~$55/10^6$/year, contradicting the general association between affluence and incidence [3]. Why might this be? Costa Rica is a unique Central American country. In the 1970s, the radical political decision was made to abandon the military and divert resources into education and medicine. Literacy rates are now higher than the USA and medical care is excellent. Over a 20 year period, parity dropped from around seven children per family to around 2.5 on average. One interpretation therefore is that these rapidly changing social circumstances significantly reduced the opportunities to natural infectious exposures in infancy.

East Germany statistics on the incidence of childhood ALL may be especially informative. Prior to 1989, the reliable data indicated a rate of ALL some one third below that of West Germany. In the period after reunification of Germany, 1992–1996, the incidence rate of ALL increased dramatically by 25% [40]. If we assume that case ascertainment did not change (which the authors of the report appear confidently assert), then clearly something 'environmental' had changed abruptly to alter the risk of ALL. The authors suggest the most plausible explanation was the dramatic change in child care that followed the collapse of the Berlin wall and the old East German communist regime in 1989. Prior to this date, effectively all 3 months plus infants attended large state-run crèches, in order that their mothers could return to work. This ceased almost immediately after 1989 with infants and toddlers being kept at home. In the context of the 'delayed infection' (or 'hygiene') hypothesis, this is precisely a circumstance that should change (increase) incidence rates of ALL.

The SARS incident in Hong Kong in 2003 provides another very unusual circumstance and 'natural experiment'. An emergency and province-wide government directive ensured that all children stayed at home rather than travelling to attend school. This embargo lasted for one year. So which prediction follows from this with respect to the 'delayed infection' model? This cohort of children with greatly reduced infectious exposure (confirmed by documenting levels of measles, chicken pox and scarlet fever) should immediately be deprived of the 'triggering' infection for ALL and rates for ALL should drop in that same year. A significant drop did indeed occur which the authors interpreted in the light of the 'delayed infection' hypothesis [41]. Another prediction however is that individuals who were infants (< 12 months) during that year should, paradoxically, have an increased risk of ALL some 3–5 years later. The data for this are not yet available.

Other infection-based hypotheses

One other infection hypothesis for childhood leukaemia has achieved significant support. This is the 'population mixing' idea of Leo Kinlen [42]. Based initially on a consideration of a cluster of cases in proximity to the nuclear reprocessing plant near Seascale village, Cumbria, United Kingdom, Kinlen's subsequent studies elsewhere in the UK consistently showed that where a sudden influx of migrant workers (engineers for the Seascale plant or, elsewhere, oil industry workers or army recruits) or rapid increases in population (rural 'new towns'), this was followed a few years later by a transient increase (av. ~two fold) in leukaemia rates (though often documented as leukaemia deaths rather from incidence rates). Some other supportive data from independent studies have been provided including our own from the Hong Kong new territories [43]. Kinlen's favoured explanation is that based on herd immunity [42]; namely that in the setting of rapid population mixing, one population infected with a 'leukaemia virus' could infect previously unexposed and non-immune individuals.

Historically, the first marked cluster of cases of ALL was in Niles, a suburb of Chicago in 1967 in which most cases attended the same school or church [44]. This transient increase in ALL coincided with an episode of streptococcal/rheumatic fever. The latest cluster, and the most marked one to date, is in the small town of Fallon, Nevada, USA, which happens to be very close to a 'top gun' naval air base. Conspiracy theories abound including that leakage of very carcinogenic jet fuel is to blame [17]. Of the 13 cases of ALL (and only one case of AML), where only ~1 was expected, most were actually born outside the area which immediately suggests that whatever environmental event has occurred, its impact on leukaemia risk was postnatal – and therefore likely in the secondary, promotional phase [17]. An abnormal response to infection is the most plausible, but unprovable, explanation. The increase of cases was during 1999–2003 and no new cases have occurred subsequently [45].

Retrospective identification of causal agents or exposures in transient clusters of leukaemia is extraordinarily difficult, if not impossible. This author's view is that the 'population mixing' and 'delayed infection' hypotheses are entirely compatible and point, jointly, to the same single explanation. Kinlen, an epidemiologist, suggests that causation involves an, as yet unidentified, transforming virus, that this applies to all forms of childhood blood cell cancer (Tab. 1), including non-Hodgkin's lymphoma, and that the critical infection may operate around the time of birth. I see no evidence to support these particular contentions despite the overall data sets providing persuasive, indirect evidence for an infectious aetiology.

Infection and other cancers

Some 15% of adult cancers worldwide involve specific infectious agents that colonise particular tissues, including DNA and RNA viruses, bacteria and parasitic worms [46]. Additionally, a large fraction of gastrointestinal cancers involve premalignant chronic inflammation, at least some of which is infection-related [47]. Infection is therefore a very significant cause of cancer, particularly in less developed countries where it accounts for some 40% of all cancer cases. Primary prophylactic trials are now in progress for some of these agents (HPV, EBV and hepatitis B/C). In the case of gastric cancer associated with *Helicobacter pylori* – and more recently other mucosal lymphomas associated with distinct bacterial species [48], antibiotic regimes have proven remarkably efficacious in resolving the tumour. More advanced or malignant gastric lymphomas are independent of bacterial/T cell drive and unresponsive to antibiotics.

In these cancers, and cancers in general, I have argued that vulnerability can be understood best from the perspective of evolutionary biology or 'Darwinian medicine' [1, 19]. Does, however, the 'delayed infection' or hygiene hypothesis apply to these cancers? I suspect not, particularly since the evidence is that risk (for Hep B/C liver cancer, EBV Burkitt's lymphoma, HTLV-1 and all T cell leukaemia/lymphoma, gastric cancer/lymphoma and *H. pylori*) is associated with early (not delayed) and chronic or persistent infection.

There is one exception, however, and that is for another blood cell cancer – Hodgkin's lymphoma (HL) in young adults (~15–40 years). Some subtypes of HL are associated with Epstein Barr virus but the majority of HL in young adults are not. The incidence trend of HD parallels childhood ALL. It is most elevated in affluent countries and there is evidence that risk may be linked to a delay or absence in exposure to common childhood infection [49]. The authors of these studies drew the analogy with polio with reference to the much earlier observation that the pathological impact of polio virus depends on social circumstances and age of exposure [50].

Prevalent diseases of children and young adults in affluent societies: the paradoxical outcome of evolutionary mismatches?

Others in this book have discussed the application of, and evidence for, the 'hygiene hypothesis' for allergies, Type 1 diabetes and multiple sclerosis. Notwithstanding that alternative explanations for the pathogenesis of these diseases and their apparent link with affluence are still tenable, there is an underlying theme in these ideas that is grounded in the evolutionary programming of the immune system.

The immune system network structure, versatility and function have been historically driven by the arms race challenge of potentially lethal infection. This has resulted in very marked allelic variation in the many genes involved in human immune systems and likely prior selection over thousands of years by endemic lethal infections and periodic plagues. Additionally, and critically as an adaptive system, much like the brain, genetic programming of the immune system at birth is not hard-wired. Rather it is set up for fine tuning by early infectious exposures, primarily *via* the agency of regulatory T cell function. It would be quite extraordinary if the otherwise very beneficial impact of depriving our infants and their naïve immune systems of this infectious educational or 'priming' experience did not come at a price.

Whether this diminished exposure in contemporary, affluent societies is *via* the respiratory, skin or gastrointestinal sites; whether it involves lessened contact with soil, domestic animals or other children; whether it involves one species of microbe (or parasitic worm) or another, the same principle may hold. It is striking that ALL, allergies and Type 1 diabetes track each other in international rates. Yet within individuals of families they do not. ALL and allergies have a reciprocal relationship in risk. There are a number of explanations for these relationships. Common risk factors (- diminished 'early' infection in affluent societies) may precipitate different pathologies dependent upon the genetic background of the individual. An allergic episode could re-set the immunological rheostat and diminish risk of ALL. Different microbes or parasites might be allied to these distinct pathologies.

If these explanations are correct and shown to be so, then the challenge is clearly for prevention. Encouraging more social contacts in the very young should be beneficial but any central or governmental edict in this direction would doubtless be seen as over-prescriptive social engineering. A more viable alternative, as actively being considered in the allergy and autoimmune fields, is prophylaxis in infancy or early life with an appropriate microbial or parasite-derived vaccine. Could a generic vaccine along these lines reduce the incidence and disease burden of all three types of disease? And if it did work out this way, would it not be the best possible endorsement of Darwinian medicine and the relevance of the evolutionary perspective on human ailments [51]?

Acknowledgements
The author's research is supported by the Leukaemia Research Fund, the Kay Kendall Leukaemia Fund and The Institute of Cancer Research.

References

1 Greaves M (2000) *Cancer. The Evolutionary Legacy*. Oxford University Press, Oxford
2 Greaves MF, Colman SM, Beard MEJ, Bradstock K, Cabrera ME, Chen P-M, Jacobs P, Lam-Po-Tang PRL, MacDougall LG, Williams CKO et al (1993) Geographical distribution of acute lymphoblastic leukaemia subtypes: second report of the collaborative group study. *Leukemia* 7: 27–34
3 Parkin DM, Stiller CA, Draper GJ, Bieber CA, Terracini B, Young JL (eds) (1988) *International incidence of childhood cancer*. IARC Scientific Publications, Lyon
4 Feltbower RG, McKinney PA, Greaves MF, Parslow RC, Bodansky HJ (2004) International parallels in leukaemia and diabetes epidemiology. *Arch Dis Child* 89: 54–56
5 Shah A, Coleman MP (2007) Increasing incidence of childhood leukaemia: a controversy re-examined. *Br J Cancer* 97: 1009–1012
6 Poynton FJ, Thursfield H, Paterson D (1922) The severe blood diseases of childhood: a series of observations from the Hospital for Sick Children, Great Ormond Street. *Br J Child Dis* XIX: 128–144
7 Preston DL, Kusumi S, Tomonaga M, Izumi S, Ron E, Kuramoto A, Kamada N, Dohy H, Matsui T, Nonaka H et al (1994) Cancer incidence in atomic bomb survivors. Part III: Leukemia, lymphoma and multiple myeloma, 1950–1987. *Radiat Res* 137 (Suppl): S68–S97
8 Greaves MF (1997) Aetiology of acute leukaemia. *Lancet* 349: 344–349
9 Greaves MF, Maia AT, Wiemels JL, Ford AM (2003) Leukemia in twins: lessons in natural history. *Blood* 102: 2321–2333
10 Greaves MF, Wiemels J (2003) Origins of chromosome translocations in childhood leukaemia. *Nature Rev Cancer* 3: 639–649
11 Greaves MF (1988) Speculations on the cause of childhood acute lymphoblastic leukemia. *Leukemia* 2: 120–125
12 Tsuzuki S, Seto M, Greaves M, Enver T (2004) Modelling first-hit functions of the t(12;21) TEL-AML1 translocation in mice. *Proc Natl Acad Sci USA* 101: 8443–8448
13 Hong D, Gupta R, Ancliffe P, Atzberger A, Brown J, Soneji S, Green J, Colman S, Piacibello W, Buckle V et al (2008) Initiating and cancer-propagating cells in TEL-AML1-associated childhood leukemia. *Science* 319: 336–339
14 Mori H, Colman SM, Xiao Z, Ford AM, Healy LE, Donaldson C, Hows JM, Navarrete C, Greaves M (2002) Chromosome translocations and covert leukemic clones are generated during normal fetal development. *Proc Natl Acad Sci USA* 99: 8242–8247
15 Mullighan CG, Goorha S, Radtke I, Miller CB, Coustan-Smith E, Dalton JD, Girtman

K, Mathew S, Ma J, Pounds SB et al (2007) Genome-wide analysis of genetic alterations in acute lymphoblastic leukaemia. *Nature* 446: 758–764
16 Bateman CM, Horsley SW, Chaplin T, Young BD, Ford AM, Kearney L, Greaves M (2008) Sequence of genetic events in *ETV6-RUNX1* positive B precursor ALL: insights from identical twins with concordant leukaemia: *Blood*. American Society of Hematology, San Francisco
17 Greaves M (2006) Infection, immune responses and the aetiology of childhood leukaemia. *Nat Rev Cancer* 6: 193–203
18 Gluckman P, Hanson M (2006) *Mismatch. Why our world no longer fits our bodies*. Oxford University Press, Oxford
19 Greaves M (2007) Darwinian medicine: a case for cancer. *Nat Rev Cancer* 7: 213–221
20 Strachan DP (2000) Family size, infection and atopy: the first decade of the "hygiene hypothesis". *Thorax* 55: S2–S10
21 Zur Hausen H (2006) *Infections causing human cancer*. Wiley-VCH, Weinheim
22 UK Childhood Cancer Study Investigators (2000) The United Kingdom Childhood Cancer Study: objectives, materials and methods. *Br J Cancer* 82: 1073–1102
23 UK Childhood Cancer Study Investigators (2000) Childhood cancer and residential proximity to power lines. *Br J Cancer* 83: 1573–1580
24 UK Childhood Cancer Study Investigators (2002) The United Kingdom Childhood Cancer Study of exposure to domestic sources of ionising radiation: 2: gamma radiation. *Br J Cancer* 86: 1727–1731
25 Gilham C, Peto J, Simpson J, Roman E, Eden TOB, Greaves MF, Alexander FE, for the UKCCS Investigators (2005) Day care in infancy and risk of childhood acute lymphoblastic leukaemia: findings from a UK case-control study. *Br Med J* 330: 1294–1297
26 Ma X, Buffler PA, Wiemels JL, Selvin S, Metayer C, Loh M, Does MB, Wiencke JK (2005) Ethnic difference in daycare attendance, early infections, and risk of childhood acute lymphoblastic leukemia. *Cancer Epidemiol Biomarkers Prev* 14: 1928–1934
27 Kamper-Jørgensen M, Woodward A, Wohlfahrt J, Benn CS, Simonsen J, Hjalgrim H, Schmiegelow K (2007) Childcare in the first 2 years of life reduces the risk of childhood acute lymphoblastic leukemia. *Leukemia* 22: 189–193
28 Louhiala PJ, Jaakkola N, Ruotsalainen R, Jaakkola JJ (1995) Form of day care and respiratory infections among Finnish children. *Am J Public Health* 85: 1109–1112
29 McNally RJQ, Eden TOB (2004) An infectious aetiology for childhood acute leukaemia: a review of the evidence. *Br J Haematol* 127: 243–263
30 Roman E, Simpson J, Ansell P, Kinsey S, Mitchell CD, McKinney PA, Birch JM, Greaves M, Eden T (2007) Childhood acute lymphoblastic leukemia and infections in the first year of life: a report from the United Kingdom Childhood Cancer Study. *Am J Epidemiol* 165: 496–504
31 Cardwell CR, McKinney PA, Patterson CC, Murray LJ (2008) Infections in early life and childhood leukaemia risk: a UK case-control study of general practitioner records. *Br J Cancer* 99: 1529–1533

32 Greaves M, Buffler PA (2009) Infections in early life and risk of childhood ALL. *Br J Cancer* 100: 863

33 Taylor GM, Dearden S, Ravetto P, Ayres M, Watson P, Hussain A, Greaves M, Alexander F, Eden OB, UKCCS Investigators (2002) Genetic susceptibility to childhood common acute lymphoblastic leukaemia is associated with polymorphic peptide-binding pocket profiles in *HLA-DPB1*0201*. *Hum Mol Genet* 11: 1585–1597

34 Josephs ZM, Gonzalez De Castro D, Johnson DC, Novosel A, Borkhardt A, Pritchard-Jones K, Greaves MF (2005) The impact of T_h1/T_h2 response variations on risk of developing childhood leukaemia: a pilot study. *Blood* 106: Abstract 849

35 Easton DF, Pooley KA, Dunning AM, Pharoah PD, Thompson D, Ballinger DG, Struewing JP, Morrison J, Field H, Luben R et al (2007) Genome-wide association study identifies novel breast cancer susceptibility loci. *Nature* 447: 1087–1093

36 Amos CI, Wu X, Broderick P, Gorlov IP, Gu J, Eisen T, Dong Q, Zhang Q, Gu X, Vijayakrishnan J et al (2008) Genome-wide association scan of tag SNPs identifies a susceptibility locus for lung cancer at 15q25.1. *Nat Genet* 40: 616–622

37 Ford A, Palmi C, Bueno C, Hong D, Cardus P, Knight D, Cazzaniga G, Enver T, Greaves M (2009) The TEL-AML1 leukaemia fusion gene dysregulates the TGFβ pathway in early B lineage progenitor cells. *J Clin Invest* 119: 826–836

38 Pasqualucci L, Bhagat G, Jankovic M, Compagno M, Smith P, Muramatsu M, Honjo T, Morse III HC, Nussenzweig MC, Dalla-Favera R (2008) AID is required for germinal center-derived lymphomagenesis. *Nat Genet* 40: 108–112

39 Mullighan CG, Miller CB, Radtke I, Phillips LA, Dalton J, Ma J, White D, Hughes TP, Le Beau MM, Pui C-H et al (2008) BCR-ABL1 lymphoblastic leukaemia is characterized by the deletion of Ikaros. *Nature* 453: 110–114

40 Spix C, Eletr D, Blettner M, Kaatsch P (2008) Temporal trends in the incidence rate of childhood cancer in Germany 1987–2004. *Int J Cancer* 122: 1859–1867

41 Li CK, Zee B, Lee J, Chik KW, Ha SY, Lee V (2007) Impact of SARS on development of childhood acute lymphoblastic leukaemia. *Leukemia* 21: 1353–1356

42 Kinlen LJ (1995) Epidemiological evidence for an infective basis in childhood leukaemia. *Br J Cancer* 71: 1–5

43 Alexander FE, Chan LC, Lam TH, Yuen P, Leung NK, Ha SY, Yuen HL, Li CK, Li CK, Lau YL et al (1997) Clustering of childhood leukaemia in Hong Kong: association with the childhood peak and common acute lymphoblastic leukaemia and with population mixing. *Br J Cancer* 75: 457–463

44 Heath Jr CW, Hasterlik RJ (1963) Leukemia among children in a suburban community. *Am J Med* 34: 796–812

45 Steinmaus C, Lu M, Todd RL, Smith AH (2004) Probability estimates for the unique childhood leukemia cluster in Fallon, Nevada, and risks near other U.S. military aviation facilities. *Environ Health Perspect* 112: 766–771

46 Goedert JJ (ed) (2000) *Infectious Causes of Cancer*. Humana Press, New Jersey

47 Balkwill F (2004) Cancer and the chemokine network. *Nat Rev Cancer* 4: 540–550

48 Isaacson PG, Du M-Q (2004) MALT lymphoma: from morphology to molecules. *Nat Rev Cancer* 4: 644–653
49 Gutensohn N, Cole P (1980) Epidemiology of Hodgkin's disease. *Sem Oncol* 7: 92–102
50 Backett EM (1957) Social patterns of antibody to poliovirus. *Lancet* i: 778–783
51 Stearns SC, Koella JC (eds) (2008) *Evolution in health and disease*. Oxford University Press, New York

Is there room for Darwinian medicine and the hygiene hypothesis in Alzheimer pathogenesis?

W. Sue T. Griffin[1], *and Robert E. Mrak*[2]

[1]Donald W. Reynolds Department of Geriatrics and the Department of Neurobiology and Developmental Science, University of Arkansas for Medical Sciences, Little Rock, Arkansas, USA, and the Geriatric Research Education and Clinical Center, Central Arkansas Veterans Healthcare System, Little Rock, Arkansas, USA
[2]Department of Pathology, University of Toledo College of Medicine, Toledo, Ohio, USA

Abstract

Improvements in modern hygiene and public health have resulted in decreased human contact with organisms associated with so-called 'dirtier' environs. These changes, in turn, have led to an appreciation of the potential importance of such 'friendly' organisms toward proper development of the human immune system. Based on this, a novel hypothesis (the hygiene hypothesis) has been formulated. This idea suggests that a paucity of exposure to environmental pathogens retards proper immune system development, and consequently decreases its ability to effectively thwart a variety of effectors with degenerative consequences, such as those associated with chronic inflammatory responses in diseases as seemingly diverse as those of the gut and the brain. In this chapter, we review current information, including the potential contribution of inheritance to development of hypotheses regarding the pathogenesis of chronic neurodegenerative diseases, especially Alzheimer's disease. We further explore ways in which the hygiene hypothesis and ideas in Darwinian medicine may play a role in the neuropathogenesis of these diseases.

Neuropathological changes in dementia

Cognitive decline, characterized by memory loss and learning disabilities, is associated with a number of age-related dementias. Clinically diagnosable dementias include Alzheimer's disease (AD), vascular dementia, dementia with Lewy bodies, Huntington's disease, and prion disease (e.g., Creutzfeldt-Jakob disease, CJD). In addition, stroke and Parkinson's disease are two neurodegenerative diseases in which cognitive decline is less prominent. Each of these diseases can present as a neuropathologically 'pure' entity or in combination with other dementia-related neuropathological changes. For example, spongiform changes and prion plaques are unique diagnostic features of prion diseases, but CJD may exist concurrent with all the characteristic neuropathological changes of AD as well as Lewy body

The Hygiene Hypothesis and Darwinian Medicine, edited by Graham A. W. Rook

dementia [1]. The neuropathological changes characteristic of AD – neuritic Aβ plaques that are the basis of neuropathological diagnosis of AD and neurofibrillary tangles that are used for staging of the disease – are not infrequently found alongside the cortical fibrillar α-synuclein-immunopositive Lewy bodies [2] characteristic of Lewy body dementia. The original neuropathological characteristics of Parkinson's disease, as described by Ikeda [3], are Lewy bodies in neurons within the substantia nigra as well as cortical areas. Many PD patients, however, show an associated dementia and show the characteristic neuropathological features of AD [4]. Pure Huntington's disease is related to inheritance of specific gene sequences encoding the huntingtin protein [5–7], which accumulates as neuronal fibrillary huntingtin [8] in Huntington's disease but has also been found in neurons of both AD and PD patients [9].

Clinically diagnosed neurodegenerative diseases, such as those mentioned above, may be confirmed by postmortem neuropathological findings of unique or shared features characteristic of the disease, such as spongiform changes in CJD (usually not accompanied by other dementia-related neuropathological changes); and Lewy bodies composed of fibrillar α-synuclein in cerebal cortical neurons in dementia with Lewy bodies (often accompanied by concurrent neuropathological changes of AD). The neuropathological changes of PD include Lewy bodies in the brain stem with or without Lewy bodies in cortical neurons. Whereas AD is characterized by both intraneuronal and extraneuronal accumulations of proteins, viz. paired helical filaments of hyperphosphorylated tau within cortical neurons and deposits of β-amyloid (Aβ) lying between neurons, other dementias such as PD and Huntington's are more often characterized by circumscribed distributions of a single protein [10]. At end stages of these diseases, the neuropathological changes suggest that there is either a disease-related increase in production, a failure in appropriate processing and elimination, or alterations in appropriate conformation ('misfolding') of an otherwise normal and essential protein. Recently, a new term has been coined for protein misfolding-associated neurodegenerative diseases: 'foldopathies' [11]. For example, in CJD the prion protein (PrP) misfolds into an abnormal prion isoform (PrPsc) present in animals with scrapie and in the PrPsc plaques in patients with CJD [12]. In Lewy body dementia and PD, α-synuclein misfolds to form intracellular Lewy bodies [13, 14]. And in AD the foldopathies are more complex – one is a 40- or 42-amino acid Aβ sequence, which folds into the β-pleated sheet structures present in Aβ plaques; another is a hyperphosphorylated tau, misfolded into paired helical filaments that comprise the neurofibrillary tangles of AD [15]. Systematic processing failures that account for the formation of these aberrantly folded proteins, and their associated dementias, are currently under intense investigation. However, as early as 1991, Braak and Braak [16] suggested that the accumulation of these misfolded proteins, whether inside or outside neurons, are the result of nucleation events in which one misfolded protein is associated with misfolding of like proteins. This makes necessary development of new ideas for therapeutic interventions not only into the genesis of misfolded proteins but also in consequent nucleation events.

Many hypotheses have been put forward regarding the genesis of production, processing, and elimination errors potentially giving rise to dementia-related neuropathological changes, and these are discussed here. We will concentrate on ways in which the hygiene hypothesis and Darwinian Medicine could impact on the development of AD, the most prevalent of these dementias, and the ways in which the principal hypotheses regarding Alzheimer pathogenesis, including the amyloid, tau, and neuroinflammatory hypotheses, may intersect with the hygiene hypothesis. The importance of protein misfolding in neuropathogenesis and the potential for the removal of misfolded proteins by immunological processes suggests that hygiene-related changes – which are proposed to result in development of a stronger immune system – could perhaps stave off development of the brain changes characteristic of AD, or act to quell them in established disease.

The amyloid hypothesis

Development of this hypothesis was largely the result of several discoveries that connected the most notable neuropathological change associated with Alzheimer's disease (AD), viz, the presence of numerous Aβ plaques, to the cognitive and behavioral changes in AD associated with learning and memory and appropriate responses to environmental cues. Alzheimer himself was first to recognize an association between the presence of Aβ plaques and cognitive changes. Alzheimer was also first to recognize the potential involvement of the small cells (activated glia) in Alzheimer pathogenesis; these cells were later shown to comprise the innate immune system of the brain (for an English translation of his original paper please see [17], and for a review of Cajal's writings regarding Alzheimer's findings and of his own neuropathological studies of AD please see [18, 19]). Alzheimer's discoveries of a relationship between Aβ plaques and glial activation have potential as a gateway for examination of the influence of the hygiene hypothesis to favor developmental strengthening of the immune system.

A further strong link between Aβ and Alzheimer pathogenesis was provided by publication of the Aβ peptide sequence [20, 21], and the subsequent sequencing of the precursor protein of β-amyloid (βAPP) gene and its mapping to chromosome 21 [22–26] . This link was confirmed by the recognition by Wisniewski and colleagues that trisomy 21 (Down's syndrome) is invariably associated with development of Alzheimer neuropathological changes at middle age [27], suggesting that inheritance of multiple copies of a normal APP gene could precipitate AD. This was later confirmed in two reports showing that inheritance of extra copies of the normal βAPP gene without duplication of other chromosome 21 genes is sufficient to cause early onset AD [28–30]. In studies of AD families, Goate and Hanger and their colleagues showed that inheritance of mutated βAPP genes is causative for AD at earlier than expected ages [31, 32]. Together these studies made clear the importance of βAPP – whether due to excessive expression or mutation of the normal gene – in Alzheimer pathogenesis. However, as βAPP mutations account for at most 2% of those with AD, neither gene loading nor βAPP mutations can explain the development of sporadic AD [33], suggesting that unknown effectors associated with neuropathological changes drive excessive expression of βAPP.

An early strategy aimed at clearing the brain of Aβ plaques was immunization against Aβ. This was quite successful in experimental animals transfected with mutated human βAPP [34, 35]. The promise of Aβ immunization to clear brain of plaques and prevent memory loss in transgenic animals [36] prompted a trial to test the effectiveness of active immunization to clear Aβ plaques and prevent memory loss in AD patients [37, 38]. Albeit a great deal of weight has been given to the idea that ridding the brain of Aβ would provide a cure for AD, to date, immunization strategies have not provided much success [39, 40]. The first trial (Elan) was halted due to several adverse events. However, among the approximately 20% of patients who showed evidence of antibody production, some were later found to be free of plaques [41]. Whether this result was beneficial or not remains uncertain [39]. For instance, a recent study determined that Aβ immunization-induced removal of Aβ plaques from brains of AD patients had little or no correlation with positive clinical outcomes, even in those patients in which immunization appeared to clear the most plaques [40, 41]. Regardless of these less-than-promising results of human immunization, there may be potential benefits from clearing Aβ plaques from the brain, and toward this goal modified immunization protocols are being devised [40]. However, doubts continue to be raised concerning whether ridding the brain of the large Aβ space-occupying lesions located among the neurons is the best strategy for combating dementia. For instance, there are a number of older individuals who have plaque densities similar to AD without cognitive impairment [42, 43], and brain amyloid load is not elevated with progression of clinical signs in some cases of early onset AD [44]. These findings and the lack of cognitive improvement following active immunization of AD patients [40, 41] suggest that factors other than, or in addition to, immunization may be a better course of action against AD [45]. In addition, these results may suggest that age-related deficits in immune responsiveness resulting from less than ideal development of the immune system could contribute to the inability of the brain to deal with the effectors of AD that set in motion neurodegenerative changes that culminate in dementia.

Immunization-related safety and other concerns constitute significant pitfalls as noted by Morgan [46]. For instance, in the Elan trial cited above, encephalitis occurred in 5% of patients and a dramatic increase in vascular amyloidosis and in cerebral microhemorrhages was noted in one patient. In addition, the cost of immunization could be viewed as 'wasted' on the 50% of the population who reach age 85 without neuropsychological changes consistent with cognitive impairment, and also wasted on the 50% of immunized individuals who remain unresponsive. Perhaps most importantly, when an antibody response was detected it did not necessarily correspond to improvements in cognitive scores [41]. Another concern about this therapeutic approach is that immunization responsiveness is irreversible. Thus, should immunization lead to adverse events such as autoimmune disease, patient suffering and caregiver stress would be exacerbated. These and other concerns [46] may be ameliorated by development of passive immunization strategies. However, it is also important to note that, in the absence of more precise knowledge of normal Aβ functions in the central nervous system, immunization strategies aimed at reducing Aβ levels over a long period of time may have unexpected adverse effects on normal brain functions. On a more hopeful note, enhancement of immune system functions, perhaps gained through adoption of strategies suggested in the hygiene hypothesis, may preclude the necessity for exogenous manipulation of the immune system, especially in already established disease.

The tau hypothesis

The tau hypothesis arose from an attempt to understand and thwart the genesis of yet another foldopathy: the paired helical filament arrangement of the hyperphosphorylated tau protein in neurofibrillary tangles present in neurons in AD brain. The association between these tangles and progression of clinically assessed cognitive decline was first demonstrated by Braak and Braak [47, 48]. Unlike Aβ, tau has important duties in normal development and throughout life. These duties include its role in formation, maintenance, and stability of the microtubules, structures in neuronal processes that have necessary structural and functional properties. Neurofibrillary tangle formation anomalies interfere with microtubule function in AD neurons, including normal flow of molecules to and from the cell body [49]. Importantly, in this interaction between tau and molecular motors [50], tau is responsible for regulating the number of engaged motors per cargo to and from synaptic regions [51]. The identification of phosphorylated tau [52] as the principle component of neurofibrillary tangles [53–55] of AD [56, 57] and its association with microtubule dissolution gave rise to a number of studies into the pathological consequences of fibrillary tangles [55, 56, 58]. For instance, tangle-bearing neurons have less polyadenylated mRNA, signifying impaired total protein synthesis [59], but more βAPP mRNA [60]. Decreasing overall protein synthetic capacity while simultaneously increasing synthesis of βAPP could be internally detrimental to neuronal function due to declining transport functions and externally detrimental to neuronal function due to accumulation of neurotoxic Aβ [61]. Further, upregulation of βAPP could lead to elevated levels of sAPP, a secreted cytokine-like fragment of βAPP, that has been shown to stimulate phosphorylation of tau both in cell lines [62] and in primary neurons [63]. This result could thus further favor formation of tangles, selective loss of synapses [64], and dysfunction and loss of neurons [65, 66]. Moreover, these tangle-related problems may be further exacerbated due to sAPP-induced microglial activation and excessive production and release of interleukin-1 (IL-1) [67–69], which increases neuronal production of phosphorylated tau, thus again favoring tangle formation [69–71].

It may seem obvious that hyperphosphylation of tau and formation of neurofibrillary tangles within neurons would contribute to the neuronal dysfunction and loss associated with dementia. However, a direct link between mutations in the tau gene and AD (similar to that between βAPP mutations and AD) has not been established. Interestingly, patients with βAPP 717 mutations [72–74] do have neurofibrillary pathology [75]. Carriers of a specific tau mutation do exhibit the characteristic features of both frontal lobe dementia and AD [76]. Tangle-bearing neurons are extremely long lived, suggesting that they may cause the insidious but prolonged decline in neuron function as AD progresses, but that they are not, in and of themselves, acutely lethal [77].

Immunization strategies similar to those directed toward Aβ plaque clearance in AD patients have been proposed and are currently being developed for clearing neurofibrillary tangles from mutant mice. Immunization with a phosphorylated tau derivative that crosses the blood brain barrier and binds to phosphorylated tau in neurons is associated with improvement in tangle-related behavioral deficits in a mutant mouse model [78]. Such tau immunization strategies may have promise in patients but may encounter problems similar to those noted with Aβ immunization strategies.

In another approach, Gozes and her colleagues [79, 80] are using an agent called NAP, an 8-amino acid peptide derived from the activity-dependent neuroprotective peptide [81] for treatment or prevention of tauopathies. Three months of treatment of mice transgenic for both mutated βAPP and tau gene sequences, at a time when both plaques and neurofibrillary tangles form in these animals' brains, resulted in reductions in both soluble and insoluble tau. Mice treated for 6 months showed similar reductions as well as improvements in cognitive or behavioral functions [80]. Potential advantages of this treatment include a less invasive route and more direct benefit to neurons, thus it may alleviate the neuron dysfunction and loss, which is seminal to disease neuropathogenesis.

Potential for drivers of both the amyloid- and tau-opathies

Inheritance of unfortunate gene mutations and/or polymorphisms are always factors in development of diseases. With regard to AD, there are several genetic mutations that are causative. For instance, as mentioned above, inheritance of specific mutations in the genes for βAPP, for presenilin-1, and for presenilin-2. In addition, individuals with trisomy of chromosome 21 (Down's syndrome) develop the characteristic neuropathological changes and diminished cognitive abilities of AD by middle age. The gene for βAPP is on chromosome 21, and the 50% increase in expression of the βAPP gene due to triplication of chromosome 21 further highlights the importance of this gene in AD pathogenesis. However, these causative genetic anomalies cannot explain the more common sporadic forms of AD. For these, variants or polymorphisms in several genes have been implicated in contributing risk for the development of AD; the most common of these is the epsilon 4 (ε4) variant of the apolipoprotein E (ApoE) gene: 50% of patients with AD have one or more such ε4 alleles. Recently, we have shown that neuronal levels of ApoE and βAPP increase in response to microglial activation and microglial IL-1 expression; such microglial IL-1 overexpression, in turn, can be induced *in vitro* by glutamate-induced neuronal stress, and *in vivo* by the wear and tear of the aging process. We also found that ApoE induces expression of βAPP, and that the ApoE ε3 variant is more effective at such induction than is ApoE ε4. These responses may be important and beneficial in limited periods of stress but, as explained above, have the potential to give rise to development of Alzheimer-type neuropathological changes when neuronal stress is prolonged, as occurs, for example, in normal aging [82].

Another set of genes that have been implicated in conferring risk for sporadic Alzheimer's disease are the IL-1A and B genes which encode IL-1α and IL-1β. For IL-1A, the gene polymorphism associated with AD lies in the gene promoter region while for IL-1B the risk-conferring polymorphism lies in the coding region. Two studies of individuals who died with AD have shown that either of these polymorphisms confers a three-fold increase in risk for AD, and that inheritance of both polymorphisms is associated with a nine-fold increase in risk [83, 84]. Another of our studies of AD patients showed that the IL-1A polymorphism is associated not only with increased risk (4.5 times) but also with earlier disease onset (7–9 years earlier) [85]. Subsequent studies have not always replicated these findings, but meta-analyses of all such studies have supported an association at least for the IL-1A gene [86, 87]. Some variation in results might be expected if these polymorphisms are dependent on co-inheritance of other presently unknown polymorphisms so that the increased risk is associated with such gene combinations rather than a single genotype. The dramatic increase in risk that we found for co-inheritance of both IL-1A and B polymorphisms supports the multigene and additive nature of risk conferrance, which is in contrast to the single gene mechanisms through which gene mutations may cause disease. In addition to linkage of AD risk to IL-1 genes, increased risk has also been associated with polymorphisms in other inflammatory genes, such as tumor necrosis factor-α (TNF-α) [88]. IL-1 dysregulation is also implicated in a variety of neural conditions that are themselves associated with increased risk for precocious development of AD or AD-like clinical and neuropathological changes [89–91]. These genetic associations imply a role for IL-1 *via* an independent, unbiased approach and thus complement the neuropathological findings that (i) IL-1 neural tissue levels are elevated in AD and AD-related conditions, (ii) microglial activation and excessive IL-1 expression are associated with Aβ plaque progression, and (iii) microglial IL-1 overexpression is associated with formation of neurofibrillary tangles.

Some demented patients have few Aβ plaques and neurofibrillary tangles while some cognitively intact individuals have many. This, together with the lack of an association between immunization-induced Aβ plaque removal and cognitive improvement or decrease in time to severe dementia [40] suggests that a better strategy against AD may be preventive measures that counteract effects of conditions associated with precocious development of neuropathological changes and dementia. These conditions include heart disease [92], epilepsy [90, 93–95], and AIDS [89, 96–98]. In each of these conditions neuroinflammatory changes – neuronal trauma, glial activation, and excessive IL-1 expression – are observed before there is full-blown AD pathology. These neuronal and glial changes are present as early as fetal and neonatal ages and continue throughout life in Down's syndrome, a certain predictor of AD [99]. The association between neuronal trauma and βAPP overexpression in primary neurons, animals, and in humans is well established [60, 91, 100–102]. The missing link explaining the universal observation of neuronal injury, βAPP overexpression, and glial activation in neurodegenerative conditions was provided by the finding that neuronal trauma-induced production of βAPP results in release of sAPP, which in turn induces glial activation and excessive expression of IL-1 [67, 103]. Findings implicating excessive production of βAPP and release of sAPP and IL-1 as inducers of the precursors of each of the neuropathological changes characteristic of AD provided a final link between neuronal trauma, glial activation, and subsequent development of AD in conditions as diverse as Down's and PD [99, 104, 105]. In addition, they were the basis for developing the cytokine hypothesis.

The cytokine hypothesis

The cytokine hypothesis states that in response to neuronal injury or insult, whether induced by effectors such as trauma, genetics, or aging or a combination of all three, microglia and astrocytes become activated [106] and overexpress and release cytokines, in particular proinflammatory IL-1 and neuritogenic S100B. In limited circumstances, such activation and increased expression of specific cytokines may be important in maintenance and repair of neurons in the normal course of aging. The gradual increase in S100B expression during normal aging [107] may occur in an effort to counteract age-related changes in synaptic and neuronal maintenance of neurons, their neurites and synaptic functions [108, 109]. Similarly, IL-1 may be elevated in response to injury in an effort to repair and maintain membranes [110–112].

The elevation of βAPP, sAPP, IL-1 and S100B that facilitates the repair of neurons in limited circumstances may, as neuronal dysfunction becomes chronic, have unintended and less salutary consequences that drive neuropathogenesis [113]. For instance, elevated levels of IL-1 (i) promote tau phosphorylation, favoring neurofibrillary tangle formation [68–70]; (ii) induce expression of α-synuclein [114], perhaps explaining comorbidities such as AD with Lewy body disease [104]; (iii) increase both the expression and activity of acetylcholinesterase, favoring excess degradation of the neurotransmitter acetylcholine [115, 116]; (iv) and exacerbate neuronal and glial expression of ApoE, which together with the greater effectiveness of ApoE3 (compared to ApoE4) in inducing neuronal expression of βAPP, may explain the negative impact of ApoE4 inheritance on AD risk [82]. Perhaps as importantly, IL-1 and TNF-α, as well as S100B [117] have been shown to act as gliotransmitters or modulators of neurotransmitter functions at synapses for regulation of long-term potentiation (LTP) [118–120], and thus regulation of learning and memory functions. Age-related [121] and stress-related (e.g., neuronal exposure to Aβ [122]) changes in LTP have long been recognized as important in learning and memory [117]. Taken together, these findings suggest that reported age- and Alzheimer-related increases in S100B, IL-1, and TNF-α may have beneficial effects in limited circumstances of neuronal stress or perhaps early in disease progression but later act as drivers of neuropathological change, thus creating a self-propagating cycle of neurodegenerative events.

A new term, 'inflammaging', has been coined to explain the age-related overexpression of inflammatory cytokines and consequent activation of innate immunity [123]. Discovery of associations between increased risk for development of AD and normal aging [124], head injury [91, 125], stroke [126, 127], epilepsy [94, 95], and viral infection [89] gave credence to the idea that inflammagenic events such as those in the cytokine cycle act as drivers of neuropathogenesis [128]. These discoveries were a departure from conventional ideas and created a sea change in our thinking about not only the reasoning behind the progressive nature of AD but also the potential for development of novel and rational therapeutic strategies for prevention or delay of disease onset. Earlier (non-AD related) use of anti-inflammatory drugs (NSAIDS) has been shown to delay predicted onset age [129–132]. However, clinical trials of NSAIDs in patients with established AD have been disappointing [133], perhaps suggesting that the drugs need to be taken prior to onset of any clinical symptoms to exert their protective effects. This should be obvious if the idea is to quell neuroinflammation by slowing the cytokine cycle, taking into account the decades necessary for development of the characteristic cognitive and neuropathological changes of AD in Down's syndrome. Encouragingly, large well-designed epidemiological studies do show that NSAID use confers impressive decreases in risk for later development of AD [129, 132, 134–137]. Such studies suggest that 'turning' microglia and astrocytes toward non-inflammatory functions would be helpful in combating the propagation of Alzheimer pathogenesis.

Ways in which hygiene-related changes in immune system development may forestall neuropathogenesis

Removal of Aβ aggregates might be facilitated by a more effective immune system such as that resulting from a hygiene-related better immune system. For example, hygiene-related enhancement of a patient's immune system might improve the chance of immunization-driven clearance of Aβ or tau aggregates, but it might at the same time increase the risk for development of autoimmune diseases. Most important in any discussion of clearing the brain of constitutively expressed proteins is our lack of knowledge regarding its normal functions; such is the case with Aβ. For example, the lack of improvement in cognitive status with removal of Aβ plaques may suggest that reducing Aβ levels adversely affects neuron functions. The functions of tau in neurons appear to be critical, including its role in microtubule assembly, stability, and transport of neurotransmitters and other proteins. Since all of these functions are essential, neurons may particularly susceptible to long-term therapeutic strategies that sequester tau into aggregates which can no longer participate in dynamic interactions with microtubules [79]. If pitfalls noted with current strategies can be circumvented, combining strategies aimed toward eliminating both Aβ and tau aggregates may be useful. Better Aβ and tau aggregate immunization strategies are currently underway and some are being assessed for effectiveness in primates [78, 138, 139]. In addition, new less invasive routes, e.g., mucosal delivery of immunogen, may have promise as they may more effectively clear aggregates as well as decrease adverse effects such as amyloidosis [140]. Combining strategies such as immunization against neuropathological aggregates with another approach such as anti-inflammatory drug treatment may also have unexpected benefits. For example, Wilcock et al. [141] showed that both NSAID treatment and Aβ immunization resulted in decreases parenchymal Aβ, but immunization was associated with increases in amyloidosis and microhemorrhage.

An important consideration for any potential therapeutic approach to alleviate problems associated with Alzheimer neuropathological change is the potential risk of harm. This, plus the risk of maintaining, rather than alleviating, an unacceptable cognitive state, raises important questions concerning not only patient rights but also caregiver, societal, and economic worries. Taken together these concerns suggest that the most rational strategies are those aimed at prevention rather than clearance of dementia-related neuropathological changes. Such strategies include those like NSAIDS, shown in large epidemiological studies to be associated with lowered risk for later development of Alzheimer dementia [129, 132, 134–137]. These studies, together with others showing that NSAIDS directly downregulate microglial activation and excessive expression of IL-1 [142] and excess production of each of the precursors of AD neuropathological change [113, 116], are consistent with the idea that quelling neuroinflammatory drivers of neuropathological change prevents or slows onset of dementia. Furthermore, the role of microglia in the brain's innate immune system suggests that increasing their power to fight neuropathological change and neuronal damage would be advantageous in preventing progression of these changes to full fledged AD. If so, this may lend credence to a beneficial effect of a hygiene-related boosted immune system.

The hygiene hypothesis states that urbanization-related decreases in contact between humans and 'friendly' organisms, which otherwise acts to improve immune regulatory mechanisms, result in weakening of immune system efforts to thwart effectors of disease. In the brain, microglia perform the regulatory mechanisms likely to be the targets of such improvement resulting from interactions between humans and such organisms at times when the immune system is developing. There is evidence that immunization of mouse models of AD results in microglial phenotype changes from an inflammatory to an more anti-inflammatory phenotype, suggesting that as Aβ plaques are cleared microglia revert to a resting, non-inflammatory state [143]. But whether switching microglia to states more consistent with challenging effectors of neurodegeneration would quell or exacerbate degeneration is unknown. The studies of Morgan and colleagues seem to suggest that microglia are more effective in plaque removal when exhibiting an inflammatory phenotype and revert to resting phenotypes only when Aβ removal reaches some unknown point [143]. Considering this idea together with the capacity of microglial activation and excessive production of IL-1 to drive neurodegenerative processes, and NSAIDS to quell such processes, one is struck by the importance of such small cells in perhaps whimsical promotion or demotion of neuropathogenesis. If the hygiene hypothesis is pertinent to neuropathogenic events, would exposure to 'friendly' organisms favor development of more aggressive or less aggressive, i.e., inflammatory or non-inflammatory, phenotypes? On the one hand, favoring development of more aggressive phenotypes would simultaneously promote Aβ plaque clearance and neuronal production of substrates necessary for formation of Aβ plaques as well as neurofibrillary tangles. On the other hand, favoring development of less aggressive phenotypes might have salutary results, such as those of NSAIDS, including decreases in neurodegenerative drivers associated with microglial activation and cytokine production. The findings and suppositions discussed in this chapter raise more questions than answers. However, one fact remains – without interventions directed at effectors of neuronal insults, whether they be mutations as in familial AD, extra βAPP gene copies as in Down's syndrome, or environmental factors such as head injury, microglia will be activated and overexpress proinflammatory cytokines. This induction of microglial activation by neuronal stress can then set in motion a self-amplifying cascade – the cytokine cycle – which, unchecked, leads to further neuronal damage. If perchance hygiene-related changes in innate immune responses in the brain quelled microglial responses, such changes should be implemented. This calls for more research into this idea as neglect of other less felicitous outcomes from exposure to dirty environments might outweigh any beneficial effects on development of neurodegenerative diseases. Such gathering of, and attention to, facts that provide the wherewithal to test alternate hypotheses – perhaps best stated by Ramon y Cajal, "*while hypotheses pass by, facts remain*" (Madrid, 16 September 1927) [144] – will be an interesting scientific trail to follow.

In conclusion, deciding between beneficial and detrimental effects of exposure of the developing human to 'friendly' organisms with regard to neuropathogenesis is not possible at this time. However, the possibility that building a better immune system – through exposure to 'friendly' organisms during immune system development – would enable microglia to thwart the insidious progression of the neuropathological changes and dementia of AD is so intriguing that it calls for further investigation to satisfy both Darwinian curiosity and a need to manipulate potential Darwinian influences on such development.

Acknowledgments

Thanks are owed to the patients and their loved ones who allowed assessment of patient cognitive abilities and neuropathological studies of patient brains. Manuscript review by Professor Steven Barger and secretarial support by Ms. Pamela Free are appreciated. Supported in part by National Institutes of Health grants AG10208, AG12411, HD37989, and NS27414.

References

1 Haraguchi T, Terada S, Ishizu H, Sakai K, Tanabe Y, Nagai T, Takata H, Nobukuni K, Ihara Y, Kitamoto T et al (2008) Coexistence of Creutzfeldt-Jakob disease, Lewy body disease, and Alzheimer's disease pathology: An autopsy case showing typical clinical features of Creutzfeldt-Jakob disease. *Neuropathology,* Epub ahead of print

2 Tashiro M, Kojima M, Kihara H, Kasai K, Kamiyoshihara T, Ueda K, Shimotakahara S (2008) Characterization of fibrillation process of alpha-synuclein at the initial stage. *Biochem Biophys Res Commun* 369: 910–914

3 Ikeda K, Ikeda S, Yoshimura T, Kato H, Namba M (1978) Idiopathic Parkinsonism with Lewy-type inclusions in cerebral cortex. A case report. *Acta Neuropathol* 41: 165–168

4 Farlow MR, Cummings J (2008) A modern hypothesis: The distinct pathologies of dementia associated with Parkinson's disease versus Alzheimer's disease. *Dement Geriatr Cogn Disord* 25: 301–308

5 Gusella JF, Wexler NS, Conneally PM, Naylor SL, Anderson MA, Tanzi RE, Watkins PC, Ottina K, Wallace MR, Sakaguchi AY et al (1983) A polymorphic DNA marker genetically linked to Huntington's disease. *Nature* 306: 234–238

6 Rubinsztein DC, Barton DE, Davison BC, Ferguson-Smith MA (1993) Analysis of the huntingtin gene reveals a trinucleotide-length polymorphism in the region of the gene that contains two CCG-rich stretches and a correlation between decreased age of onset of Huntington's disease and CAG repeat number. *Hum Mol Genet* 2: 1713–1715

7 Norremolle A, Riess O, Epplen JT, Fenger K, Hasholt L, Sorensen SA (1993) Trinucleotide repeat elongation in the Huntingtin gene in Huntington disease patients from 71 Danish families. *Hum Mol Genet* 2: 1475–1476
8 Forno LS (1992) Neuropathologic features of Parkinson's, Huntington's, and Alzheimer's diseases. *Ann NY Acad Sci* 648: 6–16
9 Singhrao SK, Thomas P, Wood JD, MacMillan JC, Neal JW, Harper PS, Jones AL (1998) Huntingtin protein colocalizes with lesions of neurodegenerative diseases: An investigation in Huntington's, Alzheimer's, and Pick's diseases. *Exp Neurol* 150: 213–222
10 Heiser V, Scherzinger E, Boeddrich A, Nordhoff E, Lurz R, Schugardt N, Lehrach H, Wanker EE (2000) Inhibition of huntingtin fibrillogenesis by specific antibodies and small molecules: implications for Huntington's disease therapy. *Proc Natl Acad Sci USA* 97: 6739–6744
11 Zilka N, Kontsekova E, Novak M (2008) Chaperone-like antibodies targeting misfolded tau protein: new vistas in the immunotherapy of neurodegenerative foldopathies. *J Alzheimers Dis* 15: 169–179
12 DeArmond SJ (2004) Discovering the mechanisms of neurodegeneration in prion diseases. *Neurochem Res* 29: 1979–1998
13 Lee VM, Goedert M, Trojanowski JQ (2001) Neurodegenerative tauopathies. *Annu Rev Neurosci* 24: 1121–1159
14 Clarimon J, Molina-Porcel L, Gomez-Isla T, Blesa R, Guardia-Laguarta C, Gonzalez-Neira A, Estorch M, Ma Grau J, Barraquer L, Roig C et al (2009) Early-onset familial lewy body dementia with extensive tauopathy: a clinical, genetic, and neuropathological study. *J Neuropathol Exp Neurol* 68: 73–82
15 Nelson PT, Braak H, Markesbery WR (2009) Neuropathology and cognitive impairment in Alzheimer disease: a complex but coherent relationship. *J Neuropathol Exp Neurol* 68: 1–14
16 Braak H, Braak E (1991) Demonstration of amyloid deposits and neurofibrillary changes in whole brain sections. *Brain Pathol* 1: 213–216
17 Alzheimer A, Stelzmann RA, Schnitzlein HN, Murtagh FR (1995) An English translation of Alzheimer's 1907 paper, "Über eine eigenartige Erkankung der Hirnrinde". *Clin Anat* 8: 429–431
18 Garcia-Marin V, Garcia-Lopez P, Freire M (2007) Cajal's contributions to the study of Alzheimer's disease. *J Alzheimers Dis* 12: 161–174
19 Ramon y Cajal S (1984) *Degeneration and Regeneration of the Nervous System.* Robert Maclehose and Co, University Press, Glasgow
20 Glenner GG, Wong CW (1984) Alzheimer's disease: initial report of the purification and characterization of a novel cerebrovascular amyloid protein. *Biochem Biophys Res Commun* 120: 885–890
21 Glenner GG, Wong CW, Quaranta V, Eanes ED (1984) The amyloid deposits in Alzheimer's disease: their nature and pathogenesis. *Appl Pathol* 2: 357–369
22 St George-Hyslop PH, Tanzi RE, Polinsky RJ, Haines JL, Nee L, Watkins PC, Myers

RH, Feldman RG, Pollen D, Drachman D et al (1987) The genetic defect causing familial Alzheimer's disease maps on chromosome 21. *Science* 235: 885–890
23 Patterson D, Gardiner K, Kao FT, Tanzi R, Watkins P, Gusella JF (1988) Mapping of the gene encoding the beta-amyloid precursor protein and its relationship to the Down syndrome region of chromosome 21. *Proc Natl Acad Sci USA* 85: 8266–8270
24 Tanzi RE, Gusella JF, Watkins PC, Bruns GA, St George-Hyslop P, Van Keuren ML, Patterson D, Pagan S, Kurnit DM, Neve RL (1987) Amyloid beta protein gene: cDNA, mRNA distribution, and genetic linkage near the Alzheimer locus. *Science* 235: 880–884
25 Goldgaber D, Lerman MI, McBride OW, Saffiotti U, Gajdusek DC (1987) Characterization and chromosomal localization of a cDNA encoding brain amyloid of Alzheimer's disease. *Science* 235: 877–880
26 Robakis NK, Ramakrishna N, Wolfe G, Wisniewski HM (1987) Molecular cloning and characterization of a cDNA encoding the cerebrovascular and the neuritic plaque amyloid peptides. *Proc Natl Acad Sci USA* 84: 4190–4194
27 Wisniewski KE, Dalton AJ, McLachlan C, Wen GY, Wisniewski HM (1985) Alzheimer's disease in Down's syndrome: clinicopathologic studies. *Neurology* 35: 957–961
28 Sachatello CR, Bivins PA, Daugherty ME, Griffin WO, Jr (1980) Diagnostic peritoneal lavage: a ten-year overview. *J Ky Med Assoc* 78: 418–422
29 Sleegers K, Brouwers N, Gijselinck I, Theuns J, Goossens D, Wauters J, Del-Favero J, Cruts M, van Duijn CM, Van Broeckhoven C (2006) APP duplication is sufficient to cause early onset Alzheimer's dementia with cerebral amyloid angiopathy. *Brain* 129: 2977–2983
30 Rovelet-Lecrux A, Hannequin D, Raux G, Le Meur N, Laquerriere A, Vital A, Dumanchin C, Feuillette S, Brice A, Vercelletto M et al (2006) APP locus duplication causes autosomal dominant early-onset Alzheimer disease with cerebral amyloid angiopathy. *Nat Genet* 38: 24–26
31 Hanger DP, Mann DM, Neary D, Anderton BH (1992) Tau pathology in a case of familial Alzheimer's disease with a valine to glycine mutation at position 717 in the amyloid precursor protein. *Neurosci Lett* 145: 178–180
32 Goate A, Chartier-Harlin MC, Mullan M, Brown J, Crawford F, Fidani L, Giuffra L, Haynes A, Irving N, James L et al (1991) Segregation of a missense mutation in the amyloid precursor protein gene with familial Alzheimer's disease. *Nature* 349: 704–706
33 Korczyn AD (2008) The amyloid cascade hypothesis. Alzheimers Dement 4: 176–178
34 Bard F, Cannon C, Barbour R, Burke RL, Games D, Grajeda H, Guido T, Hu K, Huang J, Johnson-Wood K et al (2000) Peripherally administered antibodies against amyloid beta-peptide enter the central nervous system and reduce pathology in a mouse model of Alzheimer disease. *Nat Med* 6: 916–919
35 Schenk D, Barbour R, Dunn W, Gordon G, Grajeda H, Guido T, Hu K, Huang J, Johnson-Wood K, Khan K et al (1999) Immunization with amyloid-beta attenuates Alzheimer-disease-like pathology in the PDAPP mouse. *Nature* 400: 173–177
36 Morgan D, Diamond DM, Gottschall PE, Ugen KE, Dickey C, Hardy J, Duff K, Jantzen

P, DiCarlo G, Wilcock D et al (2000) A beta peptide vaccination prevents memory loss in an animal model of Alzheimer's disease. *Nature* 408: 982–985

37 Schenk D, Seubert P, Ciccarelli RB (2001) Immunotherapy with beta-amyloid for Alzheimer's disease: a new frontier. *DNA Cell Biol* 20: 679–681

38 Masliah E, Hansen L, Adame A, Crews L, Bard F, Lee C, Seubert P, Games D, Kirby L, Schenk D (2005) Abeta vaccination effects on plaque pathology in the absence of encephalitis in Alzheimer disease. *Neurology* 64: 129–131

39 Boche D, Nicoll JA (2008) The role of the immune system in clearance of Abeta from the brain. *Brain Pathol* 18: 267–278

40 Holmes C, Boche D, Wilkinson D, Yadegarfar G, Hopkins V, Bayer A, Jones RW, Bullock R, Love S, Neal JW et al (2008) Long-term effects of Abeta42 immunisation in Alzheimer's disease: follow-up of a randomised, placebo-controlled phase I trial. *Lancet* 372: 216–223

41 Nicoll JA, Barton E, Boche D, Neal JW, Ferrer I, Thompson P, Vlachouli C, Wilkinson D, Bayer A, Games D et al (2006) Abeta species removal after abeta42 immunization. *J Neuropathol Exp Neurol* 65: 1040–1048

42 Davis DG, Schmitt FA, Wekstein DR, Markesbery WR (1999) Alzheimer neuropathologic alterations in aged cognitively normal subjects. *J Neuropathol Exp Neurol* 58: 376–388

43 (2001) Pathological correlates of late-onset dementia in a multicentre, community-based population in England and Wales. Neuropathology Group of the Medical Research Council Cognitive Function and Ageing Study (MRC CFAS) *Lancet* 357: 169–175

44 Engler H, Forsberg A, Almkvist O, Blomquist G, Larsson E, Savitcheva I, Wall A, Ringheim A, Langstrom B, Nordberg A (2006) Two-year follow-up of amyloid deposition in patients with Alzheimer's disease. *Brain* 129: 2856–2866

45 Head E, Pop V, Vasilevko V, Hill M, Saing T, Sarsoza F, Nistor M, Christie LA, Milton S, Glabe C et al (2008) A two-year study with fibrillar beta-amyloid (Abeta) immunization in aged canines: effects on cognitive function and brain Abeta. *J Neurosci* 28: 3555–3566

46 Morgan D, Landreth G, Bickford P (2008) The Promise and Perils of an Alzheimer Disease Vaccine: A Video Debate. *J Neuroimmune Pharmacol* 4: 1–3

47 Braak H, Braak E (1991) Neuropathological stageing of Alzheimer-related changes. *Acta Neuropathol (Berl)* 82: 239–259

48 Braak H, Braak E (1995) Staging of Alzheimer's disease-related neurofibrillary changes. *Neurobiol Aging* 16: 271–278; discussion 278–284

49 Cuchillo-Ibanez I, Seereeram A, Byers HL, Leung KY, Ward MA, Anderton BH, Hanger DP (2008) Phosphorylation of tau regulates its axonal transport by controlling its binding to kinesin. *Faseb J* 22: 3186–3195

50 Trinczek B, Ebneth A, Mandelkow EM, Mandelkow E (1999) Tau regulates the attachment/detachment but not the speed of motors in microtubule-dependent transport of single vesicles and organelles. *J Cell Sci* 112 (Pt 14): 2355–2367

51 Vershinin M, Carter BC, Razafsky DS, King SJ, Gross SP (2007) Multiple-motor based transport and its regulation by Tau. *Proc Natl Acad Sci USA* 104: 87–92

52 Ueda K, Masliah E, Saitoh T, Bakalis SL, Scoble H, Kosik KS (1990) Alz-50 recognizes a phosphorylated epitope of tau protein. *J Neurosci* 10: 3295–3304

53 Kosik KS (1990) Tau protein and neurodegeneration. *Mol Neurobiol* 4: 171–179

54 Kosik KS (1993) The molecular and cellular biology of tau. *Brain Pathol* 3: 39–43

55 Grundke-Iqbal I, Iqbal K, Tung YC, Quinlan M, Wisniewski HM, Binder LI (1986) Abnormal phosphorylation of the microtubule-associated protein tau (tau) in Alzheimer cytoskeletal pathology. *Proc Natl Acad Sci USA* 83: 4913–4917

56 Dayanandan R, Van Slegtenhorst M, Mack TG, Ko L, Yen SH, Leroy K, Brion JP, Anderton BH, Hutton M, Lovestone S (1999) Mutations in tau reduce its microtubule binding properties in intact cells and affect its phosphorylation. *FEBS Lett* 446: 228–232

57 Kosik KS, Shimura H (2005) Phosphorylated tau and the neurodegenerative foldopathies. *Biochim Biophys Acta* 1739: 298–310

58 Trojanowski JQ, Lee VM (1995) Phosphorylation of paired helical filament tau in Alzheimer's disease neurofibrillary lesions: focusing on phosphatases. *Faseb J* 9: 1570–1576

59 Griffin WS, Ling C, White CL, 3rd, Morrison-Bogorad M (1990) Polyadenylated messenger RNA in paired helical filament-immunoreactive neurons in Alzheimer disease. *Alzheimer Dis Assoc Disord* 4: 69–78

60 Schmechel DE, Goldgaber D, Burkhart DS, Gilbert JR, Gajdusek DC, Roses AD (1988) Cellular localization of messenger RNA encoding amyloid-beta-protein in normal tissue and in Alzheimer disease. *Alzheimer Dis Assoc Disord* 2: 96–111

61 Huang HC, Jiang ZF (2009) Accumulated amyloid-beta peptide and hyperphosphorylated tau protein: relationship and links in Alzheimer's disease. *J Alzheimers Dis* 16: 15–27

62 Greenberg SM, Koo EH, Selkoe DJ, Qiu WQ, Kosik KS (1994) Secreted beta-amyloid precursor protein stimulates mitogen-activated protein kinase and enhances tau phosphorylation. *Proc Natl Acad Sci USA* 91: 7104–7108

63 Greenberg SM, Rebeck GW, Vonsattel JP, Gomez-Isla T, Hyman BT (1995) Apolipoprotein E epsilon 4 and cerebral hemorrhage associated with amyloid angiopathy. *Ann Neurol* 38: 254–259

64 Callahan LM, Coleman PD (1995) Neurons bearing neurofibrillary tangles are responsible for selected synaptic deficits in Alzheimer's disease. *Neurobiol Aging* 16: 311–314

65 Callahan LM, Vaules WA, Coleman PD (2002) Progressive reduction of synaptophysin message in single neurons in Alzheimer disease. *J Neuropathol Exp Neurol* 61: 384–395

66 Coleman PD, Flood DG (1987) Neuron numbers and dendritic extent in normal aging and Alzheimer's disease. *Neurobiol Aging* 8: 521–545

67 Barger SW, Harmon AD (1997) Microglial activation by Alzheimer amyloid precursor protein and modulation by apolipoprotein E. *Nature* 388: 878–881

68 Li Y, Liu L, Barger SW, Griffin WS (2003) Interleukin-1 mediates pathological effects of microglia on tau phosphorylation and on synaptophysin synthesis in cortical neurons through a p38-MAPK pathway. *J Neurosci* 23: 1605–1611

69 Sheng JG, Jones RA, Zhou XQ, McGinness JM, Van Eldik LJ, Mrak RE, Griffin WS (2001) Interleukin-1 promotion of MAPK-p38 overexpression in experimental animals and in Alzheimer's disease: potential significance for tau protein phosphorylation. *Neurochem Int* 39: 341–348

70 Sheng JG, Zhu SG, Jones RA, Griffin WS, Mrak RE (2000) Interleukin-1 promotes expression and phosphorylation of neurofilament and tau proteins *in vivo*. *Exp Neurol* 163: 388–391

71 Sheng JG, Mrak RE, Griffin WS (1997) Glial-neuronal interactions in Alzheimer disease: progressive association of IL-1alpha^{+} microglia and S100beta^{+} astrocytes with neurofibrillary tangle stages. *J Neuropathol Exp Neurol* 56: 285–290

72 Hutton M, Lendon CL, Rizzu P, Baker M, Froelich S, Houlden H, Pickering-Brown S, Chakraverty S, Isaacs A, Grover A et al (1998) Association of missense and 5'-splice-site mutations in tau with the inherited dementia FTDP-17. *Nature* 393: 702–705

73 Spillantini MG, Crowther RA, Kamphorst W, Heutink P, van Swieten JC (1998) Tau pathology in two Dutch families with mutations in the microtubule-binding region of tau. *Am J Pathol* 153: 1359–1363

74 Foster NL, Wilhelmsen K, Sima AA, Jones MZ, D'Amato CJ, Gilman S (1997) Frontotemporal dementia and parkinsonism linked to chromosome 17: a consensus conference. Conference Participants. *Ann Neurol* 41: 706–715

75 Lantos PL, Luthert PJ, Hanger D, Anderton BH, Mullan M, Rossor M (1992) Familial Alzheimer's disease with the amyloid precursor protein position 717 mutation and sporadic Alzheimer's disease have the same cytoskeletal pathology. *Neurosci Lett* 137: 221–224

76 Ostojic J, Elfgren C, Passant U, Nilsson K, Gustafson L, Lannfelt L, Froelich Fabre S (2004) The tau R406W mutation causes progressive presenile dementia with bitemporal atrophy. *Dement Geriatr Cogn Disord* 17: 298–301

77 Morsch R, Simon W, Coleman PD (1999) Neurons may live for decades with neurofibrillary tangles. *J Neuropathol Exp Neurol* 58: 188–197

78 Sigurdsson EM (2008) Immunotherapy targeting pathological tau protein in Alzheimer's disease and related tauopathies. *J Alzheimers Dis* 15: 157–168

79 Gozes I, Divinski I, Piltzer I (2008) NAP and D-SAL: neuroprotection against the beta amyloid peptide (1–42). *BMC Neurosci* 9 (Suppl 3): S3

80 Matsuoka Y, Jouroukhin Y, Gray AJ, Ma L, Hirata-Fukae C, Li HF, Feng L, Lecanu L, Walker BR, Planel E et al (2008) A neuronal microtubule-interacting agent, NAPVSIPQ, reduces tau pathology and enhances cognitive function in a mouse model of Alzheimer's disease. *J Pharmacol Exp Ther* 325: 146–153

81 Gozes I, Bassan M, Zamostiano R, Pinhasov A, Davidson A, Giladi E, Perl O, Glazner GW, Brenneman DE (1999) A novel signaling molecule for neuropeptide action: activity-dependent neuroprotective protein. *Ann NY Acad Sci* 897: 125–135

82 Barger SW, DeWall KM, Liu L, Mrak RE, Griffin WS (2008) Relationships between expression of apolipoprotein E and β-amyloid precursor protein are altered in proximity to Alzheimer β-amyloid plaques; potential explanations from cell culture studies. *J Neuropathol Exp Neurol* 67: 773–783
83 Nicoll JA, Mrak RE, Graham DI, Stewart J, Wilcock G, MacGowan S, Esiri MM, Murray LS, Dewar D, Love S et al (2000) Association of interleukin-1 gene polymorphisms with Alzheimer's disease. *Ann Neurol* 47: 365–368
84 Rebeck GW (2000) Confirmation of the genetic association of interleukin-1A with early onset sporadic Alzheimer's disease. *Neurosci Lett* 293: 75–77
85 Grimaldi LM, Casadei VM, Ferri C, Veglia F, Licastro F, Annoni G, Biunno I, De Bellis G, Sorbi S, Mariani C et al (2000) Association of early-onset Alzheimer's disease with an interleukin-1alpha gene polymorphism. *Ann Neurol* 47: 361–365
86 Rainero I, Bo M, Ferrero M, Valfre W, Vaula G, Pinessi L (2004) Association between the interleukin-1alpha gene and Alzheimer's disease: a meta-analysis. *Neurobiol Aging* 25: 1293–1298
87 Combarros O, Llorca J, Sanchez-Guerra M, Infante J, Berciano J (2003) Age-dependent association between interleukin-1A (-889) genetic polymorphism and sporadic Alzheimer's disease. A meta-analysis. *J Neurol* 250: 987–989
88 Culpan D, MacGowan SH, Ford JM, Nicoll JA, Griffin WS, Dewar D, Cairns NJ, Hughes A, Kehoe PG, Wilcock GK (2003) Tumour necrosis factor-alpha gene polymorphisms and Alzheimer's disease. *Neurosci Lett* 350: 61–65
89 Stanley LC, Mrak RE, Woody RC, Perrot LJ, Zhang S, Marshak DR, Nelson SJ, Griffin WS (1994) Glial cytokines as neuropathogenic factors in HIV infection: pathogenic similarities to Alzheimer's disease. *J Neuropathol Exp Neurol* 53: 231–238
90 Sheng JG, Boop FA, Mrak RE, Griffin WS (1994) Increased neuronal beta-amyloid precursor protein expression in human temporal lobe epilepsy: association with interleukin-1 alpha immunoreactivity. *J Neurochem* 63: 1872–1879
91 Griffin WS, Sheng JG, Gentleman SM, Graham DI, Mrak RE, Roberts GW (1994) Microglial interleukin-1 alpha expression in human head injury: correlations with neuronal and neuritic beta-amyloid precursor protein expression. *Neurosci Lett* 176: 133–136
92 Sparks DL, Hunsaker JC, 3rd, Scheff SW, Kryscio RJ, Henson JL, Markesbery WR (1990) Cortical senile plaques in coronary artery disease, aging and Alzheimer's disease. *Neurobiol Aging* 11: 601–607
93 Griffith HR, Martin RC, Bambara JK, Marson DC, Faught E (2006) Older adults with epilepsy demonstrate cognitive impairments compared with patients with amnestic mild cognitive impairment. *Epilepsy Behav* 8: 161–168
94 Breteler MM, van Duijn CM, Chandra V, Fratiglioni L, Graves AB, Heyman A, Jorm AF, Kokmen E, Kondo K, Mortimer JA et al (1991) Medical history and the risk of Alzheimer's disease: a collaborative re-analysis of case-control studies. EURODEM Risk Factors Research Group. *Int J Epidemiol* 20 (Suppl 2): S36–42
95 Griffin WS, Yeralan O, Sheng JG, Boop FA, Mrak RE, Rovnaghi CR, Burnett BA, Feok-

tistova A, Van Eldik LJ (1995) Overexpression of the neurotrophic cytokine S100 beta in human temporal lobe epilepsy. *J Neurochem* 65: 228–233

96 Esiri MM, Biddolph SC, Morris CS (1998) Prevalence of Alzheimer plaques in AIDS. *J Neurol Neurosurg Psychiatry* 65: 29–33

97 Kusdra L, Rempel H, Yaffe K, Pulliam L (2000) Elevation of $CD69^+$ monocyte/macrophages in patients with Alzheimer's disease. *Immunobiology* 202: 26–33

98 Rempel HC, Pulliam L (2005) HIV-1 Tat inhibits neprilysin and elevates amyloid beta. *AIDS* 19: 127–135

99 Griffin WS, Stanley LC, Ling C, White L, MacLeod V, Perrot LJ, White CL, 3rd, Araoz C (1989) Brain interleukin 1 and S-100 immunoreactivity are elevated in Down syndrome and Alzheimer disease. *Proc Natl Acad Sci USA* 86: 7611–7615

100 McKenzie JE, Gentleman SM, Roberts GW, Graham DI, Royston MC (1994) Increased numbers of beta APP-immunoreactive neurones in the entorhinal cortex after head injury. *Neuroreport* 6: 161–164

101 Mattson MP, Scheff SW (1994) Endogenous neuroprotection factors and traumatic brain injury: mechanisms of action and implications for therapy. *J Neurotrauma* 11: 3–33

102 Sola C, Garcia-Ladona FJ, Mengod G, Probst A, Frey P, Palacios JM (1993) Increased levels of the Kunitz protease inhibitor-containing beta APP mRNAs in rat brain following neurotoxic damage. *Brain Res Mol Brain Res* 17: 41–52

103 Barger SW, Basile AS (2001) Activation of microglia by secreted amyloid precursor protein evokes release of glutamate by cystine exchange and attenuates synaptic function. *J Neurochem* 76: 846–854

104 Mrak RE, Griffin WS (2007) Common inflammatory mechanisms in Lewy body disease and Alzheimer disease. *J Neuropathol Exp Neurol* 66: 683–686

105 Griffin WS (2006) Inflammation and neurodegenerative diseases. *Am J Clin Nutr* 83: 470S-474S

106 da Cunha A, Jefferson JJ, Tyor WR, Glass JD, Jannotta FS, Vitkovic L (1993) Gliosis in human brain: relationship to size but not other properties of astrocytes. *Brain Res* 600: 161–165

107 Sheng JG, Mrak RE, Griffin WS (1998) Enlarged and phagocytic, but not primed, interleukin-1 alpha-immunoreactive microglia increase with age in normal human brain. *Acta Neuropathol* (Berl) 95: 229–234

108 Marshak DR (1990) S100 beta as a neurotrophic factor. *Prog Brain Res* 86: 169–181

109 Lewis D, Teyler TJ (1986) Anti-S-100 serum blocks long-term potentiation in the hippocampal slice. *Brain Res* 383: 159–164

110 Goldgaber D, Harris HW, Hla T, Maciag T, Donnelly RJ, Jacobsen JS, Vitek MP, Gajdusek DC (1989) Interleukin 1 regulates synthesis of amyloid beta-protein precursor mRNA in human endothelial cells. *Proc Natl Acad Sci USA* 86: 7606–7610

111 Buxbaum JD, Oishi M, Chen HI, Pinkas-Kramarski R, Jaffe EA, Gandy SE, Greengard P (1992) Cholinergic agonists and interleukin 1 regulate processing and secretion of

the Alzheimer beta/A4 amyloid protein precursor. *Proc Natl Acad Sci USA* 89: 10075–10078

112 Mattson MP (1997) Cellular actions of beta-amyloid precursor protein and its soluble and fibrillogenic derivatives. *Physiol Rev* 77: 1081–1132

113 Mrak RE, Griffin WS (2005) Glia and their cytokines in progression of neurodegeneration. *Neurobiol Aging* 26: 349–354

114 Griffin WS, Liu L, Li Y, Mrak RE, Barger SW (2006) Interleukin-1 mediates Alzheimer and Lewy body pathologies. *J Neuroinflammation* 3: 5

115 Li Y, Liu L, Kang J, Sheng JG, Barger SW, Mrak RE, Griffin WS (2000) Neuronal-glial interactions mediated by interleukin-1 enhance neuronal acetylcholinesterase activity and mRNA expression. *J Neurosci* 20: 149–155

116 Mrak RE, Griffin WS (2001) Interleukin-1, neuroinflammation, and Alzheimer's disease. *Neurobiol Aging* 22: 903–908

117 Teyler TJ, Discenna P (1984) Long-term potentiation as a candidate mnemonic device. *Brain Res* 319: 15–28

118 Stellwagen D, Beattie EC, Seo JY, Malenka RC (2005) Differential regulation of AMPA receptor and GABA receptor trafficking by tumor necrosis factor-alpha. *J Neurosci* 25: 3219–3228

119 Stellwagen D, Malenka RC (2006) Synaptic scaling mediated by glial TNF-alpha. *Nature* 440: 1054–1059

120 Lai AY, Swayze RD, El-Husseini A, Song C (2006) Interleukin-1 beta modulates AMPA receptor expression and phosphorylation in hippocampal neurons. *J Neuroimmunol* 175: 97–106

121 Jang HJ, Cho KH, Kim HS, Hahn SJ, Kim MS, Rhie DJ (2009) Age-dependent decline in supragranular long-term synaptic plasticity by increased inhibition during the critical period in the rat primary visual cortex. *J Neurophysiol* 101: 269–275

122 Schmid AW, Lynch MA, Herron CE (2008) The effects of IL-1 receptor antagonist on beta amyloid mediated depression of LTP in the rat CA1 *in vivo*. *Hippocampus*, Epub ahead of print

123 Giunta B, Fernandez F, Nikolic WV, Obregon D, Rrapo E, Town T, Tan J (2008) Inflammaging as a prodrome to Alzheimer's disease. *J Neuroinflammation* 5: 51

124 Sheng JG, Griffin WS, Royston MC, Mrak RE (1998) Distribution of interleukin-1–immunoreactive microglia in cerebral cortical layers: implications for neuritic plaque formation in Alzheimer's disease. *Neuropathol Appl Neurobiol* 24: 278–283

125 Gentleman SM, Leclercq PD, Moyes L, Graham DI, Smith C, Griffin WS, Nicoll JA (2004) Long-term intracerebral inflammatory response after traumatic brain injury. *Forensic Sci Int* 146: 97–104

126 Wainwright MS, Craft JM, Griffin WS, Marks A, Pineda J, Padgett KR, Van Eldik LJ (2004) Increased susceptibility of S100B transgenic mice to perinatal hypoxia-ischemia. *Ann Neurol* 56: 61–67

127 Mori T, Tan J, Arendash GW, Koyama N, Nojima Y, Town T (2008) Overexpression

of human S100B exacerbates brain damage and periinfarct gliosis after permanent focal ischemia. *Stroke* 39: 2114–2121
128 Griffin WS, Mrak RE (2002) Interleukin-1 in the genesis and progression of and risk for development of neuronal degeneration in Alzheimer's disease. *J Leukoc Biol* 72: 233–238
129 Breitner JC, Gau BA, Welsh KA, Plassman BL, McDonald WM, Helms MJ, Anthony JC (1994) Inverse association of anti-inflammatory treatments and Alzheimer's disease: initial results of a co-twin control study. *Neurology* 44: 227–232
130 Breitner JC, Welsh KA, Helms MJ, Gaskell PC, Gau BA, Roses AD, Pericak-Vance MA, Saunders AM (1995) Delayed onset of Alzheimer's disease with nonsteroidal anti-inflammatory and histamine H2 blocking drugs. *Neurobiol Aging* 16: 523–530
131 Breitner JC, Zandi PP (2001) Do nonsteroidal antiinflammatory drugs reduce the risk of Alzheimer's disease? *N Engl J Med* 345: 1567–1568
132 Zandi PP, Anthony JC, Hayden KM, Mehta K, Mayer L, Breitner JC (2002) Reduced incidence of AD with NSAID but not H2 receptor antagonists: the Cache County Study. *Neurology* 59: 880–886
133 Aisen PS (2008) The inflammatory hypothesis of Alzheimer disease: dead or alive? *Alzheimer Dis Assoc Disord* 22: 4–5
134 McGeer PL, Rogers J (1992) Anti-inflammatory agents as a therapeutic approach to Alzheimer's disease. *Neurology* 42: 447–449
135 Vlad SC, Miller DR, Kowall NW, Felson DT (2008) Protective effects of NSAIDs on the development of Alzheimer disease. *Neurology* 70: 1672–1677
136 Szekely CA, Breitner JC, Fitzpatrick AL, Rea TD, Psaty BM, Kuller LH, Zandi PP (2008) NSAID use and dementia risk in the Cardiovascular Health Study: role of APOE and NSAID type. *Neurology* 70: 17–24
137 in t' Veld BA, Ruitenberg A, Hofman A, Launer LJ, van Duijn CM, Stijnen T, Breteler MM, Stricker BH (2001) Nonsteroidal antiinflammatory drugs and the risk of Alzheimer's disease. *N Engl J Med* 345: 1515–1521
138 Trouche SG, Asuni A, Rouland S, Wisniewski T, Frangione B, Verdier JM, Sigurdsson EM, Mestre-Frances N (2009) Antibody response and plasma Abeta1–40 levels in young Microcebus murinus primates immunized with Abeta1–42 and its derivatives. *Vaccine* 27: 957–964
139 Asuni AA, Boutajangout A, Quartermain D, Sigurdsson EM (2007) Immunotherapy targeting pathological tau conformers in a tangle mouse model reduces brain pathology with associated functional improvements. *J Neurosci* 27: 9115–9129
140 Lemere CA, Maron R, Selkoe DJ, Weiner HL (2001) Nasal vaccination with beta-amyloid peptide for the treatment of Alzheimer's disease. *DNA Cell Biol* 20: 705–711
141 Wilcock DM, Jantzen PT, Li Q, Morgan D, Gordon MN (2007) Amyloid-beta vaccination, but not nitro-nonsteroidal anti-inflammatory drug treatment, increases vascular amyloid and microhemorrhage while both reduce parenchymal amyloid. *Neuroscience* 144: 950–960
142 Morihara T, Teter B, Yang F, Lim GP, Boudinot S, Boudinot FD, Frautschy SA, Cole GM

(2005) Ibuprofen suppresses interleukin-1beta induction of pro-amyloidogenic alpha1–antichymotrypsin to ameliorate beta-amyloid (Abeta) pathology in Alzheimer's models. *Neuropsychopharmacology* 30: 1111–1120
143 Morgan D (2006) Modulation of microglial activation state following passive immunization in amyloid depositing transgenic mice. *Neurochem Int* 49: 190–194
144 May R (1984) *Degeneration and Regeneration of the Nervous System*. Robert Maclehose and Co, University Press, Glasgow

Alternative and additional mechanisms to the hygiene hypothesis

Margo C. Honeyman and Leonard C. Harrison

Walter and Eliza Hall Institute of Medical Research, Parkville, Victoria, Australia 3052

Abstract

The rising incidence of allergic and autoimmune diseases is occurring on a background of genes selected for strong immune responses. Possible mechanisms for selection and changed environmental factors that impact on immune response genes are discussed. Reduced exposure of infants to infections is discussed elsewhere in this volume. Here we consider the role of delayed exposure to infections as well as additional factors that could promote chronic immuno-inflammatory diseases. These include changes in the amount of food consumed, dietary composition, sleep reduction and lower energy expenditure due to reduced exercise and thermoneutrality of the built environment. Any or all of these may result in obesity, which is a proinflammatory state. Increases in air pollution and psychological stress and, finally, insufficiency of vitamin D, are discussed, as these may also shift immune responsiveness towards a proinflammatory state.

Introduction

The statement by Dubos that "*the response of human beings to the conditions of the present is always conditioned by the biological remembrance of things past*" [1] expresses succinctly the evolutionary approach to understanding immune responses. Biological remembrance encompasses both genes selected by a past environment to be optimal and, in a much shorter time frame, epigenetic control of those genes. However, the rising incidence of allergic and autoimmune diseases in the developed world post-World War II (WWII), for example Type 1 diabetes (Fig. 1), challenges us to understand the nature of the environmental changes that have occurred over that time and their impact on the genome. One of these changes, reduced and/or delayed exposure to microbiota in infancy, encapsulated by the 'hygiene hypothesis', has been extensively discussed in the preceding chapters. Here we examine the evolution of major immune-response genes and their interplay with the modern environment, as alternative or additive explanations to the hygiene hypothesis.

The Hygiene Hypothesis and Darwinian Medicine, edited by Graham A. W. Rook

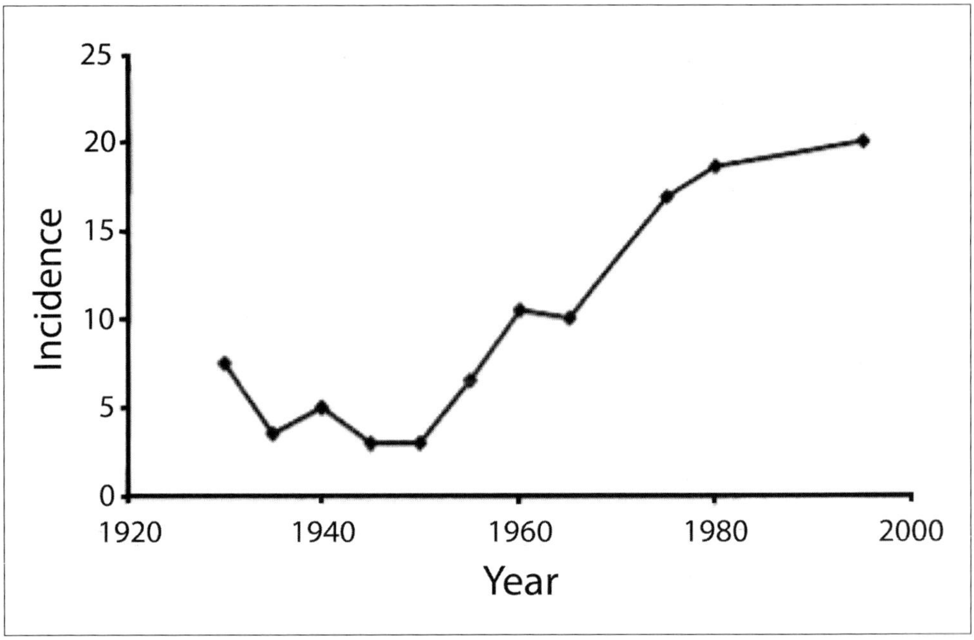

Figure 1
Incidence of Type 1 diabetes in children under age 10 in Norway, 1925–1995 (from Gale E, Diabetes, 2002 [103]).

Disease susceptibility and changes in gene penetrance

The HLA region of the human genome spans 6–8 megabases and contains approximately 150 immune response genes [2], which are the most polymorphic in the human genome. They include genes encoding Class I proteins expressed on all nucleated cells which present antigens derived from the cell cytoplasm (including viral and other exogenous proteins) to CD8 T cells (HLA-A, B) and NK cells (HLA-C, Bw4, Bw6). Proteins encoded by the Class II region (HLA-DR, -DQ, -DP) are found on antigen-presenting cells such as dendritic cells, macrophages and B cells that process exogenous proteins to generate antigenic peptides for presentation to CD4 T cells. The Class III region contains genes for proteins (e.g., MICA/B and ULBP1-3) specialised in presentation to and activation of NK and gamma delta T cells, as well as allelic variants of the cytokines TNF-α and -β, critical in the development of both innate and adaptive immune responses [3]. It also contains the gene encoding the receptor for advanced glycation end products (RAGE), important diet-derived inflammatory stimuli [4]. Thus, this region of the genome contains the greatest concentration of genes required for both innate and adaptive immune responses. While other genes for immune responses, e.g., those for the Toll-like receptors (TLRs) or for microbial pattern recognition receptors, exist elsewhere in the genome, their polymorphism is restricted, by the necessity to recognise invariant protein or carbohydrate signatures of phylogenetically distinct organisms [5]. Genes for highly phenotypically diverse T cell receptors and antibodies recombine in developing lymphocytes, but are not inherited alleles and therefore are little affected by evolutionary selection. Selection in this case takes place at the level of the developing lymphocytes. The selective pressure of infectious diseases has resulted in HLA alleles at different loci on the one chromosome (haplotype) being in linkage disequilibrium if they operate coordinately [2]. Recombination is selected against member loci of a haplotype. Combinations of alleles at HLA loci constituting advantageous haplotypes will be selected if they favour strong immune responses to common infectious agents that affect survival before the reproductive years. This is demonstrated by haplotypes that include HLA-DR3 [6]. HLA-DR3 is linked to DQ2 in several populations such as Northwest Europeans (NW Europeans, 'Caucasoids'), Negroid, northern Indian and Filipino populations, attesting to its immunological efficacy. In the NW European population, these linked genes occur as part of the extended haplotype HLA-A1-B8-Cw7-DR3-DQ2, which is associated with multiple autoimmune diseases including Type 1 diabetes, myasthenia gravis, systemic lupus erythematosus and coeliac disease, as well as with the accelerated onset of AIDS (reviewed in [2]) [7]. In Basques and Southern Europeans, this 'frozen block' of DR3-DQ2 DNA is also found as part of the extended HLA-A26-B18-Cw5-DR3-DQ2 haplotype, again associated with the development of Type 1diabetes but also with Graves' disease, while HLA-B8-DR3-DQ2 (without A1) is associated with Graves' disease in Asian Indians [8].

HLA haplotypes were selected for protection against infectious diseases. Sequence studies indicate that HLA-DR3 favours binding of peptides with a central basic amino acid [7]. The Class I alleles on the HLA-A1 and B8 haplotype are also predicted to be poor binders of many peptides that lack central basic amino acids [9]. DR3 is particularly good at presenting a naturally processed and immunodominant peptide (amino acids 31–50) from the HspX protein of *Mycobacterium tuberculosis* (TB) that is upregulated during latency [10]. Individuals with a DR3 haplotype who mount an immune response to the HspX protein would be more likely to control their TB infection. In addition, healthy individuals with DR3, rather than other HLA-DR alleles, have been shown to have higher IFN-gamma responses to *M. tuberculosis* [11]. These and other studies [7, 12, 13] suggest that TB, a disease of urban societies which remains a major killer in the world today, may have contributed to the maintenance of the DR3-DQ2 haplotype in multiple populations, driving the selection of the A1-B8-Cw7-DR3-DQ2 extended haplotype in NW Europeans. The 158kb DR3-DQ2 frozen block contained within both the A1-B8-Cw7-DR3-DQ2 and A26-B18-Cw5-DR3-DQ2 haplotypes is most likely to confer susceptibility to Type 1 diabetes. However, as the two haplotypes differ in conferring susceptibility to Graves' disease, susceptibility to Graves' disease can only be determined by genes on the haplotype outside DR3-DQ2. This suggests selection for functional interaction between the various immune response genes encoded on the extended haplotype [14]. Other haplotypes associated with autoimmune disease may have been functionally selected in a similar fashion. The haplotype HLA-A3-B7-DR15 (2)-DQB1*0602, common in Northern Europe [15], confers susceptibility to multiple sclerosis [16], but is protective against Type 1 diabetes [17]. Childhood (i.e., pre-reproductive) disease(s) against which it might confer protection are not defined but, interestingly, discovery of a highly conserved vitamin D response element (VDRE) in the promoter region of HLA-DR15 suggests selection for this haplotype might instead occur by responsiveness to vitamin D [18], particularly in northern Europe (see section on vitamin D, below) .

The gene for Slc11a1 (formerly known as NRAMP1) is also polymorphic and encodes a late endosomal/lysosomal protein that regulates iron homeostasis in macrophages, controlling activation, MHC Class II expression and antigen presentation [19]. The gene for Slc11A1 maps to chromosomal locus IDDM13, which confers susceptibility to Type 1 diabetes. Promoter polymorphisms of Slc11a1 can be associated with a strong Th1 response and resistance to intracellular pathogens such as Mycobacteria, Leishmania and *Salmonella* [20, 21]. Slc11A1 is also associated with susceptibility to diabetes in the non-obese diabetic (NOD) mouse model of Type 1 diabetes, in which it maps to Idd5.2 [22]. The common allele of Slc11A1 has also been associated with inflammation in Alzheimer's disease, with a variant allele conferring protection, but only in males [23].

Other genes that regulate innate immunity could also be subject to selection by infectious diseases for strong immune responses. These include the highly polymorphic MHC Class I chain associated (MICA) genes encoded within the HLA Class III region [24]. MICA proteins are preferentially expressed on fibroblasts and epithelial cells including enterocytes in response to cellular stress and inflammation, and are recognised by the natural killer cell receptor, NKG2D, which also occurs on CD8 and gamma delta T cells. While the MICA*A5 allele is present in the HLA-A1-B8-Cw7-DR3-DQ2 autoimmune susceptibility haplotype, the MICA*A6 allele has been shown independently to confer susceptibility to Type 1 diabetes [25].

Selection of high responder polymorphisms in TLR2 and 4 genes could explain strong innate immune responses associated with asthma [26] and atopy [27]. Signalling through these TLRs, triggered by recognition of Gram-positive and Gram-negative bacteria, Mycobacteria, Mycoplasmas, parasites and yeasts (for TLR2) and bacterial lipopolysaccharide (LPS) (for TLR4) [28], triggers production of IL-12 by dendritic cells, leading to increased IFN-gamma production by T cells and NK cells. The function of $CD4^{+}CD25^{+}$ regulatory T cells (Tregs) may be affected by TLR4 polymorphisms, as Tregs express TLR4 and are rendered more immunosuppressive upon its ligation [29].

Urbanisation, and migration from country to cities, may increase exposure to potentially fatal childhood infections, encouraging selection for genes that confer strong immunity to diseases in that environment. Migration of people out of a selecting environment could not only result in immunological naivety to newly encountered and highly transmitted infections, but also in newly unregulated immune responses which might also prove fatal. This is demonstrated by the people of the Watut Valley in Papua New Guinea, who moved from the highlands where they had lived for 15,000 years, to a coastal valley where they were exposed to malaria. Those with HLA-DR2 had hyper IgM responses to malaria, resulting in malarious splenomegaly syndrome and death of most inhabitants in some villages [30].

Thus, selection for high immune responses would be unlikely to confer risk for autoimmune and allergic diseases until changes in the modern environment encouraged penetrance of the genes for strong immune responsiveness or reduced regulation.

Analysis of the frequency of defined HLA susceptibility genotypes in Type 1 diabetes demonstrates a clear interaction between the modern environment and selected immune response genes. The increasing incidence of Type 1 diabetes between 1950–2005 can be entirely attributed to new cases with lower risk HLA genotypes, the number of the highest risk genotypes remaining unchanged [31] (Fig. 2), Thus, recently operating environmental factors must be invoked to explain the increased penetrance of genes that previously conferred only moderate risk for the development of Type 1 diabetes.

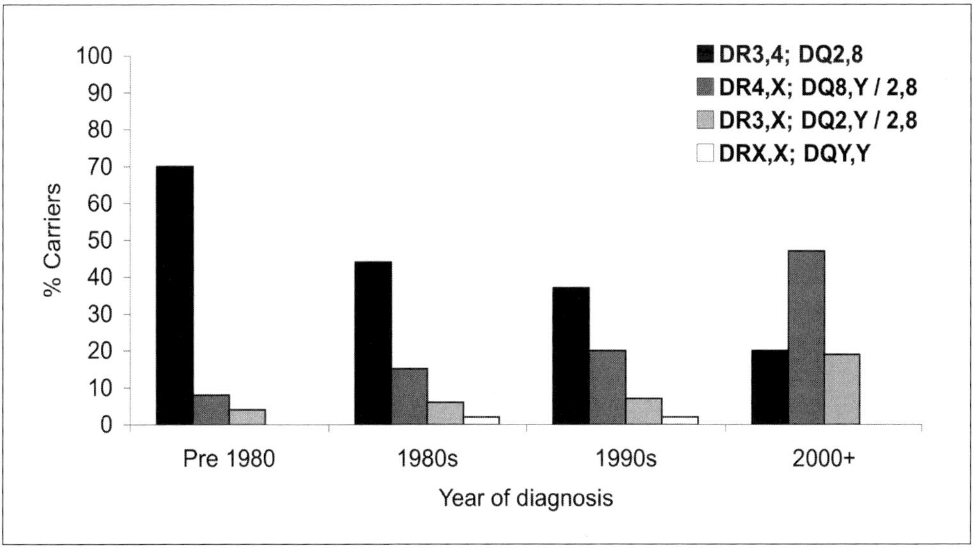

Figure 2
The frequency of HLA-DR genotypes in childhood-onset Type 1 diabetes according to year of diagnosis (where X is a non-3 or non-4 DR allele).

How could the environment result in increased penetrance of lower risk genes for chronic inflammatory diseases like Type 1 diabetes? It could promote inflammation or diminish regulation of innate and adaptive immune responses, or both. We suggest that the modern environment is 'proinflammatory', activating innate immune inflammatory pathways that promote pathology, e.g., insulin resistance in diabetes or arteriosclerosis, or promote adaptive immunity thereby the incidence of autoimmune and allergic diseases. On this background, we will consider a number of factors in the modern environment: the age at which first exposure to infection occurs, the amount and composition of food intake, pollution, hours of sleep, amount of exercise, psychological stress and sunlight exposure and vitamin D sufficiency.

Timing of infections

Opportunities for infant exposure to infection have diminished since WWII due to social changes. For example, previously the practice of placing neonates in hospital nurseries ensured swift transfer of rotavirus infection. However, with 'rooming-in' of mothers and their neonates, severe gastroenteric infection due to rotavirus is delayed, peaking at 18 months of age (R. Alexander, pers.comm.). Once home, infants were frequently infected with viruses by elder siblings in the first few months of life, but smaller family sizes result in less opportunities for such infections. During the first 9 months of life, transplacentally transmitted antibodies from the mother, and then antibodies in breast milk, coupled with antiviral molecules such as lactadherin in the milk, lead to diminished viral load and clinical symptoms but don't prevent actual infection [32] – nature's vaccination. A decrease in the frequency and duration of breast feeding has impacted on this protective period for the infant, while compromising the development of a well-regulated mucosal immune response [33]. Delay of infection to beyond this early period of antibody cover could conceivably lead to poor development of T cell regulation or an enhanced Th1 response at the time of infection on entry to day care [33].

The effect of altered timing of infection is clearly illustrated in the NOD mouse. Evidence in humans indicates that rotavirus may trigger or promote islet autoimmunity [34–37], leading to Type 1 diabetes. When NOD mice are infected with rotavirus at birth, diabetes onset is significantly delayed. In contrast, infection at weaning, at an age equivalent to when children enter day care, does not delay and may even accelerate diabetes onset (Fig. 3). Social factors and the consequent pattern of enterovirus infections may also be relevant in humans [38]. The incidence of Type 1 diabetes is very high in Finns in whom enterovirus infections are sporadic, but is very low in genetically-similar Estonians in whom enteroviruses are ubiquitous, family sizes large and infections more likely to happen during breast feeding [39].

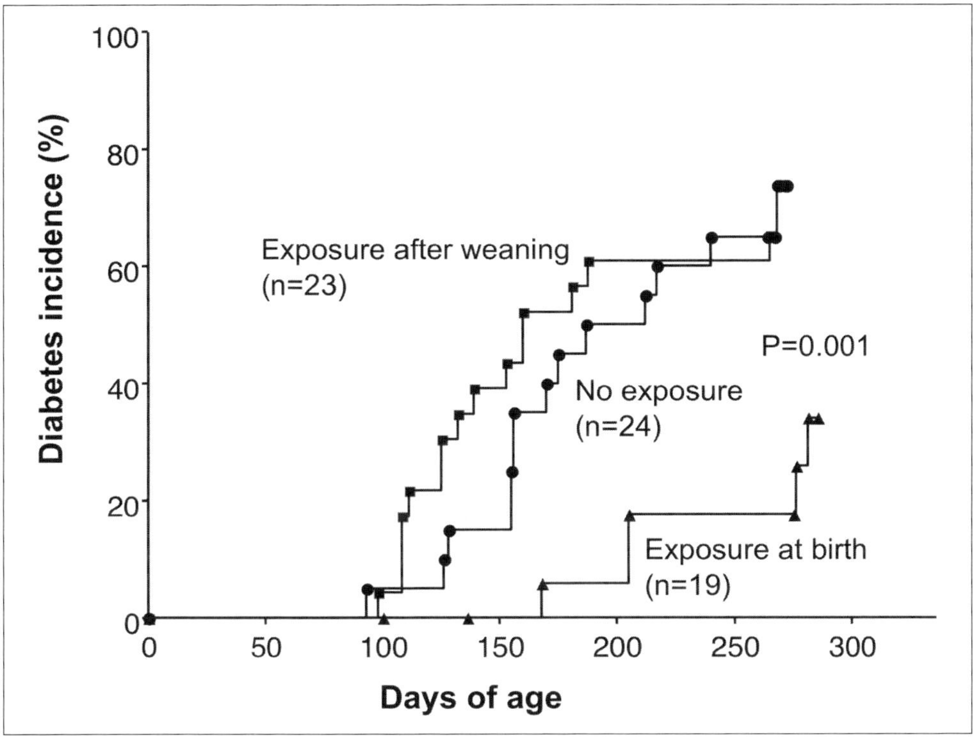

Figure 3
Incidence of diabetes in non-obese diabetic mice either unexposed or exposed at birth or weaning to murine rotavirus.

Food, obesity and energy utilisation

Total energy intake in the developed world has increased since WWII, with the body mass index (BMI) increasing in children of different ethnicities [40, 41]. However while asthma and atopy increased along with BMI this was predominantly in Caucasoid [38] not Asian [39] children in the USA, consistent with an effect of genetic background or some other undefined, ethnic-associated difference in environment. Acquisition of obesity is associated with the development of insulin resistance, which could account not only for the rising incidence of Type 2 but also Type 1 diabetes [42]. Insulin resistance per se has been shown to accelerate progression to clinical Type 1 diabetes in children with islet autoimmunity [43].

Although obesity is now recognised to be a state of low-grade chronic inflammation [44] and adipose tissue in obesity a seat of inflammation [45], many factors in the obesogenic environment are themselves proinflammatory.

The effects of dietary composition on mitochondrial free radical generation, activation of the NF-κB pathway and endoplasmic reticulum (ER) stress is a subject of considerable current interest [46–48]. Accumulation of lipid in adipocytes increases ER stress [49] leading to the production of chemoattractants such as monocyte chemoattractant protein-1, which recruit and differentiate monocytes to proinflammatory adipose tissue macrophages [50] implicated in insulin resistance [51]. Specific dietary components such as advanced glycation end products (AGEs) [52], saturated fats including trans fatty acids [53] and fructose [54] are associated with elevated circulating pro-inflammatory cytokines and insulin resistance. AGEs are non-enzymatically glycated or oxidated proteins, lipids and nucleic acids that are formed in conditions of oxidant stress including hyperglycemia. Binding of AGEs to their receptors (RAGEs), on macrophages for example, activates the NF-κB pathway and transcription of proinflammatory factors [55]. AGEs are formed in foods following cooking of sugars with fats or proteins, and are particularly conspicuous in 'fast food', much beloved by children.

Trans fatty acids (TFAs) are produced in deep fried foods through the transformation of polyunsaturated fatty acids from their natural cis form to the trans form. The process of partial hydrogenation also converts vegetable oils to semisolid fats like margarines and peanut butter that contain high concentrations of TFAs and are commonly used in bakeries [53]. Ingestion of diets high in fat, especially TFAs, but not isocaloric high fibre diets, provokes proinflammatory changes in the blood within hours [56].

The intake of fructose rose >1000% from 1970 to 1990 in the USA [57] and excessive fructose intake (>50 g/d) has been proposed as one of the underlying etiologies of metabolic syndrome and Type 2 diabetes [58]. While the primary sources of fructose are sugar (sucrose) and high fructose corn syrup, the recent trend to drink 'healthy' fruit juices can result in consumption at the one time of the equivalent of 10 fructose-containing oranges without the associated fibre. Unlike glucose, fructose does not stimulate secretion of insulin or leptin, key afferent signalling molecules in the regulation of food intake and body weight [59].

Obesity is encouraged by diminished total energy outlay due to changes to work and leisure practices, including less exercise, fewer natural play areas for children and increased TV viewing and use of computer games. It may also result from a reduction in hours of sleep. The average hours of sleep of US adults has diminished from 9 h to 7 h over the last few decades [60]. Sleep curtailment in young adults results in a constellation of metabolic and endocrine alterations, including decreased glucose tolerance, increased insulin resistance, increased evening concentrations of cortisol, increased and decreased levels, respectively, of the orexigenic hormones ghrelin leptin, and increased hunger and appetite [61]. These data suggest that improving sleep length could reduce the epidemic of obesity and chronic inflammatory diseases. Reduced energy expenditure could also result from the deliberate control of humidity and temperature in our living and working environments. Air-conditioning and heating mean that energy required in the past for thermoregulation is now diminished, even when we are inactive.

Pollution

Recently, it was reported that differentiation of CD4 T cells into Treg and Th17 cells is controlled by ligation of the aryl hydrocarbon receptor (AHR) by polycyclic aromatic hydrocarbons [62]. The AHR is upregulated specifically in Th17 cells, in the presence of TGF-β and IL-6 [62, 63], and ligation provides the signal for the production of IL-17F, IL-22 and IL-21. IL-17A and IL-17F form a heterodimeric cytokine implicated in range of inflammatory disorders [64]. Numerous candidates for endogenous ligands of the AHR have been proposed [65], including prostaglandins, bilirubin at high concentrations and modified low-density lipoprotein, but the evidence is strongest for ultraviolet photoproducts of tryptophan, the two most active of which are 6-formylindolo[3,2-b]carbazole (FICZ) and 6,12-diformylindolo[3,2-b] carbazole (dFICZ) [65]. The capacity of the AHR to recognise exogenously-derived polycyclic aromatic hydrocarbons, such as 2,3,7,8 tetrachlorodibenzo-p-dioxin (TCDD), ubiquitously present in atmospheric pollution, links Th17-mediated autoimmunity to environmental toxins [63]. Background levels of polycyclic aromatic hydrocarbons are 0.1–1 ng/m^3 but in urban environments are 0.5–3 ng/m^3, and close to some industries involving coal or coke burning levels can reach 30 ng/m^3. In urban environments, polycyclic aromatic hydrocarbons are derived domestically from tobacco smoke, cooking, and heating (coke and wood burning), as well as industrial activity. In rural environments, stubble burning (now banned in the UK), open burning of moorland and bushfires also contribute polycyclic aromatic hydrocarbons to the atmosphere. Cars are estimated to contribute 24% to measurable atmospheric polycyclic aromatic hydrocarbons, industrial combustion 24%, domestic combustion 19% and natural fires 18% (European commission working paper on polycyclic aromatic hydrocarbons, 2001). Atmospheric levels of polycyclic aromatic hydrocarbons are higher in winter, due to meteorologic conditions, greater use of heating and reduced degradation by photo-oxidation and reduction by –OH radicals. While environmental regulations have begun to diminish levels of polycyclic aromatic hydrocarbons since the late 1990s in Europe, levels continue to rise in emerging industrial giants like China [66].

The *in vivo* immunological effects of TCDD are most clearly illustrated by the disasters at Seveso, Italy in 1976 and Bhopal, India in 1984, in which victims were immunosuppressed and had chloracne [67]. *In vitro*, in murine and human cells, ligation of the AHR with FICZ upregulated genes encoding p450 enzymes such as Cyp1a1, and strongly upregulated IL-17A and F and the proinflammatory cytokine IL-22 [63]. This also occurred with an alternative polycyclic aromatic hydrocarbon, beta-naphthoflavone. However another study [62] found that while FICZ induced experimental allergic encephalomyelitis (EAE), TCDD induced Treg and suppressed EAE *in vivo*, and that resveratrol from red wine was an antagonist of the AHR. Thus, the role of polycyclic aromatic hydrocarbons in genesis of immune disorders is complex but likely to be a fruitful area of research relevant to diseases of the modern society.

Psychological stress

While there are no studies to indicate that population stress overall has increased, psycho-neuro-immunologic studies show that acute mental stress not only elicits increases in blood pressure and heart rate but also in the circulating proinflammatory IL-6 (56% rise 2 h after a stressful task) [68]. Stress has been related to both the development of asthma and atopy [69], as measured by IgE levels at 6 years of age. Parenting difficulties at 3 weeks of age correlated with 6-year measures of maternal depression and child psychological risk score, and children with asthma were rated as being at greater psychological risk. Stress is important in the evolution of atopic dermatitis [70]; it increases eosinophil infiltration, $VCAM^+$ blood vessels and epidermal thickness [71], impairs the skin barrier function and favours a shift towards Th2 cytokine responses. The stress neuropeptide substance P is released in the skin from cutaneous nerve fibres, the resultant neurogenic inflammation leading to mast cell degranulation, mast cell apoptosis and endothelial gaping [71]. Such a stress-induced process is also likely to be involved in the initiation of autoimmunity in the skin. In the early stages of alopecia areata (autoimmune hair loss), which has been associated with stress [72], the number of substance P positive nerve fibres in the skin increases, mast cells degranulate and hair follicles regress [73].

The hypothesis that maltreatment in childhood results in low glucocorticoid release that facilitates chronic low-grade inflammation later in life was tested. Significantly raised C-reactive protein (CRP) was observed in 32 year-old adults who were reported by parents 20 years previously as experiencing episodes of maltreatment. Elevated CRP was graded with maltreatment, being 1.8 times more likely in those who experienced multiple episodes, independent of co-occurring early life risks (low birth weight, low socio-economic status, low IQ), stress in adulthood and adult health and behaviour [74]. Finally, two studies show that stress affects the development of Type 1 diabetes. The total frequency of life events did not differ in children who developed Type 1 diabetes compared to controls, but the life events were more severe, e.g., actual or threatened losses within the family [75]. Serious life events, foreign origin of the mother, high parenting stress and low paternal education gave odds ratios of 2.3–1.6 for development of Type 1 diabetes-related autoantibodies at 1 year of age, independent of a family history of diabetes [76]. This implies a possible relationship with the increasing divorce rate in the developed world [77].

Vitamin D

Vitamin D, a steroid hormone previously thought to be necessary only for healthy bone development, is made in the skin by the action of ultraviolet (UV) B light (290–315 nm) on cholesterol in the skin. Humans evolved in the sunny tropics where they synthesized at least 5,000 international units (IU) of vitamin D per day, but in contemporary Northern European populations this has dropped to 200–500 IU/day (Fig. 4), resulting in a high prevalence of vitamin D deficiency. Skin pigmentation slows the synthesis of vitamin D [78], promoting selection against dark skin in populations that pre-historically travelled northward, especially to beyond latitude 35° where there are no vitamin D-producing wavelengths in winter and day lengths are short. Diminished exposure to UVB is consistent with the (northern hemisphere) north-south gradient in diseases such as multiple sclerosis [79] and Type 1 diabetes [80] which is reversed in the southern hemisphere [81], as well as with winter onset of Type 1 diabetes [82] and winter relapses in multiple sclerosis [83]. In the last 15 years, many studies have demonstrated that vitamin D is also critical for immunological health (see www.sunvitamin.org). Vitamin D facilitates killing of intracellular parasites by activating macrophage cathelicidin [78] and promotes tolerogenic dendritic cells [84] and the generation of IL-10-secreting Tregs [85]. It also improves the capacity of defective T regs to function [86]. Sun avoidance behaviour, lack of mobility of the elderly or very young, shift working, rapid migration of dark-skinned people to low UVB high latitude zones of the world, covering clothing, global dimming due to atmospheric pollution, and combinations of all these can result in deficiency of UVB and vitamin D. This is seen particularly in dark-skinned (and/or covering) populations that have migrated to live in high

Figure 4
The supply of vitamin D for which the human genome was selected was much higher than it is today.

latitude environments. Examples are Asian immigrants to the UK or African refugees in southern Australia, in whom tuberculosis re-emerges [87, 88], Latinos in the US who have a high incidence of asthma [89], the Maori of New Zealand who have a very high prevalence of Type 2 diabetes [90] and in African Americans who have earlier onset and more severe hypertension than white Americans [91].

Genetic evidence for the role of vitamin D comes from multiple studies. First, specific polymorphisms of the CYP27B1 gene encoding 1alpha hydroxylase are reported to confer susceptibility to Type 1 diabetes [92], Graves' disease and Hashimoto's thyroiditis [93]. 1 alpha hydroxylase converts circulating 25 OH D3 to the active form 1,25 $(OH)_2$ D3 in the kidney. Second, a highly conserved vitamin D response element (VDRE) has been discovered in the promoter of HLA-DRB1*1501, the HLA allele that confers susceptibility to multiple sclerosis [18] (see below). Diminishing levels of UVB exposure or deficiency of vitamin D have been linked to autoimmune diseases such as Type 1 diabetes [79, 94], multiple sclerosis [79, 81] and rheumatoid arthritis [79], and to diseases associated with low-grade inflammation such as Type 2 diabetes [95] and atherosclerosis [96]. Wheezing at age 5 is also linked to deficiency of vitamin D prenatally [97].

It may be relevant that while the incidence of Type 1 diabetes in Finland has increased to be the highest in the world the recommended daily allowance of vitamin D in Finland has dropped from 2,000 IU/day to 200 IU/day (the amount in a teaspoon of cod-liver oil sufficient to prevent rickets) over the last 40 years. Vitamin D deficiency during pregnancy has been demonstrated in Scandinavian studies of children developing Type 1 diabetes in the first year of life [98], and the offspring of pregnant women with the highest intake of dietary vitamin D had 1/5 of the chance of developing Type 1 diabetes compared to offspring of mothers with the lowest intake of vitamin D [99]. Deficiency of vitamin D during pregnancy could have effects on the developing foetus, including epigenetic effects. While our understanding of epigenetics is as yet rudimentary, the genetic effects of starvation of the mother have been shown to last at least 60 years [100]. In pregnant mice, a diet high in methyl donors (folate, vitamin B12, choline, methionine) led to offspring with increased CpG methylation of the Runx3 gene, that was heritable through several generations. Importantly, these offspring had trans-generational allergic airways disease [101], indicating that immune responses may be controlled epigenetically and inherited. Recent evidence suggests that many inflammatory genes, including TLRs, are regulated by epigenetic modifications to individual promoters. TLR2 and 4 are downregulated on monocytes in a vitamin D receptor (VDR)-dependent manner [102]. One could speculate from these studies that children of vitamin D-deficient mothers could potentially have upregulated TLR2 and 4 for their and their children's lifetimes Vitamin D deficiency would be expected to influence foetal development in many ways. In foetuses who are DRB1*1501, vitamin D deficiency resulting in impaired stimulation of the DRB1*1501 VDRE and leading to lower expression of DRB1*1501 on developing T cells in the thymus could impair negative selection of potentially autoreactive T cells, increasing the risk of DRB1*1501-associated disease such as multiple sclerosis in later life [18]. Prospective, randomised controlled trials are required to demonstrate definitively a protective effect of vitamin D on the emergent immuno-inflammatory disorders.

Epilogue

At the dawn of the new millennium, chronic non-communicable disorders, promoted by environmental changes are the major cause of morbidity and mortality worldwide. The rising tide of autoimmune and allergic diseases is accompanied by other disorders associated with chronic low-grade inflammation, including obesity, insulin resistance and Type 2 diabetes (T2D), atherosclerosis and cardiovascular disease, and arguably Alzheimer's disease and cancer. Expensive drugs for established diseases is not the long-term answer. The situation demands novel preventative and therapeutic approaches aimed in the first instance at the environment. The task is to identify in the complex environmental matrix those factors that impact most adversely on health and to institute public health and political solutions at national and international levels. Swift action is required not only in developed societies but worldwide to avoid the pitfalls of modernity.

Acknowledgements
This work was funded by Program (516700) and Infrastructure (361646) grants from the National Health and Medical Research Council of Australia (NHMRC). LCH is a Senior Principal Research Fellow of the NHMRC. The authors wish to acknowledge Dr Ashton Embry for Figure 4.

Competing interests
The authors declare no competing interests.

References

1 Dubos R, Savage D, Schaedler R (1966) Biological Freudianism. Lasting effects of early environmental influences. *Pediatrics* 38: 789–800
2 Traherne JA, Horton R, Roberts AN, Miretti MM, Hurles ME, Stewart CA, Ashurst JL, Atrazhev AM, Coggill P, Palmer S et al (2006) Genetic analysis of completely sequenced disease-associated MHC haplotypes identifies shuffling of segments in recent human history. *PLoS Genet* 2: e9

3 Shiina T, Hosomichi K, Inoko H, Kulski JK (2009) The HLA genomic loci map: expression, interaction, diversity and disease. *J Hum Genet* 54: 15–39

4 Schmidt AM, Stern DM (2001) Receptor for age (RAGE) is a gene within the major histocompatibility class III region: implications for host response mechanisms in homeostasis and chronic disease. *Front Biosci* 6: D1151–1160

5 Ferrer-Admetlla A, Bosch E, Sikora M, Marques-Bonet T, Ramirez-Soriano A, Muntasell A, Navarro A, Lazarus R, Calafell F, Bertranpetit J et al (2008) Balancing selection is the main force shaping the evolution of innate immunity genes. *J Immunol* 181: 1315–1322

6 Geluk A, Elferink DG, Slierendregt BL, van Meijgaarden KE, de Vries RR, Ottenhoff TH, Bontrop RE (1993) Evolutionary conservation of major histocompatibility complex-DR/peptide/T cell interactions in primates. *J Exp Med* 177: 979–987

7 Geluk A, Van Meijgaarden KE, Janson AA, Drijfhout JW, Meloen RH, De Vries RR, Ottenhoff TH (1992) Functional analysis of DR17(DR3)-restricted mycobacterial T cell epitopes reveals DR17-binding motif and enables the design of allele-specific competitor peptides. *J Immunol* 149: 2864–2871

8 Tandon N, Mehra NK, Taneja V, Vaidya MC, Kochupillai N (1990) HLA antigens in Asian Indian patients with Graves' disease. *Clin Endocrinol (Oxf)* 33: 21–26

9 Rammensee H, Bachmann J, Emmerich NP, Bachor OA, Stevanovic S (1999) SYFPEITHI: database for MHC ligands and peptide motifs. *Immunogenetics* 50: 213–219

10 Geluk A, Lin MY, van Meijgaarden KE, Leyten EM, Franken KL, Ottenhoff TH, Klein MR (2007) T-cell recognition of the HspX protein of *Mycobacterium tuberculosis* correlates with latent *M. tuberculosis* infection but not with *M. bovis* BCG vaccination. *Infect Immun* 75: 2914–2921

11 Vidyarani M, Selvaraj P, Prabhu Anand S, Jawahar MS, Adhilakshmi AR, Narayanan PR (2006) Interferon gamma (IFN-gamma) & interleukin-4 (IL-4) gene variants & cytokine levels in pulmonary tuberculosis. *Indian J Med Res* 124: 403–410

12 Geluk A, van Meijgaarden KE, de Vries RR, Sette A, Ottenhoff TH (1997) A DR17–restricted T cell epitope from a secreted *Mycobacterium tuberculosis* antigen only binds to DR17 molecules at neutral pH. *Eur J Immunol* 27: 842–847

13 Posch PE, Hurley CK, Geluk A, Ottenhoff TH (1996) The impact of DR3 microvariation on peptide binding: the combinations of specific DR beta residues critical to binding differ for different peptides. *Hum Immunol* 49: 96–105

14 Geluk A, van Meijgaarden KE, Franken KL, Drijfhout JW, D'Souza S, Necker A, Huygen K, Ottenhoff TH (2000) Identification of major epitopes of *Mycobacterium tuberculosis* AG85B that are recognized by HLA-A*0201-restricted $CD8^+$ T cells in HLA-transgenic mice and humans. *J Immunol* 165: 6463–6471

15 Horton R, Gibson R, Coggill P, Miretti M, Allcock RJ, Almeida J, Forbes S, Gilbert JG, Halls K, Harrow JL et al (2008) Variation analysis and gene annotation of eight MHC haplotypes: the MHC Haplotype Project. *Immunogenetics* 60: 1–18

16 Ahmad T, Neville M, Marshall SE, Armuzzi A, Mulcahy-Hawes K, Crawshaw J, Sato H, Ling KL, Barnardo M, Goldthorpe S et al (2003) Haplotype-specific linkage disequilib-

rium patterns define the genetic topography of the human MHC. *Hum Mol Genet* 12: 647–656
17 Honeyman MC, Harrison LC, Drummond B, Colman PG, Tait BD (1995) Analysis of families at risk for insulin-dependent diabetes mellitus reveals that HLA antigens influence progression to clinical disease. *Mol Med* 1: 576–582
18 Ramagopalan SV, Maugeri NJ, Handunnetthi L, Lincoln MR, Orton SM, Dyment DA, Deluca GC, Herrera BM, Chao MJ, Sadovnick AD et al (2009) Expression of the multiple sclerosis-associated MHC class II Allele HLA-DRB1*1501 is regulated by vitamin D. *PLoS Genet* 5: e1000369
19 Stober CB, Brode S, White JK, Popoff JF, Blackwell JM (2007) Slc11a1, formerly Nramp1, is expressed in dendritic cells and influences major histocompatibility complex class II expression and antigen-presenting cell function. *Infect Immun* 75: 5059–5067
20 Kaye PM, Blackwell JM (1989) Lsh, antigen presentation and the development of CMI. *Res Immunol* 140: 810–815; discussion 815–822
21 Soo SS, Villarreal-Ramos B, Anjam Khan CM, Hormaeche CE, Blackwell JM (1998) Genetic control of immune response to recombinant antigens carried by an attenuated *Salmonella typhimurium* vaccine strain: Nramp1 influences T-helper subset responses and protection against leishmanial challenge. *Infect Immun* 66: 1910–1917
22 Morahan G, Huang D, Tait BD, Colman PG, Harrison LC (1996) Markers on distal chromosome 2q linked to insulin-dependent diabetes mellitus. *Science* 272: 1811–1813
23 Jamieson SE, White JK, Howson JM, Pask R, Smith AN, Brayne C, Evans JG, Xuereb J, Cairns NJ, Rubinsztein DC et al (2005) Candidate gene association study of solute carrier family 11a members 1 (SLC11A1) and 2 (SLC11A2) genes in Alzheimer's disease. *Neurosci Lett* 374: 124–128
24 Spies T, Bresnahan M, Bahram S, Arnold D, Blanck G, Mellins E, Pious D, DeMars R (2008) A gene in the human major histocompatibility complex class II region controlling the class I antigen presentation pathway. 1990. *J Immunol* 180: 2737–2740
25 Alizadeh BZ, Eerligh P, van der Slik AR, Shastry A, Zhernakova A, Valdigem G, Bruining JG, Sanjeevi CB, Wijmenga C, Roep BO et al (2007) MICA marks additional risk factors for Type 1 diabetes on extended HLA haplotypes: an association and meta-analysis. *Mol Immunol* 44: 2806–2812
26 Eder W, Klimecki W, Yu L, von Mutius E, Riedler J, Braun-Fahrlander C, Nowak D, Martinez FD (2004) Toll-like receptor 2 as a major gene for asthma in children of European farmers. *J Allergy Clin Immunol* 113: 482–488
27 Yang IA, Barton SJ, Rorke S, Cakebread JA, Keith TP, Clough JB, Holgate ST, Holloway JW (2004) Toll-like receptor 4 polymorphism and severity of atopy in asthmatics. *Genes Immun* 5: 41–45
28 Krishnan J, Selvarajoo K, Tsuchiya M, Lee G, Choi S (2007) Toll-like receptor signal transduction. *Exp Mol Med* 39: 421–438
29 Dai J, Liu B, Ngoi SM, Sun S, Vella AT, Li Z (2007) TLR4 hyperresponsiveness *via* cell surface expression of heat shock protein gp96 potentiates suppressive function of regulatory T cells. *J Immunol* 178: 3219–3225

30 Crane GG (1986) Hyperreactive malarious splenomegaly (tropical splenomegaly syndrome). *Parasitol Today* 2: 4–9
31 Fourlanos S, Varney MD, Tait BD, Morahan G, Honeyman MC, Colman PG, Harrison LC (2008) The rising incidence of type 1 diabetes is accounted for by cases with lower-risk human leukocyte antigen genotypes. *Diabetes Care* 31: 1546–1549
32 Morrow AL, Ruiz-Palacios GM, Jiang X, Newburg DS (2005) Human-milk glycans that inhibit pathogen binding protect breast-feeding infants against infectious diarrhea. *J Nutr* 135: 1304–1307
33 Harrison LC, Honeyman MC (1999) Cow's milk and type 1 diabetes: the real debate is about mucosal immune function. *Diabetes* 48: 1501–1507
34 Honeyman MC, Stone NL, Harrison LC (1998) T-cell epitopes in type 1 diabetes autoantigen tyrosine phosphatase IA-2: potential for mimicry with rotavirus and other environmental agents. *Mol Med* 4: 231–239
35 Honeyman MC, Coulson BS, Harrison LC (2000) A novel subtype of type 1 diabetes mellitus. *N Engl J Med* 342: 1835; author reply 1837
36 Honeyman MC, Coulson BS, Stone NL, Gellert SA, Goldwater PN, Steele CE, Couper JJ, Tait BD, Colman PG, Harrison LC (2000) Association between rotavirus infection and pancreatic islet autoimmunity in children at risk of developing type 1 diabetes. *Diabetes* 49: 1319–1324
37 Coulson BS, Witterick PD, Tan Y, Hewish MJ, Mountford JN, Harrison LC, Honeyman MC (2002) Growth of rotaviruses in primary pancreatic cells. *J Virol* 76: 9537–9544
38 Hyoty H, Hiltunen M, Knip M, Laakkonen M, Vahasalo P, Karjalainen J, Koskela P, Roivainen M, Leinikki P, Hovi T et al (1995) A prospective study of the role of coxsackie B and other enterovirus infections in the pathogenesis of IDDM. Childhood Diabetes in Finland (DiMe) Study Group. *Diabetes* 44: 652–657
39 Juhela S, Hyoty H, Uibo R, Meriste SH, Uibo O, Lonnrot M, Halminen M, Simell O, Ilonen J (1999) Comparison of enterovirus-specific cellular immunity in two populations of young children vaccinated with inactivated or live poliovirus vaccines. *Clin Exp Immunol* 117: 100–105
40 von Mutius E, Schwartz J, Neas LM, Dockery D, Weiss ST (2001) Relation of body mass index to asthma and atopy in children: the National Health and Nutrition Examination Study III. *Thorax* 56: 835–838
41 Henkin S, Brugge D, Bermudez OI, Gao X (2008) A case-control study of body mass index and asthma in Asian children. *Ann Allergy Asthma Immunol* 100: 447–451
42 Fourlanos S, Harrison LC, Colman PG (2008) The accelerator hypothesis and increasing incidence of type 1 diabetes. *Curr Opin Endocrinol Diabetes Obes* 15: 321–325
43 Fourlanos S, Narendran P, Byrnes GB, Colman PG, Harrison LC (2004) Insulin resistance is a risk factor for progression to type 1 diabetes. *Diabetologia* 47: 1661–1667
44 Bluher M (2008) The inflammatory process of adipose tissue. *Pediatr Endocrinol Rev* 6: 24–31
45 Berg AH, Scherer PE (2005) Adipose tissue, inflammation, and cardiovascular disease. *Circ Res* 96: 939–949

46 Wellen KE, Hotamisligil GS (2005) Inflammation, stress, and diabetes. *J Clin Invest* 115: 1111–1119
47 Makowski L, Hotamisligil GS (2004) Fatty acid binding proteins--the evolutionary crossroads of inflammatory and metabolic responses. *J Nutr* 134: 2464S-2468S
48 Ozcan U, Cao Q, Yilmaz E, Lee AH, Iwakoshi NN, Ozdelen E, Tuncman G, Gorgun C, Glimcher LH, Hotamisligil GS (2004) Endoplasmic reticulum stress links obesity, insulin action, and type 2 diabetes. *Science* 306: 457–461
49 Lionetti L, Mollica MP, Lombardi A, Cavaliere G, Gifuni G, Barletta A (2009) From chronic overnutrition to insulin resistance: The role of fat-storing capacity and inflammation. *Nutr Metab Cardiovasc Dis* 19: 146–152
50 Weisberg SP, McCann D, Desai M, Rosenbaum M, Leibel RL, Ferrante AW, Jr (2003) Obesity is associated with macrophage accumulation in adipose tissue. *J Clin Invest* 112: 1796–1808
51 Hevener AL, Olefsky JM, Reichart D, Nguyen MT, Bandyopadyhay G, Leung HY, Watt MJ, Benner C, Febbraio MA, Nguyen AK et al (2007) Macrophage PPAR gamma is required for normal skeletal muscle and hepatic insulin sensitivity and full antidiabetic effects of thiazolidinediones. *J Clin Invest* 117: 1658–1669
52 Yamagishi S, Ueda S, Okuda S (2007) Food-derived advanced glycation end products (AGEs): a novel therapeutic target for various disorders. *Curr Pharm Des* 13: 2832–2836
53 Odegaard AO, Pereira MA (2006) Trans fatty acids, insulin resistance, and type 2 diabetes. *Nutr Rev* 64: 364–372
54 Scarpace PJ, Zhang Y (2008) Leptin resistance: a prediposing factor for diet-induced obesity. *Am J Physiol Regul Integr Comp Physiol* 296: R493–500
55 Lin L (2006) RAGE on the Toll Road? Cell *Mol Immunol* 3: 351–358
56 Esposito K, Nappo F, Giugliano F, Di Palo C, Ciotola M, Barbieri M, Paolisso G, Giugliano D (2003) Meal modulation of circulating interleukin 18 and adiponectin concentrations in healthy subjects and in patients with type 2 diabetes mellitus. *Am J Clin Nutr* 78: 1135–1140
57 Bray GA, Nielsen SJ, Popkin BM (2004) Consumption of high-fructose corn syrup in beverages may play a role in the epidemic of obesity. *Am J Clin Nutr* 79: 537–543
58 Sanchez-Lozada LG, Le M, Segal M, Johnson RJ (2008) How safe is fructose for persons with or without diabetes? *Am J Clin Nutr* 88: 1189–1190
59 Zimmermann MB, Aeberli I (2008) Dietary determinants of subclinical inflammation, dyslipidemia and components of the metabolic syndrome in overweight children: a review. *Int J Obes (Lond)* 32 (Suppl) 6: S11–18
60 Patel SR, Hu FB (2008) Short sleep duration and weight gain: a systematic review. *Obesity (Silver Spring)* 16: 643–653
61 Van Cauter E, Knutson KL (2008) Sleep and the epidemic of obesity in children and adults. *Eur J Endocrinol* 159 (Suppl 1): S59–66
62 Quintana FJ, Basso AS, Iglesias AH, Korn T, Farez MF, Bettelli E, Caccamo M, Oukka

M, Weiner HL (2008) Control of T(reg) and T(H)17 cell differentiation by the aryl hydrocarbon receptor. *Nature* 453: 65–71

63 Veldhoen M, Hirota K, Westendorf AM, Buer J, Dumoutier L, Renauld JC, Stockinger B (2008) The aryl hydrocarbon receptor links TH17-cell-mediated autoimmunity to environmental toxins. *Nature* 453: 106–109

64 Fouser LA, Wright JF, Dunussi-Joannopoulos K, Collins M (2008) Th17 cytokines and their emerging roles in inflammation and autoimmunity. *Immunol Rev* 226: 87–102

65 Nguyen LP, Bradfield CA (2008) The search for endogenous activators of the aryl hydrocarbon receptor. *Chem Res Toxicol* 21: 102–116

66 Ma J, Kannan K, Cheng J, Horii Y, Wu Q, Wang W (2008) Concentrations, profiles, and estimated human exposures for polychlorinated dibenzo-p-dioxins and dibenzofurans from electronic waste recycling facilities and a chemical industrial complex in Eastern China. *Environ Sci Technol* 42: 8252–8259

67 North CM, Kim BS, Snyder N, Crawford RB, Holsapple MP, Kaminski NE (2009) TCDD-mediated suppression of the *in vitro* anti-sheep erythrocyte IgM antibody forming cell response is reversed by interferon-gamma. *Toxicol Sci* 107: 85–92

68 Steptoe A, Willemsen G, Owen N, Flower L, Mohamed-Ali V (2001) Acute mental stress elicits delayed increases in circulating inflammatory cytokine levels. *Clin Sci (Lond)* 101: 185–192

69 Klinnert MD, Nelson HS, Price MR, Adinoff AD, Leung DY, Mrazek DA (2001) Onset and persistence of childhood asthma: predictors from infancy. *Pediatrics* 108: E69

70 Arndt J, Smith N, Tausk F (2008) Stress and atopic dermatitis. *Curr Allergy Asthma Rep* 8: 312–317

71 Pavlovic S, Daniltchenko M, Tobin DJ, Hagen E, Hunt SP, Klapp BF, Arck PC, Peters EM (2008) Further exploring the brain-skin connection: stress worsens dermatitis *via* substance P-dependent neurogenic inflammation in mice. *J Invest Dermatol* 128: 434–446

72 Willemsen R, Vanderlinden J, Roseeuw D, Haentjens P (2008) Increased history of childhood and lifetime traumatic events among adults with alopecia areata. *J Am Acad Dermatol* 60: 388–393

73 Siebenhaar F, Sharov AA, Peters EM, Sharova TY, Syska W, Mardaryev AN, Freyschmidt-Paul P, Sundberg JP, Maurer M, Botchkarev VA (2007) Substance P as an immunomodulatory neuropeptide in a mouse model for autoimmune hair loss (alopecia areata) *J Invest Dermatol* 127: 1489–1497

74 Danese A, Pariante CM, Caspi A, Taylor A, Poulton R (2007) Childhood maltreatment predicts adult inflammation in a life-course study. *Proc Natl Acad Sci USA* 104: 1319–1324

75 Hagglof B, Blom L, Dahlquist G, Lonnberg G, Sahlin B (1991) The Swedish childhood diabetes study: indications of severe psychological stress as a risk factor for type 1 (insulin-dependent) diabetes mellitus in childhood. *Diabetologia* 34: 579–583

76 Sepa A, Wahlberg J, Vaarala O, Frodi A, Ludvigsson J (2005) Psychological stress may induce diabetes-related autoimmunity in infancy. *Diabetes Care* 28: 290–295

77 Shamsie J (1985) Family breakdown and its effects on emotional disorders in children. *Can J Psychiatry* 30: 281–287
78 Liu PT, Stenger S, Li H, Wenzel L, Tan BH, Krutzik SR, Ochoa MT, Schauber J, Wu K, Meinken C et al (2006) Toll-like receptor triggering of a vitamin D-mediated human antimicrobial response. *Science* 311: 1770–1773
79 Ebers GC, Sadovnick AD (1993) The geographic distribution of multiple sclerosis: a review. *Neuroepidemiology* 12: 1–5
80 Nystrom L, Dahlquist G, Ostman J, Wall S, Arnqvist H, Blohme G, Lithner F, Littorin B, Schersten B, Wibell L (1992) Risk of developing insulin-dependent diabetes mellitus (IDDM) before 35 years of age: indications of climatological determinants for age at onset. *Int J Epidemiol* 21: 352–358
81 Ponsonby AL, Lucas RM, van der Mei IA (2005) UVR, vitamin D and three autoimmune diseases – multiple sclerosis, type 1 diabetes, rheumatoid arthritis. *Photochem Photobiol* 81: 1267–1275
82 Laron Z (2002) Incidence and seasonality of type 1 diabetes mellitus – what now? *J Pediatr Endocrinol Metab* 15: 573–575
83 Embry AF, Snowdon LR, Vieth R (2000) Vitamin D and seasonal fluctuations of gadolinium-enhancing magnetic resonance imaging lesions in multiple sclerosis. *Ann Neurol* 48: 271–272
84 Adorini L, Giarratana N, Penna G (2004) Pharmacological induction of tolerogenic dendritic cells and regulatory T cells. *Semin Immunol* 16: 127–134
85 Barrat FJ, Cua DJ, Boonstra A, Richards DF, Crain C, Savelkoul HF, de Waal-Malefyt R, Coffman RL, Hawrylowicz CM, O'Garra A (2002) *In vitro* generation of interleukin 10-producing regulatory CD4(+) T cells is induced by immunosuppressive drugs and inhibited by T helper type 1 (Th1)- and Th2-inducing cytokines. *J Exp Med* 195: 603–616
86 Xystrakis E, Kusumakar S, Boswell S, Peek E, Urry Z, Richards DF, Adikibi T, Pridgeon C, Dallman M, Loke TK et al (2006) Reversing the defective induction of IL-10-secreting regulatory T cells in glucocorticoid-resistant asthma patients. *J Clin Invest* 116: 146–155
87 Yesudian PD, Berry JL, Wiles S, Hoyle S, Young DB, Haylett AK, Rhodes LE, Davies P (2008) The effect of ultraviolet B-induced vitamin D levels on host resistance to *Mycobacterium tuberculosis*: a pilot study in immigrant Asian adults living in the United Kingdom. *Photodermatol Photoimmunol Photomed* 24: 97–98
88 Gibney KB, MacGregor L, Leder K, Torresi J, Marshall C, Ebeling PR, Biggs BA (2008) Vitamin D deficiency is associated with tuberculosis and latent tuberculosis infection in immigrants from sub-Saharan Africa. *Clin Infect Dis* 46: 443–446
89 Salari K, Burchard EG (2007) Latino populations: a unique opportunity for epidemiological research of asthma. *Paediatr Perinat Epidemiol* 21 (Suppl 3): 15–22
90 Sundborn G, Metcalf PA, Gentles D, Scragg RK, Schaaf D, Dyall L, Black P, Jackson R (2008) Ethnic differences in cardiovascular disease risk factors and diabetes status for

Pacific ethnic groups and Europeans in the Diabetes Heart and Health Survey (DHAH) 2002–2003, Auckland New Zealand. *N Z Med J* 121: 28–39

91 Scragg R, Sowers M, Bell C (2007) Serum 25-hydroxyvitamin D, ethnicity, and blood pressure in the Third National Health and Nutrition Examination Survey. *Am J Hypertens* 20: 713–719

92 Bailey R, Cooper JD, Zeitels L, Smyth DJ, Yang JH, Walker NM, Hypponen E, Dunger DB, Ramos-Lopez E, Badenhoop K et al (2007) Association of the vitamin D metabolism gene CYP27B1 with type 1 diabetes. *Diabetes* 56: 2616–2621

93 Yang J, Xiong F (2008) [Relevance of CYP27B1 gene promoter polymorphism to autoimmune thyroid diseases]. *Nan Fang Yi Ke Da Xue Xue Bao* 28: 606–608

94 Mathieu C, Badenhoop K (2005) Vitamin D and type 1 diabetes mellitus: state of the art. *Trends Endocrinol Metab* 16: 261–266

95 Scragg R, Sowers M, Bell C (2004) Serum 25-hydroxyvitamin D, diabetes, and ethnicity in the Third National Health and Nutrition Examination Survey. *Diabetes Care* 27: 2813–2818

96 Giovannucci E, Liu Y, Hollis BW, Rimm EB (2008) 25-hydroxyvitamin D and risk of myocardial infarction in men: a prospective study. *Arch Intern Med* 168: 1174–1180

97 Devereux G, Litonjua AA, Turner SW, Craig LC, McNeill G, Martindale S, Helms PJ, Seaton A, Weiss ST (2007) Maternal vitamin D intake during pregnancy and early childhood wheezing. *Am J Clin Nutr* 85: 853–859

98 Hypponen E, Laara E, Reunanen A, Jarvelin MR, Virtanen SM (2001) Intake of vitamin D and risk of type 1 diabetes: a birth-cohort study. *Lancet* 358: 1500–1503

99 Stene LC, Joner G (2003) Use of cod liver oil during the first year of life is associated with lower risk of childhood-onset type 1 diabetes: a large, population-based, case-control study. *Am J Clin Nutr* 78: 1128–1134

100 Heijmans BT, Tobi EW, Stein AD, Putter H, Blauw GJ, Susser ES, Slagboom PE, Lumey LH (2008) Persistent epigenetic differences associated with prenatal exposure to famine in humans. *Proc Natl Acad Sci USA* 105: 17046–17049

101 Hollingsworth JW, Maruoka S, Boon K, Garantziotis S, Li Z, Tomfohr J, Bailey N, Potts EN, Whitehead G, Brass DM et al (2008) *In utero* supplementation with methyl donors enhances allergic airway disease in mice. *J Clin Invest* 118: 3462–3469

102 Sadeghi K, Wessner B, Laggner U, Ploder M, Tamandl D, Friedl J, Zugel U, Steinmeyer A, Pollak A, Roth E et al (2006) Vitamin D3 downregulates monocyte TLR expression and triggers hyporesponsiveness to pathogen-associated molecular patterns. *Eur J Immunol* 36: 361–370

103 Gale EA (2002) The rise of childhood type 1 diabetes in the 20th century. *Diabetes* 51: 3353–3361

Index

The PIR-Series
Progress in Inflammation Research

Homepage: www.birkhauser.ch

Up-to-date information on the latest developments in the pathology, mechanisms and therapy of inflammatory disease are provided in this monograph series. Areas covered include vascular responses, skin inflammation, pain, neuroinflammation, arthritis cartilage and bone, airways inflammation and asthma, allergy, cytokines and inflammatory mediators, cell signalling, and recent advances in drug therapy. Each volume is edited by acknowledged experts providing succinct overviews on specific topics intended to inform and explain. The series is of interest to academic and industrial biomedical researchers, drug development personnel and rheumatologists, allergists, pathologists, dermatologists and other clinicians requiring regular scientific updates.

Available volumes:

T Cells in Arthritis, P. Miossec, W. van den Berg, G. Firestein (Editors), 1998
Medicinal Fatty Acids, J. Kremer (Editor), 1998
Cytokines in Severe Sepsis and Septic Shock, H. Redl, G. Schlag (Editors), 1999
Cytokines and Pain, L. Watkins, S. Maier (Editors), 1999
Pain and Neurogenic Inflammation, S.D. Brain, P. Moore (Editors), 1999
Apoptosis and Inflammation, J.D. Winkler (Editor), 1999
Novel Inhibitors of Leukotrienes, G. Folco, B. Samuelsson, R.C. Murphy (Editors), 1999
Metalloproteinases as Targets for Anti-Inflammatory Drugs,
K.M.K. Bottomley, D. Bradshaw, J.S. Nixon (Editors), 1999
Gene Therapy in Inflammatory Diseases, C.H. Evans, P. Robbins (Editors), 2000
Cellular Mechanisms in Airways Inflammation, C. Page, K. Banner, D. Spina (Editors), 2000
Inflammatory and Infectious Basis of Atherosclerosis, J.L. Mehta (Editor), 2001
Neuroinflammatory Mechanisms in Alzheimer's Disease. Basic and Clinical Research,
J. Rogers (Editor), 2001
Inflammation and Stroke, G.Z. Feuerstein (Editor), 2001
NMDA Antagonists as Potential Analgesic Drugs,
D.J.S. Sirinathsinghji, R.G. Hill (Editors), 2002
Mechanisms and Mediators of Neuropathic pain, A.B. Malmberg, S.R. Chaplan (Editors), 2002
Bone Morphogenetic Proteins. From Laboratory to Clinical Practice,
S. Vukicevic, K.T. Sampath (Editors), 2002
The Hereditary Basis of Allergic Diseases, J. Holloway, S. Holgate (Editors), 2002
Inflammation and Cardiac Diseases, G.Z. Feuerstein, P. Libby, D.L. Mann (Editors), 2003
Mind over Matter – Regulation of Peripheral Inflammation by the CNS,
M. Schäfer, C. Stein (Editors), 2003
Heat Shock Proteins and Inflammation, W. van Eden (Editor), 2003
Pharmacotherapy of Gastrointestinal Inflammation, A. Guglietta (Editor), 2004
Arachidonate Remodeling and Inflammation, A.N. Fonteh, R.L. Wykle (Editors), 2004
Recent Advances in Pathophysiology of COPD, P.J. Barnes, T.T. Hansel (Editors), 2004
Cytokines and Joint Injury, W.B. van den Berg, P. Miossec (Editors), 2004

Cancer and Inflammation, D.W. Morgan, U. Forssmann, M.T. Nakada (Editors), 2004
Bone Morphogenetic Proteins: Bone Regeneration and Beyond, S. Vukicevic, K.T. Sampath (Editors), 2004
Antibiotics as Anti-Inflammatory and Immunomodulatory Agents, B.K. Rubin, J. Tamaoki (Editors), 2005
Antirheumatic Therapy: Actions and Outcomes, R.O. Day, D.E. Furst, P.L.C.M. van Riel, B. Bresnihan (Editors), 2005
Regulatory T-Cells in Inflammation, L. Taams, A.N. Akbar, M.H.M Wauben (Editors), 2005
Sodium Channels, Pain, and Analgesia, K. Coward, M. Baker (Editors), 2005
Turning up the Heat on Pain: TRPV1 Receptors in Pain and Inflammation, A.B Malmberg, K.R. Bley (Editors), 2005
The NPY Family of Peptides in Immune Disorders, Inflammation, Angiogenesis and Cancer, Z. Zukowska, G.Z. Feuerstein (Editors), 2005
Toll-like Receptors in Inflammation, L.A.J. O'Neill, E. Brint (Editors), 2005
Complement and Kidney Disease, P.F. Zipfel (Editor), 2006
Chemokine Biology – Basic Research and Clinical Application, Volume 1: Immunobiology of Chemokines, B. Moser, G.L. Letts, K. Neote (Editors), 2006
The Hereditary Basis of Rheumatic Diseases, R. Holmdahl (Editor), 2006
Lymphocyte Trafficking in Health and Disease, R. Badolato, S. Sozzani (Editors), 2006
In Vivo *Models of Inflammation, 2nd Edition, Volume I*, C.S. Stevenson, L.A. Marshall, D.W. Morgan (Editors), 2006
In Vivo *Models of Inflammation, 2nd Edition, Volume II*, C.S. Stevenson, L.A. Marshall, D.W. Morgan (Editors), 2006
Chemokine Biology – Basic Research and Clinical Application. Volume II: Pathophysiology of Chemokines, K. Neote, G.L. Letts, B. Moser (Editors), 2007
Adhesion Molecules: Function and Inhibition, K. Ley (Editor), 2007
The Immune Synapse as a Novel Target for Therapy, L. Graca (Editor), 2008
The Resolution of Inflammation, A.G. Rossi, D.A. Sawatzky (Editors), 2008
Bone Morphogenetic Proteins: From Local to Systemic Therapeutics, S. Vukicevic, K.T. Sampath (Editors), 2008
Angiogenesis in Inflammation: Mechanisms and Clinical Correlates, M.P. Seed, D.A. Walsh (Editors), 2008
Matrix Metalloproteinases in Tissue Remodelling and Inflammation, V. Lagente, E. Boichot (Editors), 2008
Microarrays in Inflammation, A. Bosio, B. Gerstmayer (Editors), 2009
Th 17 Cells: Role in Inflammation and Autoimmune Disease, B. Ryffel, F. Di Padova (Editors), 2009
New Therapeutic Targets in Rheumatoid Arthritis, P.P. Tak (Editor), 2009